Human Factors of Systems and Technology

hfes europe chapter

Edited by

D. De Waard
N. Merat
A.H. Jamson
Y. Barnard
O.M.J. Carsten

2012, Shaker Publishing

Europe Chapter of the Human Factors and Ergonomics Society

http://hfes-europe.org

© Copyright Shaker 2012

All rights reserved. No part of this publication may be reproduced, stored in a retrieval system, or transmitted, in any form or by any means, electronic, mechanical, photocopying, recording or otherwise, without the prior permission of the publishers.

Printed in The Netherlands.

D. de Waard, N. Merat, A.H. Jamson, Y. Barnard, and O.M.J. Carsten (Eds.).

Human Factors of Systems and Technology

ISBN 978-90-423-0416-1

Shaker Publishing BV
St. Maartenslaan 26
6221 AX Maastricht
Tel.: +31 43 3500424
Fax: +31 43 3255090
http://www.shaker.nl

Contents

Preface 9
   Dick de Waard, Natasha Merat, Hamish Jamson, Yvonne Barnard,
   & Oliver Carsten

The Ballad of Blind Jack Metcalf 11
   Ian Duhig

Automation 13

Developing a method for measuring Situation Awareness in rail signalling 15
   David Golightly, Sarah Sharples, John Wilson, & Emma Lowe

User trust and acceptance of real time rail planning tools 27
   Rebecca Charles, Nora Balfe, Sarah Sharples, John Wilson,
   & Theresa Clarke

Connect 4? The compatibility of driver, motorcyclist, cyclist, and pedestrian
situation awareness 37
   Paul M. Salmon, Kristie Young, & Miranda Cornelissen

Development of a model predicting the use of automated decision aids 51
   Rebecca Wiczorek & Linda Onnasch

The effects of preliminary information about adaptive cruise control on trust
and the mental model of the system: a matched-sample longitudinal driving
simulator study 63
   Matthias Beggiato & Josef F. Krems

The impact of a large-screen projection of the technical process on shared
mental models and team performance in a furnace control room 75
   Vera Hagemann, Annette Kluge, & Björn Badura

Exploring the acceptance of mobile technologies using walking interviews 91
   Frances Hodgson, Yvonne Barnard, Mike Bradley, & Ashley D. Lloyd

Aviation 103

Controller-Pilot communication in a Multiple-Airport-Control scenario 105
   Nora Wittbrodt & Adeline Nelsiana Chandra

Air Traffic Controller assistance systems for attention direction: comparing
visual, acoustical, and tactile feedback 119
   Maik Friedrich, Bernhard Weber, SimonSchätzle, Hendrik Oberheid,
   Carsten Preusche, & Barbara Deml

The Eurofighter Typhoon cockpit assessment process 129
   Suzanne Broadbent

Introduction of ramp-LOSA at KLM Ground Services 139
   Robert J. de Boer, Bekir Koncak, Robbin Habekotté,
   & Gert-Jan van Hilten

A phonetic approach for detecting sleepiness from speech in simulated
Air Traffic Controller-communication 147
   Jarek Krajewski, Thomas Schnupp, Christian Heinze,
   Sebastian Schnieder, Tom Laufenberg, David Sommer, & Martin Golz

Surface Transportation 157

The Lane Change Test: United Kingdom results from a multi-laboratory
calibration study 159
    Terry C. Lansdown
Electric vehicles: an eco-friendly mode of transport which induces changes in
driving behaviour 171
    Elodie Labeye, Myriam Hugot, Michael Regan, & Corinne Brusque
Interaction between driver and infotainment system using a touchpad
with haptic feedback 181
    Andreas Blattner, Roland Spies, Klaus Bengler, & Werner Hamberger
Behaviour of deck officers with new assistance systems in
the maritime domain 189
    Albert Kircher, Fulko van Westrenen, Håkan Söderberg,
    & Margareta Lützhöft
Investigating visually distracted driver reactions in rear-end crashes and
near crashes based on 100-car study data 201
    Henrik Lind, Selpi, & Marco Dozza
Contributing factors to driving errors in trucking industry: drivers'
individual, task and organisational attributes 213
    Ya Li & Kenji Itoh
Don't be upset! Can cars regulate anger by communication? 223
    Sabine Wollstädter, Hans-Rüdiger Pfister, Mark Vollrath, & Rainer Höger
Precision of congestion warnings: Do drivers really need warnings with precise
information about the position of the congestion tail 235
    Ingo Totzke, Frederik Naujoks, Dominik Mühlbacher, & Hans-Peter Krüger
The existence and impact of the Psychological Refractory Period effect in
the driving environment 249
    Daryl Hibberd, Samantha Jamson, & Oliver Carsten

Vulnerable Road Users 263

Drivers' visual behaviour at cycle crossings 265
    Carmen Kettwich & Carina Fors
Towards understanding hazard perception abilities among child-pedestrians 277
    Anat Meir, Tal Oron-Gilad, Avinoam Borowsky, & Yisrael Parmet

Skills and Remote Control 291

On-shore supervision of off-shore gas production - Human Factors challenges 293
    Ruud N. Pikaar, Renske B. Landman, Niels de Groot, & Leen de Graaf
Cognitive Task Analysis – a relevant method for the development of
simulation training in surgery 307
    Norman Geissler, Anke Hoffmeier, Susanne Kotzsch, Stephanie Trapp,
    Nadine Riemenschneider, & Werner Korb
Cognitive performance limitations in operating rooms 317
    Nicki Marquardt, Kristian Gerstmeyer, Christian Treffenstädt,
    & Ricarda Gades-Büttrich

Collecting battlefield information using a multimodal personal digital assistant 327
   Stas Simon Krupenia, Mathilde Cuizinaud, Tijmen Muller, & Anja H. van der Hulst

MODELLING 341

Developing a unified model of driving behaviour for cars and trains 343
   Björn Peters, Anna Vadeby, Åsa Forsman, & Andreas Tapani

Implementation of a Unified Model of Driver into numerical algorithms for a predictive simulation of behaviour in different transportation contexts 359
   Magnus Hjälmdahl, Aladino Amantini, & Pietro C. Cacciabue

How is surrounding traffic complexity related to driver workload? 373
   Evona Teh, Samantha Jamson, & Oliver Carsten

Technologies to support socially connected journeys: Designing to encourage user acceptance and utilisation 385
   Sarah Sharples, David Golightly, Caroline Leygue, Claire O'Malley, James Goulding, & Ben Bedwell

Acknowledgement to reviewers 397

Preface

Dick de Waard[1], Natasha Merat[2], Hamish Jamson[2],
Yvonne Barnard[2], & Oliver Carsten[2]
[1] University of Groningen, the Netherlands
[2] ITS University of Leeds, UK

In October 2011 the HFES Europe Chapter annual conference took place in Leeds, England. Papers included in this book were presented at that three day meeting. As is its tradition, the conference roamed around multiple aspects of human factors in one single set of plenary sessions, enabling interaction among researchers and PhD students using a variety of investigative techniques to examine different domains. It is this opportunity for interaction that gives the conference series its unique flavour. Conference papers ranged from the emotional impact of music on driving to the use of PDAs to collect battlefield information. Once again the rapid-fire talking poster session was a highlight.

On the first day of the conference, Mr Ian Duhig, a Leeds-based poet was commissioned by the Europe Chapter to read his poem "The Blind Ballad of Jack Metcalf" to the conference delegates. Ian is a Leeds University Graduate and has written six books of poetry, won the Forward Best Poem Prize once, the National Poetry Competition twice, and has been shortlisted for the T.S. Eliot Prize three times. He is also a Fellow of the Royal Society of Literature. His poem about John Metcalf (1717–1810), also known as Blind Jack of Knaresborough or Blind Jack Metcalf, is based on the story of a professional road builder from the North of England. Blind from the age of six, due to a smallpox infection, John Metcalf started life as a fiddler, but became a major road builder and civil engineer, building one of the main roads which currently goes from Knaresborough through to the centre of Leeds. The poem is the opening contribution of this book.

All papers in this book are covered by the broad title of the book "Human Factors of Systems and Technology". Presentations are further grouped into the following chapters: Automation / Situation Awareness, Transportation (Aviation, Surface Transportation and Vulnerable Road Users), Skills and remote control, and Modelling.

We would like to thank all reviewers who helped to review the manuscripts for this book; their names are listed as always on the final page. Finally, we are also very grateful for the support we received by the European Office of Aerospace Research and Development of the USAF, under Award No. FA8655-11-1-5056.

In D. de Waard, N. Merat, A.H. Jamson, Y. Barnard, and O.M.J. Carsten (Eds.) (2012). *Human Factors of Systems and Technology* (pp. 9). Maastricht, the Netherlands: Shaker Publishing.

The Ballad of Blind Jack Metcalf

(1717-1810)
By Ian Duhig
October, 2011

Verse by the numbers, numbered years
summing up loved dead;
small fingers feeling headstone faces -
all young Blind Jack read.

A man, he read behind their words
how men and women felt,
like faces, suits and numbers stamped
on tavern cards he dealt.

Sharp dealer, traffic was his gift,
in fish and flesh he'd trade;
a soldier, smuggler, fiddler, guide -
roadmaker, when that paid.

He spun his tales then webs of tar
as dark as all he saw;
some swore Jack trailed a sulphur smell,
who laid down his own law.

Still dark in bronze on Market Square,
he hears the traffic snarling:
some might sing of Bonnie Princes -
this song is Jack's darling.

His waywiser beside his bench,
around his metalled hat
his secret tale's a road of braille,
and what it tells is that...

In D. de Waard, N. Merat, A.H. Jamson, Y. Barnard, and O.M.J. Carsten (Eds.) (2012). *Human Factors of Systems and Technology* (pp. 11). Maastricht, the Netherlands: Shaker Publishing.

Automation

Developing a method for measuring Situation Awareness in rail signalling

David Golightly[1], Sarah Sharples[1], John Wilson[1,2], & Emma Lowe[2]
[1]Centre for Rail Human Factors, University of Nottingham
[2]Network Rail Ergonomics Specialist Team
UK

Abstract

This paper presents work to develop methods for measuring Situation Awareness in rail signalling, based on the Situation Present Assessment Technique (SPAM) and the Situation Awareness Rating Technique (SART). The paper describes the rationale and design features of the methods used, before presenting the outcomes of an exploratory trial using a simulated workstation and experienced signallers. While the results found no differences due to the independent variable (automation versus no automation), the process of developing and using the methods highlighted a number of features of real-time probe measures in the rail control domain, and points to further validation of the SPAM approach.

Introduction

The following paper presents an exploratory study of measuring Situation Awareness (SA) for the domain of rail signalling. Rail signalling is a key role in the safe and punctual operation of the railways. The signaller (sometimes a 'dispatcher' outside of Great Britain) is responsible for monitoring the progress of trains within a given territory. This understanding serves as the basis for setting signals and points in order for trains to proceed safely and in accordance with the timetable. Signalling is typical of dynamic, safety-critical control domains where SA has been proposed as a relevant construct (Endsley, 1995a) but, to date, has received relatively little attention in comparison to domains such as ATC or driving (Golightly et al., 2010). Having a good understanding of both the elements of SA, and the processes used to build and maintain SA, could be of benefit in defining requirements for future signaling interfaces, in assessment, or in training. Building a more complete understanding of SA in signalling is closely linked to developing appropriate methods for examining SA. Such methods can be used as tools to explore what constitutes SA for signalling, and how this may change based on factors such as different interfaces, levels of experience, workstation layout, or the introduction of automation.

The following work had the aim of understanding the viability of a method, and specifically a quantitative *measure*, of SA for rail signaling. This fitted within a larger programme of work to investigate the relevance of SA for rail signalling

In D. de Waard, N. Merat, A.H. Jamson, Y. Barnard, and O.M.J. Carsten (Eds.) (2012). *Human Factors of Systems and Technology* (pp. 15 - 26). Maastricht, the Netherlands: Shaker Publishing.

(Golightly et al., 2009, 2010). The work presented is exploratory; it was a first iteration of methods using a small, though highly specialised, sample of active and experienced signallers. The work had a secondary aim of capturing any relevant insight into the nature of SA in signalling, particularly with regard to automation. The sample would be too small to draw substantial conclusions at this stage, but may highlight potential characteristics of signalling worthy of further investigation.

Background

Rail signalling requires the operator, or 'signaller', to have an accurate understanding of the state of the system that he or she controls, in order to make correct, timely decisions and take effective action. This understanding needs to include train, track, signal status and routes set. It also requires knowledge of the performance parameters that the system should be working within (e.g. the timetable), local and geographical factors that might impact regulation, knowledge of any system failures or restrictions and, where appropriate, what decisions any automation may be making (Golightly et al., 2010). Broadly, this understanding can be referred to as 'Situation Awareness'.

While a number of definitions exist, many describe Situation Awareness (SA) as the understanding of the elements in a dynamic system relevant to the operator's goals. Rather than SA being merely the awareness of these elements, theories such as Endsley's (1995a) and Durso and Dattel's (2004) emphasise the synthesised comprehension of these elements in combination. Such theories also emphasise that operators do not just perceive elements in terms of what they mean now, but also to predict the implications of the situation in the short-term future. Endsley, for example, describes SA in terms of three levels; level 1 – perception, level 2 – comprehension, and level 3, prediction. To use a signalling example, having accurate SA requires identifying that there are two trains on a workstation, understanding what their relationship is now (e.g. they are heading towards the same set of points) and in the short term future (e.g. potential conflict leading to delay).

There are alternative views on SA. For example it may be more useful to view SA as strategies of attention rather than just as a state of knowledge (Adams et al., 1995), or that SA is distributed between actors and artefacts, rather than a product or process of an individual. Others question SA as a construct altogether (Dekker & Hollnagel, 2004). In the long-term, it is possible these alternatives may prove relevant to our understanding of SA in rail signalling. The prevalence, however, of the view that SA is, at least in part, related to a state of knowledge means this perspective is a practical starting point for any investigation of SA that might lead to design guidance for systems, training approaches and evaluation methods.

Observational and ethnographic approaches are useful in understanding how SA is constructed and maintained in a control setting. For example, qualitative methods have highlighted the importance of strategies in SA such as 'active overhearing' of events in the rail control room (Roth et al., 2003). One useful output for a method of SA is that it should be possible to identify changes that occur due to factors such as technology or process re-design. It is useful if this output is quantitative (and therefore can support empirical comparison between conditions), and there have been efforts to capture this kind of data. Such data might be objective, in terms of understanding the elements that are relevant to SA, or subjective, in terms of understanding operator's own experience of SA. Several methods exist – these involve using queries to probe operator's current SA (e.g. Situation Awareness Global Assessment Technique (SAGAT, Endsley 1995b) or Situation Present Assessment Method (SPAM, Durso & Dattel, 2004)), structuring output from observations and other elicitation to identify key constructs (e.g. social network analyses, see Salmon and Young, this volume), or through inventories and survey tools (e.g. Situation Awareness Rating Technique, see Jones, 2000). The aim of the work presented in the rest of this paper is to test variants of two of these methods – SPAM and SART, in the rail control setting.

Selection of methods

The choice of the methods described in this paper was influenced by a number of factors. First, Network Rail, Great Britain's rail infrastructure provider and sponsor of this work, were seeking a method that could be used to explore changes in SA, particularly with regard to the implementation of new technology, changes in workload, and with the introduction of new training procedures. The method had to be practical, lightweight, and could be used with little or no adaptation of standard signalling processes and activities, and thereby retaining a high level of validity.

Additionally, the method should be suitable for use both by HF practitioners for investigative and empirical work, and also by signalling supervisors to make more day-to-day assessments of performance in real-time operations. A SAGAT-type measure had been trialed in the past with signallers (Wilson et al., 2001). SAGAT (Endsley, 1995b) requires the scenario or environment to be removed from the participant's view while queries are presented probing their current state of SA. In signalling, this requires either the use of a simulator, or a relief signaller to step in and continue to control the panel while the participant responds to queries. The aim of the current study was to find a method that could be used with a simulator, as reported in this paper, but could also be used during actual signalling operations with minimal impact.

As well as these practical constraints, other factors informed the choice of method. One argument is that the display should not be treated as separate from the operator's cognition. When using displays, operators do not have to retain everything they see in either working or long-term memory. Instead, it may be more efficient to engage in 'display-based problem solving', leaving the information in the external representation until it is needed (Larkin, 1989). In the case of SA, the question may not be one of 'does the operator have knowledge of what is happening' but more one of 'do they know where to get that knowledge from, should they need it?' (Durso & Dattel, 2004). By removing the display during queries, such as SAGAT (Endsley, 1995b), an artificial situation is created that is undermining the signaller's ability to answer queries effectively. In line with this, a pilot to develop a measure of signaller SA based on freeze probes had found that signaller recall of train positions was highly variable, prone to individual differences and unrelated to variables such as workload or performance (Golightly et al., 2009).

Therefore, the decision was made to trial an adaption of the Situation Present Assessment Measure (SPAM) approach (Durso & Dattell, 2004). As with SAGAT, task analysis is used to develop queries to probe an operator's awareness of the state of elements in the system. Table 1 presents an example set of queries. Unlike SAGAT, where questions are asked during freezes in the simulation, SPAM queries are put to the operator while the display is still visible to them, allowing them to draw on the display for reference. Along with response accuracy as a measure of SA, response latency is also recorded, with longer latencies indicating a need for the operator to search the display for information that is not currently active in working memory.

Another complimentary approach is through studying operator's subjective sense of their own SA, which may be important for ensuring trust with new forms of equipment or procedure. Subjective measures are easy to use in real-world settings as they can be quick and therefore relatively unobtrusive (Jones, 2000), offering the capability to use them during real-time operations without interrupting the signaller. The Situation Awareness Rating Technique (SART) (Taylor, 1990, cited in Jones, 2000) is the most common of the subjective measures, based around 10 constructs that have been identified as important to the experience of SA. Each of these constructs is presented as a Likert-type scale in a 10-item survey.

These two approaches, SPAM and SART, were modified for an exploratory study in the rail domain. While both of these methods should be appropriate for use during real signalling operations, the first step was to run a trial in a simulated signalling environment. Many of the signalling centres in Great Britain have high fidelity simulators, offering accurate recreation of train patterns and events using equivalent equipment to that used by signallers.

As noted above, one important output of a measure of SA is that is should be able to detect changes due to factors such as technology. An independent variable was therefore required to test the sensitivity of the two proposed measures in different conditions. Automation is one characteristic of control that has been shown to influence SA and SA-related strategies. For example, Kaber and Endsley (2004) found SA to be highest when there was an intermediate level of automation in a simulated control task, as opposed to either complete automation or low levels of automation. Using a simulated Air Traffic Control task, Vu et al. (2011) found that SA was highest when controllers were responsible for maintaining separation, as opposed to when it was controlled by an automated system. In the Great Britain rail context, automation takes the form of Automatic Route Setting (ARS), which selects and sets optimal routes, based on the timetable and surrounding traffic, for trains arriving on a workstation. In previous work to examine differences in behaviours due to rail automation, monitoring of workstations has been found to be less prevalent where ARS has been implemented, with potential decrements in SA (Sharples et al., 2011). Therefore, automation had the potential to influence signallers to the degree that it could affect SA, and is also easily controlled in the simulation setting described above. Thus, automation was chosen as an independent variable to test the sensitivity of the two candidate SA measures.

Method

Participants and location

Participants were six working signallers at an Integrated Electronic Control Centre (IECC) on the GB East Coast mainline. All participants were qualified for the grade of workstation being simulated, with at least 2 years of service (approximate mean of service = 10 years). All participants regularly controlled traffic on the workstation being simulated. All signallers were currently on duty on the days the study took place, and participated in the study during their working hours while being covered by a relief.

Apparatus – simulation

The IECC was equipped with a fully simulated workstation in an adjoining room to the main control area. The simulator is capable of recreating any workstation in the main control room, with (for the purposes of this study) identical equipment to a live workstation, including main displays and inputs, and simulation of supporting information systems. Also in the simulation area, but screened off from the simulator, is the simulation control desk. From here, the investigator was able to run a scenario, set up additional events such as infrastructure failures, pause the scenario, and contact the signaller via a telephone.

The particular scenario chosen was based on a prior familiarization visit to the signalling control centre being studied, including observations of all workstations at the centre and discussion with signallers of factors such as difficulty, traffic patterns and complexity of the infrastructure. The workstation that was finally selected for the purpose of the study was graded as 7 on the 1 to 9 scale (9 as most difficult) used by Network Rail for grading workstation difficulty. The workstation was selected as it involved a mix of traffic (mainline express, stopping services, freight and branch services) without being too complex for the investigator to generate meaningful and realistic queries. The scenario was based on the timetable at 5pm (i.e. during peak traffic) to give the scenario an additional level of difficulty.

Table 1. Set A of the queries used in the trials. Type of query (current or future) is indicated in brackets

1. Will Darlington platform 1 be free in 5 minutes time? (Future)
2. Is train 1A39 running to time? (Current)
3. Will there be a train arriving on your panel from Gateshead in the next 5 minutes? (Future)
4. How many platforms are occupied at Darlington? (Current)
5. How many trains are on the workstation? (Current)
6. Does 1E45 call at Durham? (Future)
7. Are any trains due to take the Saltburn branch? (Future)
8. Is train 1S19 ahead of train 1E45? (Current)

Apparatus – material

Three types of material were developed for the study. The first comprised two sets of equivalent queries. Queries were developed through workshops with Subject

Matter Experts, observations in signaling control rooms, and interviews with signallers. This led to a generic set of queries that could be adapted to the timetable and infrastructure for any workstation. Queries were split into two types, those that queried the current state of the system and those that queried the future state of the system, and queried signallers on factors such as trains currently on workstations, future calling points, current and future conformance of trains to timetable, and presence of trains on peripheral parts of the workstation such as branchlines.

These generic queries were then edited to match the specific infrastructure and train pattern for the scenario being used in the current study. This required examining the simulation scenario and identifying specific trains, routes or timetable features that would be relevant at approximately three minute intervals (the points at which queries would be presented to the participants). This lead to two equivalent sets of queries that were similar enough to be comparable across trials, but not require the investigator to ask the same question. Query set A is presented in Table 1.

The second set of material comprised an amended version of SART with terms modified to make them relevant to the rail signalling domain (e.g. referring to 'trains' rather than 'aircraft').

The third set of material comprised printed versions of the track layout diagrams for that workstation. Participants used these diagrams after each trial to mark the location and train identifier information for any recalled trains. (The train identifier, or 'headcode', is a four-digit alphanumeric code unique to any train on the timetable for that workstation on a given day – for example, '1A33' could be an express service leaving London at 10:20am heading to Edinburgh and calling at all major stations).

Procedure

Each participant was briefed on the aims and procedure for the study. They were then asked to control the scenario as they would normally during real-time operations. The scenario was run twice – once with Automatic Route Setting (ARS) switched on, and once without (the order was balanced across participants).

At three-minute intervals the participant was asked a query from either Set A or Set B. Queries were designed to be meaningful at that particular point in the scenario. This interval was chosen as it was felt to represent minimum intrusion while affording the most number of queries (eight) in the 25-minute trial. Participants were not informed of how frequently questions would be asked, and the exact time between queries varied so that that time of the query could not be anticipated. At five-minute intervals, participants were asked to report their subjective mental workload using the Integrated Workload Scale (IWS), a nine-point scale that has been validated for rail signalling (Pickup et al., 2005). On 25 minutes, the scenario was stopped, screens were switched off, and the participant was asked to recall the position of as many trains as they could on a track diagram, and complete the SART.

The scenario was then repeated in the alternate automation condition, and using the alternative query set. At the end, the participant was asked for general impressions, including the realism of the simulation. Queries were varied slightly between trials to minimise learning effects.

Design and measures

The basic experimental design was 1 x 2 within subjects, with automation (ARS on or off) as the independent variable. Dependent variables included

- SPAM – time to respond (seconds), and accuracy (%) averaged over the trial; additional analyses were performed to compare SPAM time to respond between current- and future-orientated queries,
- Post-task recall – number of trains recalled, accuracy of position, number of correct characters recalled from train identifier, for each trial (all calculated as % of maximum possible score),
- IWS score (averaged over trial),
- SART score at the end of each trial,
- Objective measure of performance – calculated by simulator using measures such as minutes delay and trains held at red, or proceeding under caution.

Results

Table 2 shows means across the independent variables for the measures listed above. Repeated measures t-tests were performed to compare means for time to respond (df = 5, t = 1.45, NS), response accuracy (df = 5, t = -0.39, NS), time to respond for current orientated queries (df = 5, t = -0.18, NS), and time for future orientated queries (df = 5, t = 0.28, NS). Repeated measures t-test were also performed to compare means for post-trail recall of number of trains (df = 5, t = 0.26, NS), accuracy of positions (df = 5, t= 0.43, NS), and number of correct characters recalled from the train identifier (df = 5, t = 1.15, NS). Finally, repeated measures t-tests were carried out to compare means for IWS score (df = 5, t = -1.23, NS), and SART (df = 5, t = 1.19, NS).

To ensure there were no ordering effects, all of the dependent variables were tested using trial order (trial 1 vs trial 2) as an independent variable. There were no significant effects.

In line with a similar study to validate SA measures (Salmon et al., 2009), workload, SART, SPAM response time and performance variables were correlated with each other. Due to data loss there were only 10 data points for performance, with 12 data points for other variables. Calculations of significance were adjusted accordingly, to reflect differing degrees of freedom. Of those correlations, only the correlation between performance and SPAM response reached significance ($p < 0.05$ – see table 3).

Table 2. Means, and standard deviations for dependent variables

	ARS	Non-ARS
SPAM – time to respond (secs)	5.02 (1.9)	4.16 (2.5)
SPAM – accuracy (%)	93.4 (10.7)	95.8 (6.5)
SPAM – time to respond (Current orientated queries)(secs)	5.51 (3.26)	5.79 (3.41)
SPAM – time to respond (Future orientated queries)(secs)	3.70 (1.53)	3.45 (2.32)
Recall – no of trains (%)	88.5 (15.7)	87.1 (10.8)
Recall – train location accuracy (%)	77.6 (23.6)	75.2 (22.6)
Recall – train identifier accuracy (%)	78.7 (22.9)	75.2 (22.6)
IWS	1.93 (0.57)	2.2 (0.73)
SART (%)	76.6 (7.4)	72.2 (10.0)

Table 3. Correlations for SA, workload and performance measures (* $p < 0.05$)

	SART	SPAM	IWS	Performance
SART	-	0.27	-0.29	0.27
SPAM	0.27	-	-0.42	**-0.79***
IWS	-0.29	-0.42	-	0.27
Performance	0.27	**-0.79*-**	0.27	-

Discussion

The aim of this study had been to explore SPAM- and SART-type measures for examining rail SA. There is an obvious limitation with the study being exploratory and therefore participant numbers are small, though the sample came from a specific and highly relevant population (as opposed to having larger numbers of participants, but from a non-expert population). While acknowledging this limitation, there were a number of interesting facets of the data, and experiences of designing and running the study, that highlight the potential value and difficulties in using SA measures for empirical investigation for signalling, pointing to future directions for such measures.

The analysis of the results revealed almost no significant differences for any of the dependent variables. Therefore, there is little evidence from this trial that our particular formulations of SPAM or SART have proved effective as a measure. It is also worth noting that recall of positions and train identifiers at the end of the trial period was far from perfect, and variable (scores ranging from 30% up to 100%), suggesting that freeze probe measures are not likely to be useful, at least for understanding SA of routine signalling. The correlation between performance and SPAM is encouraging. It suggests that decrements in SA are related to decrements in performance, and while the sample size is too small to draw this as a conclusion, it suggests that a more substantial repeat of this study is worthwhile. Data relating to SART is inconclusive at this stage.

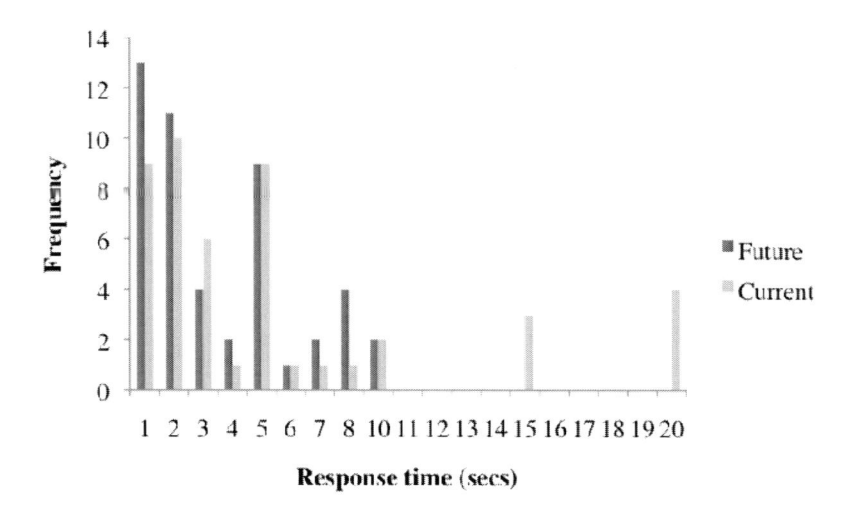

Figure 1. Distribution of query response times

Also, a review of the distribution of time (Figure 1), and observation at the time of the study, suggests groupings in the data – one group where participants are giving near instantaneous responses, a second group of response times at around 5 seconds where information may need scanning or accessing from a secondary system, but the source of that information is known, and a third set of data points at around 10 to 20 seconds where information is not known and needs to be searched from relevant data sources. It is possible that there is a standard profile of times and distributions of times that could be expected for scenarios. Different levels of expertise might lead to a redistribution of peaks from near zero to longer response times (i.e. more experienced signallers will have responses near zero, whereas less experienced signallers will have a more high response times). Shifts in response times might also indicate priorities in how the signaller is controlling their workstation. In this study, longer times were in response to queries confirming whether trains were running to time, which had to be searched from a specific system. This finding may indicate that awareness of strategic performance against the timetable is less important for signalling than train pattern information, and that signallers are engaged at a much more tactical level. Determining different profiles, and factors causing profile shifts, are worthy of further exploration with a larger sample.

One reason why comparison of means proved so inconclusive at this stage is that automation appears to have been an insufficient factor to lead to differences in SA. This may be closely linked to the issue of using expert signallers. While the knowledge of experienced signallers gave the study validity, it meant that participants were comfortable with dealing with a wide range of conditions and disruption with relative ease. Comments of participants after the study indicated that they had no difficulty dealing with the scenarios with or without automation. Also, ARS is not regarded as being 100% reliable, and therefore signallers may be maintaining awareness of the system in order to monitor the performance of ARS. In this respect, unreliable automation may be equivalent to partial automation as demonstrated by Kaber and Endsley (2004). This is worthy of further empirical investigation with a larger sample. Also, these results suggest that using automation for empirical comparisons in rail signalling for factors such as SA or workload is not straightforward, and will need careful consideration in future studies.

An additional complexity experienced in the study was the successful application of queries in a task like signalling. Deriving these queries is far from trivial; defining any kind of challenging probe takes time and has to be highly scenario-specific. Importantly, this is as much about the timing of the query as it is about the content. There is a brief window where the elements under examination in the query move from being irrelevant, to being in the focus of decision making, and then out into the display (e.g. visually indicated to the signaller as a set route with different colouring). Getting the timing correct to ask questions that are challenging needs expert knowledge of the domain, otherwise accuracy in particular will reach a ceiling (all participants scored 100% accuracy on at least one trial).

A further difficulty is that it does not make sense to ask more than one query at a time, and asking too regularly runs the risk of participants rehearsing and preparing their responses. However, asking only one query every 3 minutes leads to only 8 queries per scenario. Even with much longer trial times, the scope for extensive probing of signaller SA is limited. The probe approach should therefore be viewed as a way of 'sampling' SA rather than a means to get a complete 'audit' of SA, unless it is a very specific aspect of the domain under examination. One question that will need to be addressed if SPAM is taken forward is whether asking queries had an effect on performance through prompting the operator to look at salient aspects of the display, though previous validations with the similar SAGAT approach have shown this not to be the case (Endsley, 1995b).

Conclusion

The aim of this work was to understand whether a measure of signaller SA had value, and how to proceed. The evidence so far is that the SPAM approach does hold promise, particularly with regard to the distribution of response times, and trying to develop profiles associated with different signalling scenarios, technology or levels of expertise. However, there are difficulties in using experienced signallers, and with a factor such as automation, in order to differentiate between test conditions. Also, developing rules for the content and timing of appropriate queries needs further work. So far, work with SART has proved inconclusive.

It is interesting to note that the potential success of SPAM reinforces the relevance of strategy (how and where signallers are finding information using the display and support systems). In line with that, it would be useful to continue with more observational and elicitation-based approaches similar to that carried out by Roth et al. (2003), to further understand the relevance of strategies for SA in signalling. Also, eye-tracking technology is maturing to the point where it can be used unobtrusively during tasks such as signalling. There is also some tentative evidence in this study for automation having unexpectedly little influence on SA, and for decreased relevance of strategic timetable information. This needs validation with a larger sample.

References

Adams, M. J., Tenney, Y. J., & Pew, R. P. (1995). Situation awareness and the cognitive management of complex systems. *Human Factors, 37*, 85-104.

Dekker, S. & Hollnagel, E. (2004). Human factors and folk models. *Cognition, Technology and Work, 6*, 79-86.

Durso, F. T., & Dattell, A. R. (2004). SPAM: The real-time assessment of SA. In S. Banbury and S. Tremblay (Eds.) *A cognitive approach to situation awareness: theory and application.* (pp. 137-154). Aldershot, UK: Ashgate.

Endsley, M.R. (1995a). Towards a Theory of Situation Awareness. *Human Factors, 37*, 32-64.

Endsley, M. R. (1995b). Measurement of Situation Awareness in Dynamic Systems. *Human Factors, 37*, 65-84.

Golightly, D., Balfe, N., Sharples, S., & Lowe, E. (2009). Measuring Situation Awareness in Rail Signalling. Presented at the *Conference in Rail Human Factors, 2009, Lille, France.*

Golightly, D., Wilson, J. R., Lowe, E., & Sharples, S. (2010). The role of situation awareness for understanding signalling and control in rail operations. *Theoretical Issues in Ergonomics Science, 11*, 84-98

Jones, D.G. (2000). Subjective Measures of Situation Awareness. In M.R. Endsley and D.J. Garland (Eds.) *Situation Awareness: Analysis and Measurement* (pp 113-128). Mahwah, NJ.: Lawrence Erlbaum Associates.

Kaber, D. B., & Endlsey, M. R. (2004). The effects of level of automation of adaptive automation based on human performance, situation awareness and workload in a dynamic control task. *Theoretical Issues in Ergonomic Science, 5*, 113-153.

Larkin, J. H. (1989). Display-based problem solving. In D. Klahr and K. Kotovsky, (Eds.) *Complex Information Processing: The impact of Herbert A. Simon.* (pp. 319-341). Hillsdale NJ: Erlbaum.

Pickup, L., Wilson, J.R., Norris, B.J., Mitchell, L. & Morrisroe, G. (2005). The Integrated Workload Scale (IWS): A new self report tool to assess railway signaller workload. *Applied Ergonomics, 36*, 681-693.

Roth, E. M., Multer, J., & Rasler, T. (2003). Shared Situation Awareness as a Contributor to High Reliability Performance in Railroad Operations. *Organization Studies, 27*, 967-87.

Salmon, P. M., Stanton, N. A., Walker, G. H., Jenkins, D., Ladva, D., Rafferty, L., & Young, M (2009). Measuring situation awareness in complex systems: comparison of measures study. *International Journal of Industrial Ergonomics, 39*, 490-500.

Salmon, P., & Young, K. (2012) Connect 4? The compatibility of driver, motorcyclist, cyclist, and pedestrian situation awareness. In D. de Waard, N. Merat, A.H. Jamson, Y. Barnard, and O.M.J. Carsten (Eds.) *Human Factors of Systems and Technology* (pp. 37-50). Maastricht, the Netherlands: Shaker Publishing.

Sharples, S., Millen, L., Golightly, D., & Balfe, N. (2011). The impact of automation on rail signalling operations. *Proceedings of the Institution of Mechanical Engineers, Part F: Journal of Rail and Rapid Transit, 244*, 1-13.

Vu, K-P. L., Strybel, T. Z., Battiste, V., Lachter, J., Dao, A-Q.V., Brandt, S., Ligda, S. & Johnson, J. (2012). Pilot Performance in Trajectory-Based Operations Under Concepts of Operation That Vary Separation Responsibility Across Pilots, Air Traffic Controllers, and Automation. *International Journal of Human-Computer Interaction, 28*, 107-118

Wilson, J.R., Cordiner, L.A., Nichols, S.C., Norton, L., Bristol, N., Clarke, T. and Roberts, S. (2001). On the right track: Systematic implementation of ergonomics in railway network control. *Cognition, Technology and Work, 3*, 238-252.

User trust and acceptance of real time rail planning tools

Rebecca Charles[1], Nora Balfe[2], Sarah Sharples[1], John Wilson[2], & Theresa Clarke[2]
[1]*University of Nottingham*
[2]*Network Rail Ergonomics Team*
UK

Abstract

Regulation of train movements on busy routes can be extremely complex and there are frequent delays and alterations that require information to be easily accessible and accurate. The Train Graph is a tool that has been introduced as an aid to signalling supervisors and control staff to assist with this task. In this paper, we discuss our current understanding of technology acceptance and then discuss a field study carried out during the pilot implementation of this technology. The results indicate that people in different job roles perceived the graph differently. The results also revealed that the Train Graph has not yet been readily accepted by all users, and reasons for this are explored.

Introduction

Network Rail (NR) owns, runs, maintains and continuously improves 20,000 miles of track, 1100 signal boxes and 2500 stations in the UK. Signallers are responsible for the setting of train routes and their key objectives are to keep the railway running safely and on time. Overseeing the signal box and the signallers within it, Shift Signalling Managers (SSMs) have a clear overview of the area the box controls and are also the first point of contact within the box. Information on train running is critical to the effective management of the train service. By using databases and other computerised tools, information can be accessed quickly and easily and allow the signallers and SSMs to make their decisions accurately.

The Train Graph is a tool that shows the routes taken by trains using existing information from a data management system, known as TRUST, and displays a line-based representation of the train's path with reference to time and station. When trains are running late, the Train Graph (TG) will flag up potential future conflicts (i.e. two trains expected to reach the same place at the same time) and also provide regulating options to aid the signaller in their decisions. The Train Graph has initially been rolled out to SSMs and Train Running Specialists (TRSs) along the East Coast Mainline to assist with regulating tasks. The TRSs are responsible for the regulation of the entire route while the SSMs are responsible for their area of control (i.e. the area covered by their signal box). The TRS is a relatively new role within the company and due to the large area they control, there are considerably fewer TRS's than SSM's. Currently there are only three in the country. Following initial training given at the end of 2010, the Train Graph was introduced to five signal boxes and one control centre along the East Coast Mainline (a high speed route joining London and Edinburgh via York). This study was based on the pilot implementation of the Train Graph. The train graph continues to be developed before full implementation.

The main issue faced when introducing new or replacement technologies into established environments is whether it will be willingly accepted and effectively used by users (Venkatesh, et al. 2003), especially when the current systems (in this case TRUST and another data source, CCF) are well used. TRUST is used as an information source, providing detailed information about each train, including delay information; whereas CCF is a real-time map based display of the trains' location. There is much literature within the Information Systems field that aims to measure and predict the acceptance of a new technology within the context of a working environment using quantitative methods. One of the most well known and replicated of these is the Technology Acceptance Model (TAM; Davis, 1989). This has been adapted and added to in the past 20 years (see Venkatesh, Morris et al. 2003 for a review) but the basis remains the same: the uptake and acceptance of any technology can be determined by perceived usefulness and perceived ease of use, with perceived usefulness found to have the greatest influence (Davis, 1989). Although TAM provides feedback about general usefulness and ease of use, it does not provide specific feedback about the artefact itself. Roger's (2003) Innovation Diffusion Theory (IDT) however, specifies different characteristics of a system that are believed to determine the acceptance and rate of adoption, and therefore can provide more identifiable feedback for use within a design process. These characteristics are; relative advantage, compatibility, complexity, trialability and observability (see figure 2). The first three have been shown to have the greatest influence on uptake, these being the characteristics that directly compare the technology to existing systems. If users perceive the new technology as advantageous over existing ones, compatible with their existing ideas and values and perceive the system to be easy to understand and use then uptake is more likely. Of course, this is easier to measure or report if a new system is replacing an existing one; if the new system is offering novel functionality or it is likely or intended to change the way in which a job or task is completed, these characteristics may be harder to apply. Trialability and observability are characteristics more concerned with the roll out of a new technology. The additions to TAM, known as 'TAM2' made by Venkatesh and Davis (2000) suggest key forces underlying judgments of perceived usefulness that are complementary to the IDT (see figure 1).

Within the framework of the TAM, users have been found to perceive the technology easy to use if they have prior experience of the technology (Gefen, et al. 2003) (the technology in this case being existing planning tools). The existing technology however, in this case will remain in place even when the TG is introduced. This means that use of the TG is not compulsory and the users are not being forced or required to use it. This is in some ways useful from an analytical perspective (if the use of a technology is non-compulsory then it can be assumed that use of the technology is an indicator of its successful adoption, and, in terms of TAM, perceived usefulness or ease of use) but may also be a disadvantage as users may be less willing to persist with learning how to use the technology or overcoming initial barriers to use, and thus may be quick to reject or ignore the new system. However, the TG is also not completely voluntary as it is a work related tool. Most examples of technology in the literature are tools such as internet banking or on-line shopping sites. The TG will also not present any new data to the user; it will just display data differently – therefore the change to the task is potentially in the form of supporting reasoning or decision making with existing data sets, rather than enhancing the cognitive tasks by providing new data.

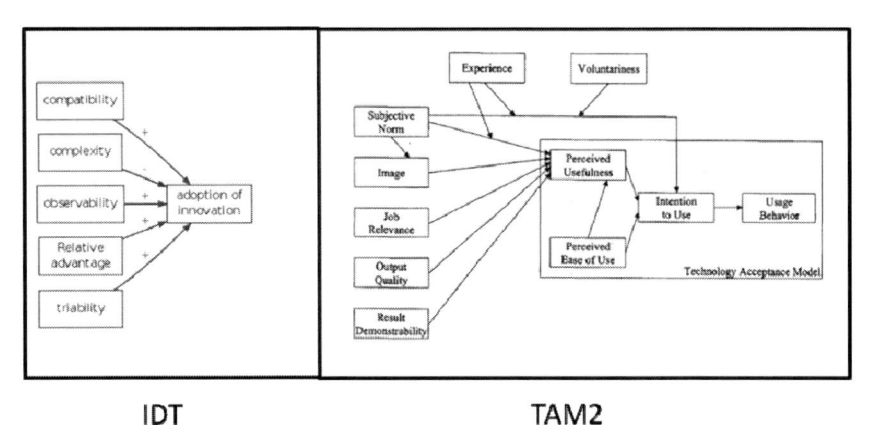

Figure 1. Representations of the IDT and TAM2

If the users have prior knowledge and experience of a system, as rail staff have experience of the railway system, they also have existing mental models of how the new technology should work. When introducing new technologies, Zhang and Xu (2011) argue that the users existing mental models need to be modified or restructured in order to continue to guide the users interaction with it. If the new technology does not fit the existing mental model, it can lead to frustration for the operator and will affect uptake and adoption (d'Apollonia, et al. 2004). Existing familiarity however has been shown to lead to increased trust with a system and can also lead to an increased belief that the technology is easy to use (Gefen, et al. 2003).

The purpose of this paper is to report initial findings from an investigative study on the acceptance of TG with its users after the first stage of introduction. This is the first part of a larger body of work currently being undertaken.

Method

Six sites were visited, and ten users (nine SSMs and one TRS) were asked general questions relating to the use of the Train Graph based around acceptance, usage and usability of the TG. The sites were chosen to be as different from one another as possible, in terms of operating systems, number of staff and level of traffic. Each interview took between 20 minutes to 1 hour and took place while the operators were carrying out their other duties. This allowed an element of opportunistic observation to be undertaken also. In addition, users were asked to report any specific issues they had with the Train Graph and questions about the interface and the general layout of the tool.

Results

Data was collected in the form of transcripts (when recording was permitted) and written notes. Due to constraints of the opportunistic nature of the data collection, the interviews were coded by site rather than individual and a total of six transcripts were analysed. These were coded and analysed in NVIVO, using qualitative thematic analysis techniques (Hayes, 2000). Each transcript was coded three times ensuring all data had been coded sufficiently. Initially, 20 themes were identified. These were then divided into groups using a card sort with three Human Factors experts in order to structure the data for reporting. Two top level themes were identified; existing job role and the Train Graph tool itself.

Table 1. Frequency of themes

	Interviews (X if yes)	Observations (X if yes)	Freq (comment)	% coverage (TRS Role)	% coverage (SSM Role)
Existing Job Role					
Signaller Qualities					
experience	X		31	7.6	5.6
knowledge	X		36	8.5	6.6
role	X	X	24	5.9	4.3
Physical attributes of signalling task					
infrastructure	X	X	14	2.5	2.8
existing technology	X	X	36	6.8	7.1
Signalling task					
signaller actions	X	X	26	4.2	5.3
Signaller strategies	X	X	22	3.4	4.6
communication routes	X	X	18	4.2	3.3
current task	X	X	33	4.2	7.1
Train Graph					
Use of Train Graph					
accessability	X		11	1.7	2.3
functionality	X		48	10.2	9.2
usefulness	X		51	5.1	11.5
relative advantage	X		30	4.2	6.4
usage	X	X	34	2.5	7.9
acceptance	X		10	2.5	1.8
Data Quality					
input quality	X		11	5.9	1.0
output quality	X	X	32	7.6	5.9
trust of TG	X		18	5.9	2.8
Interaction with Train Graph					
usability	X	X	12	3.4	2.0
interface	X	X	14	3.4	2.5

Table 1 shows the frequency of the themes and also identifies the areas that have been enriched through participant observation. In addition the table shows the frequency of themes based on role. Since more SSMs were interviewed than TRSs the percentage coverage per role rather than frequency for these two roles has been shown. The results are expanded in more detail below, using the structure presented in Table 1, concentrating mainly on the Train Graph.

Existing job role

The main role of the SSM is to manage the signal box, and ensure that the routes are being set correctly. Actually setting the route is the job of the signallers under the SSMs supervision. Each signal box is only responsible for a small section of railway. The TRS oversees the entire route, in this case the East Coast Mainline. It is the TRS's job to ensure that delays are kept to a minimum along the entire route and the signal boxes are coordinating with one another. Most of the SSMs interviewed saw the TG as being a tool most beneficial to aiding the job role of the TRS and felt that their area of control was too small to benefit. SSMs controlling older, more manual boxes that contain large displays of the control area felt particularly in control of their areas: *"You are constantly aware of the trains that are in your area, so the conflicts that the train graph flags up, you are already aware of."* Around half also acknowledged that the TRS lead their decision making process as they *"make a decision that is based on the bigger picture, so we try our best to carry out their request."*

Existing technology was a theme that emerged strongly. All of the interviewees mentioned CCF and TRUST at least once and were confident about the ability of these systems to assist them in carrying out their tasks. CCF is currently used as more of a tracking tool, and TRUST is used to enquire about a specific train. One SSM commented: *"looking at CCF you are actually looking at the infrastructure."* The phrase "actually looking at the infrastructure" strongly implies that the user's mental model suggests that the actual state of the infrastructure is directly reflected in the representation used in CCF, implying trust in the information sources used to construct the CCF representation. This is a layout that is familiar and can be *"glanced at quickly to gain information regarding delays."* The same SSM commented that *"TRUST required more interpretation."* Interestingly, the TRS found TRUST to be more beneficial to their needs and easier to use than CCF.

When reading and using data was mentioned, it was nearly always followed up with a reference to the experience and learned knowledge of the operator: *"The SSM knows the routes of all the trains and potential conflicts within his area from experience."* Many situations would present themselves in many different guises and the operator would use their knowledge and experience to guide the strategies that were used to deal with it.

Train graph – use

Usefulness (or lack thereof) emerged as the most frequent theme, specifically among the SSMs. All of the SSMs felt that in its current stage of development, the TG was not useful to them. Many felt that the TG did not add anything to the existing tools and was not really relevant to them: *"It doesn't add anything over CCF. You can see the delays coming on CCF."* This is an interesting comment as in its current form CCF does not directly represent predicted delays but presents the data in a form that allows the operator to infer them; the new TG display explicitly presents the predicted conflict, but this is not perceived by the operators as a significant benefit. Many however, did acknowledge that the tool would be beneficial to the TRS. The TRS felt it added *"additional clout to our decision making process. Especially since all of the same information is visible to the SSMs. They are able to see exactly what we are talking about, as we are talking to them."*

With regard to the functionality of the graph, three of the SSMs mentioned the way in which the graph shows conflicts: "Only shows one train conflicting another and predicts trains as first come first serve. Some conflicts would never be resolved in this manner." However, the TRS found this function extremely useful. This may suggest that they have knowledge of a larger area, but not as detailed as the SSMs so their route knowledge is not as strong. The TRS is also a relatively newly created post, so the TRS will not have the years of 'on the job' learned knowledge and experience that many of the SSMs have. Both TRS and SSMs all shared the view that the TG add ons (such as being able to access TRUST data quickly) were a good feature.

One SSM commented that the Train Graph *"made no sense. It doesn't resemble the infrastructure. CCF is easier to understand"*. It may however be the fact that the CCF is reported as being easier to understand because it is familiar, rather than because it enables the task to be completed in a cognitively more simple manner. CCF was referred to several times by the SSMs as being *"easy to glance at"* but the TRS preferred TRUST to acquire regulating information. In contrast, the TRS also found train graph easy to use and interact with. This may be because the layout of the graph fits better with the TRS's view of the railway (north to south expanded area) rather than the compact area the SSMs deal with.

Train graph – data quality

The quality of the information that TG provided was a frequently mentioned topic, specifically for TRSs. All of the interviewees made a reference to the graph showing conflicts that did not exist: *"It tends to show conflicts that are not there, or won't happen and from experience we know they won't happen."* One SSM also stated that *"the Graph is sometimes wrong. It shows conflicts between trains on fast line and trains on slow line."* Although not explicitly stated, these errors lead to mis-trust of the system and a need to query every conflict box and check each detail. The TRS was more trusting of the system, although they did still acknowledge that there were certain aspects that were not accurate. They were also more aware of the quality of data being fed to the graph: *"Had a freight train that was duplicated three times, so the graph was showing conflicts with trains that did not exist. It's only as good as the information it's provided with."*

TraingGraph - interaction

The interface was generally well received by all of the operators. One SSM commented that "it was easy to read once it had been cut down and simplified" and that "you can glance at it and it will tell you where the likely conflicts will be. To find that out in TRUST would take ages." The TRS did however state that the graph can be extremely difficult to read when densely populated: "Down near Kings Cross where you've got all the local trains coming in, it's really difficult to read, especially during peak times."

One SSM in particular found the TG extremely difficult to use at first: "There was just too much information there. It made no sense. Then I worked out how to filter out certain things and it made it easier."

Discussion

The TAM and IDT have been applied in many different situations and industries investigating the uptake and usage behaviour of both mandatory and voluntary technology. In this instance, the technology is not mandatory (Brown, et al. 2002), nor is it completely voluntary: Users still have access to the original systems but the train graph is a work related tool. The key tenet of TAM is that if a technology is perceived to be useful, it will be used. The results in this case clearly support this theory, with usefulness, or lack thereof being the most frequently mentioned topic amongst both of the roles and appearing to have a direct affect on use of the TG.

Exploring the drivers behind perceived usefulness in the updated TAM2 model, the cognitive instrumental processes (job relevance, output quality, result demonstrability and perceived ease of use) were found by Venkatesh et al. (2000) to have a direct influence on perceived usefulness in mandatory settings, and for the most part voluntary systems as well. The findings from the current study follow this theory, and also show some interesting findings regarding usage behavior with regard to role. Both the SSMs and the TRS shared the opinion that the TG was more relevant to the TRS role. The results also clearly show that output quality was an important topic for the TRS; more so than the SSMs. This interactive effect between job relevance and output quality was also demonstrated by Venkatesh et al. (2000).

Output quality, although mentioned more frequently by the TRS, was a topic discussed in depth by both job roles. In this instance, the information is fed to the Train Graph directly from TRUST, a system all of the users are familiar with and use daily. Prior experience of TRUST, in this case was not seen to have much influence on the level of trust users have with the Train Graph. Even though most of the information is the same and certain aspects of the incorrect information seen in TG is visible on the TRUST system currently, the Train Graph makes this wrong information more visible to the user, so the system is not trusted. Although unreliable data may be an external issue, the impact means that the acceptance and benefit of Train Graph is reduced. However, Davis' (1992) theory suggests that if the output quality was to improve, then the level of perceived usefulness would also increase.

Although the IDT can provide more identifiable feedback with regard to the artifact itself, specific features of the technology are not taken into account in either model when considering usage behavior. In the present study, the users demonstrated an increased level of acceptance and use when they were able to customize the interface. It could be argued that this aspect of customization could be placed within the complexity characteristic of the IDT (Rogers, 2003), or directly influence ease of use within TAM2 (Venkatesh et al., 2000). However, as the TG is essentially a new interface for existing technology, specific design characteristics emerge as an important factor in use in this case and should be further explored.

The operators showed a distinct lack of trust in the information that the train graph displayed (output). Overall the results show that generally the TRS interviewed found the Train Graph useful, whereas the SSMs did not. The TRS also had no issues with the ease of use of the tool, whereas the SSMs did. This is consistent with basic tenets of the TAM which suggests that the more a user perceives the technology to be useful, the more the user believes it is easy to use (Venkatesh et al., 2003).

The SSMs in particular mentioned CCF frequently during the interviews and used this as a primary source of information. The TRS primarily used CCF in a list format and TRUST. This may indicate that both groups are using different experiences to base their learning of the new technology on (Orlikowski, 2008). The TRS are using TRUST to make sense of Train Graph, whereas by the SSMs using CCF to base their learning on, they may be also using the existing mental model of this system to learn the new technology. Consistent with Zhang's (2011) theory, the train graph is more consistent with the TRS's existing mental models of existing systems and layout of the railway they use, so limited cognitive effort is required to adapt to the Train Graph. The SSMs however have a different view; a more compact extended view of a small section of railway. This view also agrees with Rogers' theory, specifically compatibility; existing ideas and values when applied to a new technology can be seen to drive adoption rate.

Conclusions

The findings are for the most part, supported by either IDT or TAM2. The research presented here showed strong similarities with existing findings (Venkatesh & Davis, 2000), specifically regarding the impact of output quality and job relevance on perceived usefulness, and the results showed clear differences between the two roles. Relative advantage was also seen to have an effect on the uptake of TG, as did compatibility. The parts of the models not investigated here were the more social factors – Subjective norm and image. These will need to be investigated in a further study. The ability to customize the interface had a significant effect on whether a user found the TG useful or not. Specific design characteristics are not sufficiently dealt with in the current models, but will need further investigation.

Further work

The Train Graph continues to be developed and enhanced ahead of a national roll-out. A similar study will be carried out following the upgrade of the software so that comparisons can be drawn. This should indicate whether improvements in areas such as user interfaces and accuracy of information have made any difference in how the users view the Train Graph. By using TAM and IDT as a basis for further study with a detailed observation study and a questionnaire, the concept of acceptance will be further studied, with an emphasis on how different design features influence uptake.

References

Brown, S.A., Massey, A.P., Montoya-Weiss, M.M., & Burkman, J.R. (2002). Do I really have to? User acceptance of mandated technology. *European Journal of Information Systems, 11*, 283-295.

D'Apollonia, S.T., Charles, E.S., & Boyd, G.M. (2004). Acquisition of complex systemic thinking. Mental models of evolution. *Educational Research and Evaluation, 10*, 499-521.

Davis, F.D. (1989). Perceived usefulness, perceived ease of use, and user acceptance of information technology. *MIS quarterly, 13*, 319-340.

Gefen, D., Karahanna, E., & Straub, D.W. (2003). Inexperience and experience with online stores: The importance of TAM and trust. Engineering Management. *IEEE Transactions on engineering management, 50*, 307-321.

Hayes, N. (2000). *Doing psychological research*. London: Taylor & Francis.

Orlikowski, W.J. (2008). Using technology and constituting structures: A practice lens for studying technology in organizations. *Resources, Co-Evolution and Artifacts, 3*, 255-305.

Rogers, E.M. (2003). *Diffusion of innovations*. New York: Free Press.

Venkatesh, V. & Davis, F.D. (2000). A theoretical extension of the technology acceptance model: Four longitudinal field studies. *Management science*, 186-204.

Venkatesh, V., Morris, M.G., Davis, G.B., & Davis, F.D. (2003). User Acceptance of Information Technology: Toward a Unified View. *Management Information Systems Quarterly, 27*, 425-478.

Zhang, W. (2011). Do I have to learn something new? Mental models and the acceptance of replacement technologies. *Behaviour and Information Technology, 30*, 201 - 211.

Connect 4? The compatibility of driver, motorcyclist, cyclist, and pedestrian situation awareness

Paul M. Salmon, Kristie Young, & Miranda Cornelissen
Monash University Accident Research Centre, Victoria
Australia

Abstract

Compatibility between different road users' situation awareness is critical to safe and efficient interactions between them. This paper presents an exploratory proof of concept on-road study conducted to explore situation awareness across four road user groups: drivers, motorcyclists, cyclists, and pedestrians. The aim was to test the assumption that different road users interpret the same road situations differently and to explore the extent to which these interpretations are compatible with one another. Participants from each group negotiated a pre-defined route either on foot (e.g. pedestrians) or using an instrumented car/motorcycle/bicycle. Based on verbal protocols provided en-route, a network analysis procedure was used to describe and analyse participants' situation awareness. This revealed differences both in the content and structure of each road user groups' situation awareness, along with evidence of incompatibilities at intersections. The implications of this are discussed along with potential initiatives for enhancing compatibility between different road users.

Introduction

Each form of road transportation (e.g. driving, motorcycling, cycling) requires different physical and cognitive tasks for safe and efficient performance. Evidence suggests that distinct road users, such as drivers and motorcyclists, interpret the same road situations differently (e.g. Shahar et al., 2010; Walker et al., 2011). This is perhaps not surprising; however, for safe interactions between road users, some degree of compatibility between their situational interpretations is required (Salmon et al., 2011; Walker et al., 2011). In situations in which awareness across road users becomes uncoupled, conflicts between them are likely; for example, 'right of way' accidents between cars and motorcycles (e.g. Pai, 2009) represent instances where one road user is not aware of the other. The concept of Situation Awareness (SA), which accounts for how humans understand 'what is going on' (Endsley, 1995), offers one approach to investigate different road users' understanding of the same road situations. Despite this, to date SA has received scant attention in the road transport context (Salmon et al., 2011).

In this paper we present the findings from an exploratory on-road study of SA across four different road user groups: drivers, motorcyclists, cyclists, and pedestrians. The aims were, first, to test on-road the assumption that different road users interpret the same road situations differently, and second, to explore the degree of compatibility between each road user groups' SA in different road contexts.

Situation awareness on the road: definition and assessment

Road user SA can be defined as activated knowledge, regarding road user tasks, at a specific point in time. This knowledge encompasses the relationships between road user goals and behaviours, vehicles, the road environment and infrastructure. Recent evidence suggests different road users sample the environment differently and perceive and interpret the same road situations differently, leading to differences in awareness across road users (e.g. Walker et al., 2011). These differences exist both within road user groups (i.e. drivers have differing views to one another based on goals, experience, etc.) and across different road user groups (i.e. drivers have differing views to motorcyclists).

The extent to which road users' SA differs is contingent upon various factors, including the tasks that they have to perform (e.g. driving a car versus riding a motorcycle), the design of the road environment (e.g. infrastructure, signage, road markings), and their experience and goals. A key component of road users' SA, however, are genotype and phenotype schema (Stanton et al., 2010); genotype schema represent schema held in the mind of individuals which contain prototypical responses to specific situations, whereas phenotype schema represent the state specific activated schema which is brought to bear in a particular situation. These schema drive, and determine the content of, road user SA, since they direct exploration in the world (i.e. sampling of the environment), which in turn directs behaviour, which in turn modifies schema and so on (e.g. Neisser, 1976). Different road users possess different schema for the same road situations and Walker et al. (2011) argue that schema-related issues are likely to be at the root of incompatibilities between road users. Put simply, if phenotype schema are brought to bear which do not incorporate other road users (and thus do not direct sampling of other road users), or the wrong schema are activated, then the interaction between road users is likely to be problematic. This is likely to include issues such as road users failing to see the others (i.e. looked but failed to see errors) or failing to comprehend how other road users are likely to behave. Viewing road user behaviour from a schema perspective brings into question the extent to which road systems currently support awareness across different road users; for example, the extent to which road environments are designed for all road users, as opposed to one road user group in isolation (e.g. drivers), is questionable.

The aim of the present study was to investigate, via on-road study, the differences in, and level of compatibility between, driver, motorcyclist, cyclist and pedestrian SA. Participant SA during an on-road study was modelled and analysed using a network analysis procedure known as propositional networks (Salmon et al., 2009). Using this approach, SA 'networks' are constructed using data derived from the Verbal Protocol Analysis (VPA) method, which involves participants 'thinking aloud' as they perform tasks, with the resulting verbal transcript being subjected to content analysis procedures to make inferences regarding the content of situation awareness. The content analysis procedure is used to identify keywords or concepts and the relationships between them, which enables propositional networks to be constructed. The resulting network is taken to be a representation of SA. Mathematical analysis of the networks is then used to interrogate their content and structure (e.g. Salmon et al., 2009; Walker et al., 2011), which enables comparison of SA across different actors. An overview of the network analysis process is presented in Figure 1.

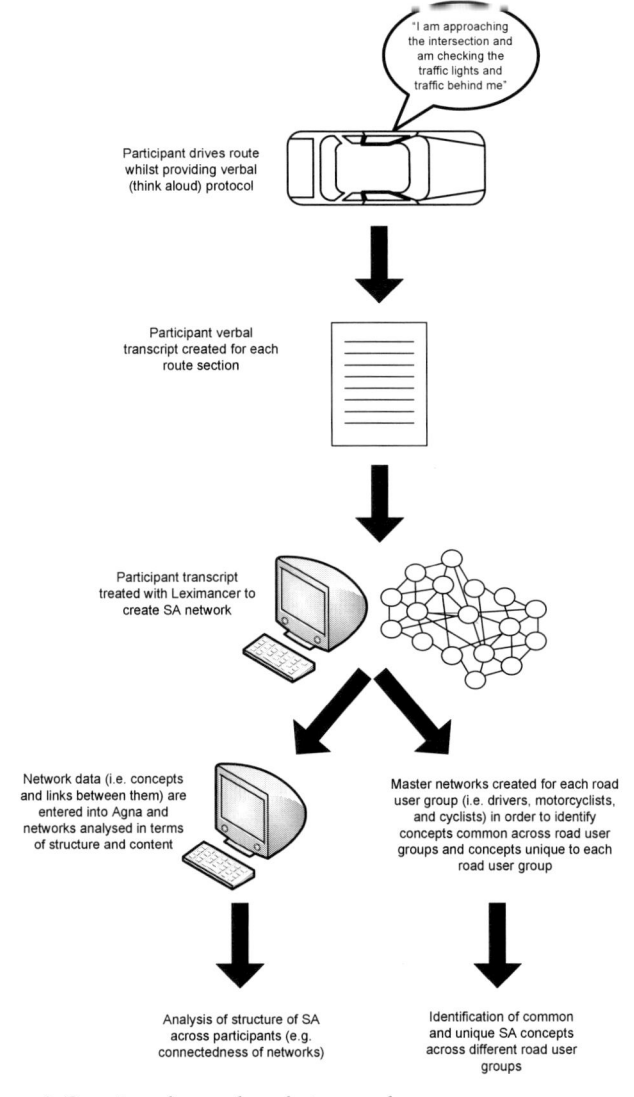

Figure 1. Overview of network analysis procedure

Based on previous research (e.g. Shahar et al., 2010; Walker et al., 2011), the hypothesis for the proof of concept study was that participants from the different road user groups would interpret similar road situations differently. Specifically, the knowledge underpinning SA would be different, both in terms of content and structure, across the road user groups studied.

Methodology

Design

The study used an exploratory semi-naturalistic paradigm whereby participants either drove an instrumented vehicle around a pre-defined urban route or walked specific sections of the route on foot. All participants provided concurrent verbal protocols as they negotiated the route. Network analysis was used to describe and analyse participant SA.

Participants

Twenty participants (12 male, 8 female) aged 21-50 years (mean = 35.31, SD = 8.19), including 5 participants from each road user group, took part in the study. Participants were recruited through a weekly on-line university newsletter and were compensated for their time. Prior to commencing the study ethics approval was formally granted by the Monash Human Ethics Committee.

Materials

A demographic questionnaire was completed using pen and paper. The study used a 16.5km urban route (including 1.5km practice route) located in the south-eastern suburbs of Melbourne. Four distinct route sections formed the basis for the analysis of road user SA: the entire route, intersections, arterial roads, and a shopping strip. The *entire route* was 15km long and comprised a mix of arterial roads (50, 60 and 80km/h speed limits), residential roads (50km/h speed limit), and university campus private roads (40km/h speed limit). Seven intersections along the route were used for the *intersections* analysis component. These comprised four fully signalised intersections (i.e. all turns controlled by traffic lights), two partially signalised (i.e. some but not all turns controlled by traffic lights) intersections, and one non-signalised intersection. The *arterial roads* component comprised approximately 6.2kms of arterial roads along the route. These had three lanes and an 80km/h speed limit. Finally, a *shopping strip* section of the route was also used. The shopping strip section was approximately half a kilometre in length, has a 60km/h speed limit, and has shops and car parking spaces running parallel to the road on either side.

Drivers drove the route in an instrumented 2004 Holden Calais sedan equipped to collect various vehicle and driver-related data. A Dictaphone was used to record drivers' verbal protocols. Motorcyclists rode the route using their own motorcycle fitted with an Oregon Scientific ATC9K portable camera, which, depending on motorcycle model was fixed either to the handlebars or front headlight assembly. The ATC9K records the visual scene, speed and distance travelled (via GPS). A microphone was fitted inside each rider's motorcycle helmet to record their verbal protocols. Cyclists cycled the route using their own bicycles. To record the cycling visual scene and the cyclist verbal protocols, the ATC9K portable camera was fitted to the cyclists' helmets. Pedestrians walked three of the intersections along the route whilst wearing spy sunglasses capable of recording high definition video and audio. All verbal protocols were transcribed using Microsoft Word. For data analysis, the Leximancer™ content analysis software and Agna™ network analysis software tools were used.

Procedure

To control for traffic conditions, all trials took place at the same pre-defined times on weekdays (10am or 2pm Monday to Friday). Prior to the study these times were subject to pilot testing to confirm the presence of similar traffic conditions. Upon completion of an informed consent form and demographic questionnaire, participants were briefed on the research and its aims. After a short VPA training session, participants were shown the study route and were given time to memorise it. Whilst motorcyclist/cyclist participants were practising the VPA method and familiarising themselves with the route, a technician fitted the ATC9K camera to their motorcycle or cycling helmet. For drivers/motorcyclists/cyclists, when comfortable with the VPA procedure and route, they were taken to their vehicle and asked to prepare themselves for the test. They were then given a demonstration of the video and audio recording equipment, which was also set to record at this point. Following this, the experimenter instructed the participant to begin the practice route. Participants were instructed to stop at the end of the practice route if they had any problems or further questions. If they felt comfortable to proceed, they were asked to continue onto the test route. For the drivers, an experimenter was located in the vehicle and provided route directions if necessary. For the motorcyclists and cyclists, an experimenter followed behind (in a car for the motorcyclists, on a bicycle for the cyclists) ready to intervene if the participants strayed off route. For the pedestrians, once comfortable with the VPA procedure and route, they were taken by car (driven by the experimenter) to the first intersection and instructed to negotiate the intersection in order to reach a set point on the road following the intersection. Once the participant reached this point, they were picked up by the experimenter and driven to the next intersection. This process was repeated until all three intersections had been successfully negotiated.

Participants' verbal protocols were transcribed verbatim using Microsoft Word. For data reduction purposes, extracts of each participant's verbal transcript for each route section were taken. The extracts were taken based on set points in the road environment (e.g. beginning and end of arterial roads). The verbal transcripts were then treated with the Leximancer™ content analysis software. Leximancer™ automates the content analysis procedure by processing verbal transcript data through five stages: conversion of raw text data, concept identification, thesaurus learning, concept location, and mapping (i.e. creation of network). This led to the creation of a series of networks for each participant. These were then entered into the Agna network analysis software program for content and structural analysis purposes.

Results

Network analysis

The networks were analysed both quantitatively and qualitatively. The quantitative analysis involved using the *density* and *diameter* metrics to analyse the structure and content of each network. The qualitative analysis involved identifying the concepts common across all road users' networks and those unique to each road user group. A brief description of each metric is given below, followed by the results derived from each form of analysis.

Network structure: density and diameter

Network density represents the level of interconnectivity of the network in terms of links between concepts. The formula is presented in Formula 1 below (adapted from Walker et al., 2011).

$$\text{Network Density} = \frac{2e}{n(n-1)}$$

Where:
e = number of links in network
n = number of information elements in network

Formula 1. Network density

Density is expressed as a value between 0 and 1, with 0 representing a network with no connections between concepts, and 1 representing a network in which every concept is connected to every other concept (Kakimoto et al., 2006; cited in Walker et al., 2011). For SA assessments, higher levels of interconnectivity indicate an enhanced, richer level of SA since there are more links between concepts. Poorer SA is embodied by a lower level of interconnectivity between concepts, since the concepts underpinning SA are not well integrated. The mean density values for participants' networks overall and for the different route sections are presented in Figure 2.

Drivers' network density was greater overall (0.80 compared to 0.62 for motorcyclists and 0.69 for cyclists), at the intersections (0.83 compared to 0.55 for motorcyclists, 0.60 for cyclists, and 0.59 for pedestrians), and along the arterial roads (0.79 compared to 0.51 for motorcyclists and 0.44 for cyclists). Motorcyclists' had more dense networks along the shopping strip (0.95 compared to 0.89 for drivers and 0.69 for cyclists). These results suggest that driver SA, when expressed as concepts and the relationships between concepts, was more interconnected for the overall route, at the intersections, and along the arterial roads. This means that the SA concepts within driver's networks were more connected to one another than in the other participant's networks'. Along the shopping strip, motorcyclists' SA networks were more interconnected than drivers and cyclists.

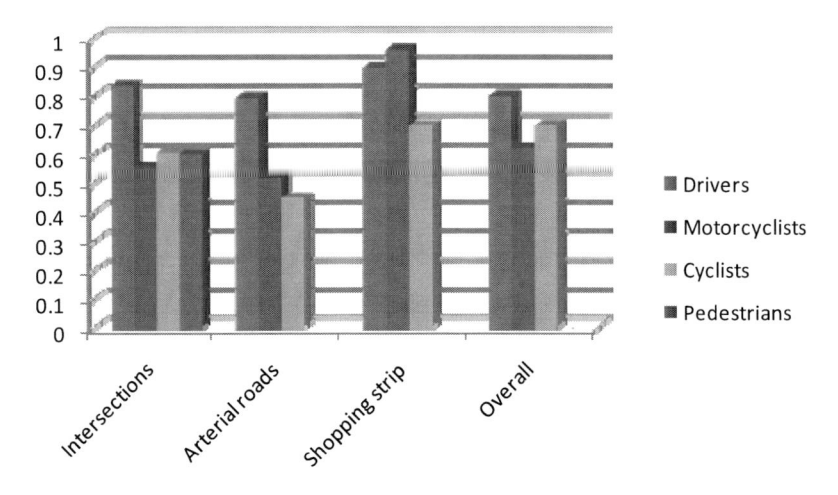

Figure 2. Mean density values for each road user group

Diameter is used to analyse the connections between concepts within networks and also the paths between the concepts (Walker et al., 2011). The formula is presented below (adapted from Walker et al., 2011).

Diameter = $\max_{uy} d(n_i, n_j)$

Where
$d(n_i, n_j)$ = largest number of concepts which must be traversed in order to travel from one concept to another

Formula 2. Network diameter

Greater diameter values are indicative of longer paths through the network. Denser networks therefore have smaller values since the routes through the network are shorter and more direct. With regard to SA, lower diameter scores are indicative of better SA, whereas higher diameter scores are indicative of a model of the situation comprising more concepts but with less links present between them (Walker et al., 2011). The mean diameter values for the participants' networks overall and at the different roadway sections are presented in Figure 3.

The cyclist group had a greater mean diameter overall, at the intersections, and along the arterial roads and shopping strip. The drivers, motorcyclists and pedestrians achieved the same mean diameter values at the intersections; however, the drivers had lower mean diameter values overall and along the shopping strip. These results suggest that overall and across all three route sections cyclist networks comprised more concepts but with fewer connections between them compared to drivers and motorcyclists. This is indicative of taking more content from the road environment but making fewer connections between this content.

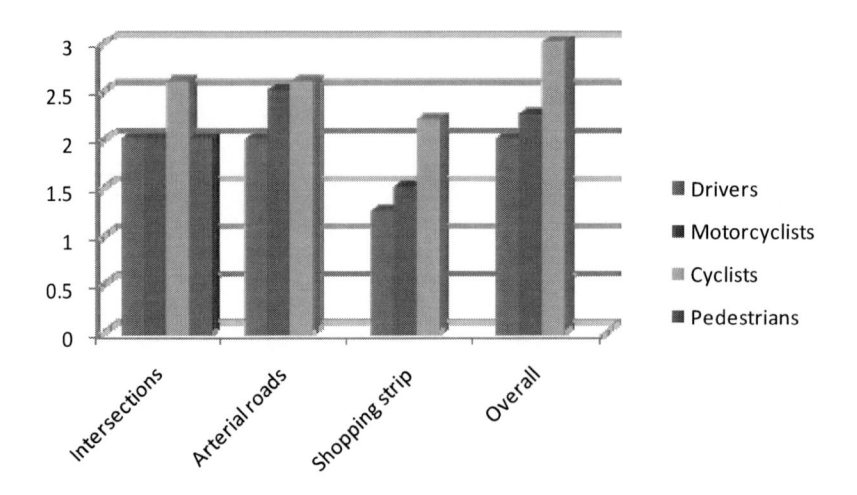

Figure 3. Mean diameter values for each road user group

Network content: common and unique concepts

The analysis of network content involved qualitative interrogation of the networks. This involved creating 'master' networks for each road user group; these were created by combining all concepts within each road user group together. This process led to four master networks being created for each road user group (apart from the pedestrians who had an 'intersection' master network only): an overall master network, an intersections master network, an arterial roads master network, and a shopping strip master network. Unique concepts (i.e. present only in one road user groups master network) and common concepts (i.e. that were present in all of the road user groups' master networks) were then identified for the overall route and for each route section.

Unique and common concepts

Table 1 shows the number of concepts within each master network along with the number and percentage of unique and common concepts.

Table 1. Common and unique concepts from master networks

	Intersections	Arterial roads	Shopping strip	Overall route
Drivers (unique concepts expressed as total number and % of drivers master network)				
Number of concepts	36	35	25	58
Unique concepts	9 (25%)	14 (40%)	7 (28%)	16 (27.6%)
Motorcyclists (unique concepts expressed as total number and % of motorcyclist master network)				
Number of concepts	50	43	19	63
Unique concepts	19 (25%)	22 (51.1%)	8 (42.1%)	20 (31.7%)
Cyclists (unique concepts expressed as total number and % of cyclist master network)				
Number of concepts	50	58	31	59
Unique concepts	21 (42%)	33 (56.8%)	16 (51.6%)	21 (35.6%)

Pedestrians (unique concepts expressed as total number and % of cyclist master network)				
Number of concepts	56	-	-	-
Unique concepts	28 (50%)	-	-	-
Common concepts (expressed as total number and % of combined master networks)				
Concepts common across all road user groups	12 (25%)	14 (30.8%)	8 (32%)	19 (31.6%)

Table 1 shows that, for the overall route and specific route sections, only between a quarter and a third of all concepts were common across the road user groups. That is, of all of the concepts underpinning SA, no more than a third were the same across all road user groups for the overall route, at the intersections, and along the arterial roads and shopping strip. Conversely, between a quarter and up to 57% of the concepts were unique to each road user group for the overall route and the three specific route sections. These finding suggests that, although around a third of concepts were similar across the master networks, SA whilst negotiating the entire route and each route section was different across the road user groups studied.

Distributed SA protagonists' argue that within collaborative sociotechnical systems, it is the level of compatibility between different portions of awareness that determines performance efficiency; that is, the extent to which different understandings of the situation connect together determines how well the system functions (Salmon et al., 2009). In systems where different situational understandings are incompatible and do not connect, performance breakdowns occur. In the road safety context then, it is important to understand how compatible different road user's awareness is. Does it connect together to support safe interactions between road users? Or are there incompatibilities which create conflicts between road users?

To explore the compatibility between the SA held by each distinct form of road user, the analysis now proceeds with a qualitative assessment of the unique and common concepts identified. This involved an assessment of the degree of compatibility between the four road user groups SA based on a review of the unique and common concepts identified. This was based on analyse judgement, for example, the presence of a 'filter' concept in the motorcyclists' SA network along with the absence of a 'side' (of car) concept in the drivers' network indicates that the two situational understandings are incompatible, since a motorcyclist filtering up the traffic queue may not be seen by a driver whose awareness does not include the traffic located to the 'side of their vehicle. The unique concepts, along with the concepts common across all road user groups, are presented in Tables 2 – 5 for intersections, arterial roads, shopping strip and overall, respectively. Following each Table discussion is presented on the extent to which the unique and common concepts identified suggest compatibilities and incompatibilities between the four road user groups.

Table 2. Shared and unique concepts at intersections

Shared across all road users	Unique to drivers	Unique to motorcyclists	Unique to cyclists	Unique to pedestrians
Behind	Follow	Taking	Hook	Parked
Traffic lights	Speed camera	Sit	Path	Attention
Look	Brake	Filter	Indicate	Street(s)
Traffic	Driving	Signals	Ready	Walk/walking
Green	Pedestrians	Gear	Clear	Work
Turning	Change (lights)	Turning arrow	Area	Trying
Road names	Aware	Sequence	Footpath	Looking
Road		Mirrors	Busy	Check
Intersection		Slowing	Line	Driveway
Cars		Opportunity	Heading	Crossing
Car		Coming	Service lane	Button
Lane		Indicating	Opposite	People
		Checking	Stick	Stand
		Cross		Man
				Now
				Flashing
				Approaching
				Waiting
				Island
				Fine
	Non relevant concepts: stay, ones	Non relevant concepts: Aren't, bloke, cause, take, doing	Non relevant concepts: We've, given, probably, spin, goes, haven't, normally, means	Non relevant concepts: supposed, wondering, case, further, anyway

As shown in Table 2, at intersections the concepts common across all road user groups were mainly related to the area behind the road users (e.g. 'behind'), the road (e.g. 'lane; 'road' and 'road name') and intersection itself (e.g. 'intersection'), surrounding traffic (e.g. car, cars, traffic) and the traffic lights and their status (e.g. traffic lights, green). There is evidence of compatibility between drivers and pedestrians, with drivers' having a 'pedestrians' concept and pedestrians having 'car', 'cars' and 'traffic' concepts. Evidence of incompatibilities is provided by the motorcyclists' 'filter' concept, which reflects their filtering behaviour at intersections (i.e. movement up between lanes of traffic to the front of the intersection), and the lack of a 'left/right side' concept present in the drivers' master network. Although drivers were focussed on the traffic approaching from behind (e.g. 'traffic' and 'behind' concepts), they were not focussed on the areas along either side of the vehicle. This suggests that, for the intersections studied, the drivers' schema was not anticipatory with regard to motorcyclists and cyclists filtering up the traffic queue. In addition, the motorcyclists concept 'signals' refers to the behaviours that they undertake to ensure that other road users were aware of their presence at the intersection (e.g. wiggling the bike, flashing brake lights). The lack of 'front' or 'side' concepts and presence of 'car/cars', 'follow' 'traffic lights' and 'change (lights)' concepts for drivers suggests that, regardless of motorcyclists' actions, the drivers schema were not directing sampling in the area of the intersection that motorcyclists would occupy and thus drivers may miss these signals. These findings suggest that, at intersections, there are likely to be instances where drivers are not aware of motorcyclists filtering up the traffic queue. Last minute lane changes by drivers could potentially lead to conflicts between them and filtering motorcyclists.

Table 3. Common and unique concepts along arterial roads sections

Freeway sections							
Shared across all road users		Unique to drivers		Unique to motorcyclists		Unique to cyclists	
Lane	Sign	Limits	80km/h	Riding	Speed	Door	Doors
Front	Lights	Pulled out	Rearview	Checking	Mirrors	Space	Room
Traffic	Road	Cameras	Looked	Slowing	Overtaking	Wide	Passed
Turning	Cars	Road names	Brake	Indicate	Time	Safety	Parked cars
Thinking	Car	Vehicle	Hand	Mirror	Head	Coming	Slip
		Change	Prepare	Wheel	Slow	Service lane	Look
				Trucks	People	Intersection	Footpath
				Full	Bike	Clear	Eye
				Blind		Roads	Lanes
						Pedestrians	Cruising
						Place	Lanes
						Keeping left	
Non relevant concepts: he's, I'm, sure, theres		Non relevant concepts: Hoping, I've		Non relevant concepts: Trying, probably, idea, means, looks		Non relevant concepts: Notoriously, chained, hurry, dumps, earlier, stick, round, ones, going, place	

Table 3 shows the unique and common concepts found for the arterial road sections. Common concepts here are related to the road itself (e.g. 'lane', 'road', 'sign'), other traffic (e.g. 'cars', 'car', 'traffic') and other traffic 'turning', the area in 'front' of the road users, traffic 'lights' and road users' thought processes (e.g. 'thinking'). There is evidence of compatibility, with drivers being focussed ahead, behind (e.g. 'rear-view') and also to the side of the vehicle (e.g. the concept 'hand' from left and right 'hand' side). The motorcyclists' unique concepts on the other hand are reflective of increased manoeuvrability and overtaking manoeuvres, with concepts such as 'overtaking', 'indicating', 'checking', 'mirrors' and 'mirror'. Importantly, they are focussed on other road users in their path (e.g. common concepts 'front', 'traffic' 'car' and 'cars'), and drivers are also focussed on other road users approaching from behind and to the side of their vehicle. Although both are negotiating the arterial roads differently, their SA is compatible since both are on the lookout for each other. There is also evidence of compatibility between the cyclists and other road users, since the cyclists' master network includes concepts regarding their position and space on the road (e.g. 'space', 'room') and other road users 'coming' towards them. This is in conjunction with drivers/motorcyclists' focus on objects in 'front' of them. Whilst drivers and motorcyclists are on the lookout for vehicles ahead of them, the cyclists are focussed on vehicles coming towards them from behind and also the amount of space between them and passing vehicles.

Table 4. Common and unique concepts along shopping strip section

Shopping strip section			
Shared across all road users	Unique to drivers	Unique to motorcyclists	Unique to cyclists
Car, Cars, Turning, Pedestrians, Ahead, Front, Behind, Lights	Brake, Driving, Straight, Yellow light, Sides, Slowly	Bus, Indicating, Change (lights), Watch, Make sure	Moving, Service lane, People, Giving, Time, Left hand side, Fast, Look, Doors, Door, Taxi, Eye (contact), Space, around (go)
	Non relevant concepts: Intend	Non relevant concepts: Guys, plenty, doing	Non relevant concepts: Important, erm,

Table 4 shows the unique and common concepts for the shopping strip section of the route studied. Common concepts here include other traffic (e.g. 'car', 'cars'), 'pedestrians', the area in front (e.g. 'ahead', 'front') and 'behind' of the road users, traffic 'lights' and traffic 'turning'. Again there is evidence of compatibility; all road users are focussed on the traffic in front and behind, and drivers are focussed on traffic either 'side' of the vehicle. This is compatible with the motorcyclist and cyclists filtering behaviours along the shopping strip; as the traffic moves along the strip in a slow queue, the motorcyclists and cyclists moved along the shopping strip by manoeuvring in and out of the traffic, both on the inside and outside of other vehicles (e.g. 'indicating' concept in the motorcycle network and 'go around' concept in the cyclists network). Whilst doing so they are focussed in 'front' and 'ahead' and, importantly, the drivers schema is anticipatory, directing attention not only ahead and behind, but also to either side of the vehicle.

Discussion

The aim of this exploratory study was to test the assumption that different road users interpret the same road situations differently, and to investigate the level of compatibility between SA across distinct road user groups. The analysis demonstrates that, during the on-road study, driver, motorcyclist, cyclist, and pedestrian's SA was different, both in terms of structure and content, when faced with similar road situations. With regard to network structure, the analysis suggests that the drivers' SA was more interconnected, in terms of relationships between concepts extracted from the driving situation for the overall route, the intersections, and the arterial road sections. For the shopping strip part of the route, motorcyclists' SA was more interconnected.

The hypothesis that the different road users would interpret similar road situations differently was further supported through the analysis of network content, which showed that SA was different across the road user groups studied. Over a quarter of concepts underpinning SA were unique to drivers, both for the overall route and for the three route sections. The amount of SA concepts unique to motorcyclists ranged from between a quarter and up to half of the concepts for the overall route and for the three route sections, whereas for cyclists the number of unique SA concepts ranged from between a third and over half of all concepts for the overall route and for the three route sections. For the pedestrians, half of the concepts underpinning SA whilst negotiating intersections were unique compared to the other road users. The assumption that different road users interpret the same situations differently is further supported by the fact that no more than a third of SA concepts were common across all road users for the overall route and three route sections. It is therefore concluded that, even when faced with similar road situations, the content of driver, motorcyclist, cyclist and pedestrian SA is different.

If SA differs across different road users, the pressing question is whether or not these different situational interpretations are compatible? That is, are distinct road users likely to be aware of other road users' presence and likely behaviour in the event that they occupy the same road situation? The compatibility analysis revealed instances of both compatibility and incompatibility along the route studied. On the arterial roads, evidence of compatibility was found between drivers and motorcyclists, with drivers being focussed ahead, behind and also to the sides of their vehicle. This was highly compatible with motorcyclists overtaking and manoeuvring in and out of traffic on the arterial roads. Evidence of a highly anticipatory schema, incorporating various observations and checking behaviours was also found for the motorcyclists. In short, drivers are on the look out for vehicles in front and approaching from behind and both sides, whereas motorcyclists are highly observant of events ahead of them whilst manoeuvring through traffic. The findings also suggest compatibility between the cyclists and other road users on the arterial roads, with the cyclists focussed on their own position and space between them and other passing vehicles, along with vehicles approaching from behind, and the drivers and motorcyclists focussing on objects ahead and to the sides. Along the shopping strip evidence of compatibility was also found; all road users were focussed on the traffic in front and behind; however, importantly drivers held up in traffic queues were focussed on traffic approaching on either side of the vehicle, which is compatible with motorcyclist and cyclists filtering behaviours (manoeuvring up through the traffic queue via the sides of slow moving vehicles). Finally, some evidence of compatibility between drivers and pedestrians was found at the intersections, with drivers' being on the look out for pedestrians and vice versa.

Evidence of incompatibility between the road users was identified at the intersections studied. Specifically, the findings point to incompatibilities between where drivers are focussing their attention (i.e. traffic behind) and motorcyclists' and cyclists' lane filtering behaviours. Despite the high likelihood that other road users will move up through traffic queues, drivers' schema orients their attention to traffic approaching directly behind them, the traffic lights, and the traffic directly in-front of them. The analysis demonstrated that the drivers studied were not focussed on traffic to the sides of them, which is likely to be problematic in the presence of filtering motorcyclists and cyclists travelling in the left hand lane.

A logical next step in this research program is to investigate means of enhancing the compatibility between different road users' SA. Potential avenues for this purpose include the use of cross mode training (whereby different road users receive training in how other road users interpret road situations; Maguzzi et al., 2006), road design (e.g. filter lanes for motorcyclists, signage warning of the likelihood of motorcyclists filtering) and educational road safety campaigns (e.g. encouraging drivers to focus on the sides of their vehicle at intersections).

References

Crundall, D., Crundall, E., Clarke, D., & Shahar, A. (2012). Why do car drivers fail to give way to motorcyclists at t-junctions? *Accident Analysis and Prevention, 44*, 88-96.

Endsley, M.R. (1995). Towards a theory of situation Awareness in dynamic systems, *Human Factors, 37*, 32 - 64.

Magazzù, D., Comelli, M., & Marinoni, A. (2006). Are car drivers holding a motorcycle licence less responsible for motorcycle-car crash occurrence? A non-parametric approach. *Accident Analysis & Prevention, 38*, 365–370.

Neisser, U. (1976). *Cognition and reality: principles and implications of cognitive psychology*. San Francisco: Freeman.

Pai, C.W. (2009). Motorcyclist injury severity in angle crashes at T-junctions: identifying significant factors and analysing what made motorists fail to yield to motorcycles. *Safety Science, 47*, 1097-1106.

Salmon, P.M., Stanton, N.A., & Young, K.L. (2011). Situation awareness on the road: review, theoretical and methodological issues, and future directions. *Theoretical Issues in Ergonomics Science. I-First, 1 - 24*

Salmon, P.M., Stanton, N.A., Walker, G.H., & Jenkins, D. P. (2009). *Distributed situation awareness: advances in theory, measurement and application to teamwork*. Ashgate, Aldershot, UK.

Shahar, A., Poulter, D., Clarke, D., & Crundall, D. (2010). Motorcyclists' and car drivers' responses to hazards. *Transportation Research Part F: Traffic Psychology and Behaviour, 13*, 243-254.

Stanton, N.A., Salmon, P.M., Walker, G.H., & Jenkins, D.P. (2010). Is situation awareness all in the mind? *Theoretical Issues in Ergonomics Science, 11*, 29-40.

Stanton, N.A., Walker, G.H., Young, M.S., Kazi, T., Salmon, P.M. (2007). Changing drivers' minds: the evaluation of an advanced driver coaching system. *Ergonomics, 50*, 1209-1234.

Walker, G.H., Stanton, N.A., Salmon, P.M. (2011). Cognitive compatibility of motorcyclists and car drivers. *Accident Analysis & Prevention, 43*, 878-888

Walker, G. H., Stanton, N. A., Kazi, T. A., Salmon, P. M., & Jenkins, D. P. (2009). Does advanced driver training improve situation awareness? *Applied Ergonomics, 40*, 678-687.

Development of a model predicting the use of automated decision aids

Rebecca Wiczorek & Linda Onnasch
Berlin Institute of Technology
Berlin, Germany

Abstract

Common theories of operators' use of assistance systems often involve the concept of trust. It is assumed that trust guides operators' behaviour. They develop a certain level of trust towards the system based on the characteristics of the automation. This amount of trust should be appropriate in order to use the automation properly. However, some studies did not find empirical evidence for this mediator role of trust. A model which is taking these contradictory findings into account and which is able to predict the usage of an assistance system is missing so far. Based on a narrative analysis of five experimental studies investigating the effect of decision aids on human behavior, we have developed a model predicting the use of decision aids. The model does not only focus on the technical components and the level of trust, but also takes the perceived risk of the anticipated output and the perceived interpretability of the situation, i.e. uncertainty into account. The model assumes that risk and interpretability have a strong influence on using an assistance system. Furthermore it is assumed that they moderate the mediator role of trust. Practical implications are made, regarding the importance of multiple sources of information to reduce operators' uncertainty. Empirical studies will show the influence of interpretability and risk on operators' behaviour and operators' trust towards the system.

Introduction

The automation of functions prior performed by human operators is continuously increasing especially in safety related work domains such as aviation or process industry. The use of automation can enhance productivity as well as safety because of faster and more precise operation methods even under dangerous conditions. However, as the implementation of automation changes operators' work environment and practices, it also leads to new challenges in regard to operator-automation interaction. A broad range of research deals with the different problems arising from operators' new role as supervisors (Sheridan, 1997) of the human-machine system. One aspect that seems to play an important role for the efficient and safe interaction of humans with sophisticated technical systems is operators' trust in the automation they use.

In D. de Waard, N. Merat, A.H. Jamson, Y. Barnard, and O.M.J. Carsten (Eds.) (2012). *Human Factors of Systems and Technology* (pp. 51 - 61). Maastricht, the Netherlands: Shaker Publishing.

The concept of human-automation trust derives from the concept of human-human trust (e.g. Lee & Moray, 1992; Muir, 1994; Lee & See, 2004; Madhavan & Wiegmann, 2004). In the wide field of trust research different definitions of trust are used. These definitions mainly differ regarding the nature of trust. It may be seen as a belief, an attitude, a behavioural intention, or the behaviour itself. Dependent on the diverse definitions, the concept of what trust actually is, how it develops, and how it may affect behaviour are rather different. In this paper, it is assumed that trust is an attitude, which is based on characteristics of the automation and affects operators' behaviour in terms of automation use. It is therefore defined as:" […] the attitude that an agent will help achieve an individual's goal in a situation characterized by uncertainty and vulnerability" (p.54; Lee & See, 2004).

Trust is seen as one of the fundamental elements (Madhavan, Wiegmann & Lacson, 2006) of human-machine interaction, which guides operators' reliance on the automation (Lee & See, 2004). A lack of trust can lead to operators' disuse of automation, whereas over-trust has the potential to result in misuse of automation in terms of overreliance (Parasuramann & Riley, 1997).

Several models (e.g. Muir, 1994, Madhavan & Wiegmann, 2004; Lee & See, 2004) describe the development of trust and its evolution over time. Most of them postulate that trust is build on the basis of certain system characteristics. That can be, for example the reliability, utility, or transparency (e.g. Lee & Moray, 1992; Muir, 1994) of the technical agent. The level of trust on its part guides operators' behaviour i.e. the way and frequency of automation use. The frequent use influences trust again; therefore, trust develops and may change over time. Lee & See (2004) point out, that there is a need for appropriate trust, because it would lead to appropriate automation use. Appropriate trust is defined as dependent of automation characteristics. One of the most important characteristics is the reliability of the automation. Poor reliability leads to less trust and less automation use, whereas high reliable systems are trusted more by the operators, who in consequence make more use of the automation.

Several studies have shown the importance of trust in human-machine interaction in different domains such as aviation, automotive or process industry (e.g. Lee & Moray, 1992; Riley, 1996; Madhavan et al., 2006; Lee & Lees, 2007). For example, Madhavan et al. (2006) found that when automation errors occurred in a task which participants perceived as easy, the resulting trust ratings were lower compared to their expressed trust in an automation that failed in a more difficult task. The degree of trust was a better predictor of system use than was the system's reliability. However, other studies led to contradictory findings as they failed to find a mediating effect of trust on the relation between system characteristics and participants' use of the systems (Bustamante, 2009, Wiczorek & Manzey, 2010).

Common models (e.g. Muir, 1994, Madhavan & Wiegmann, 2004; Lee & See, 2004) lack the ability to explain these contradictory findings. Most of the trust models focus more on the development of trust and its evolution over time rather than on the actual effect of trust on behaviour. Only a few models exist, which incorporate the prediction of automation use as for example the model of Parasuraman & Mouloua (1996). This model embodies thirteen factors which influence operators' behaviour directly and indirectly. The model has undoubted theoretical value, but does not seem applicable for an empirical investigation of the effects of trust on behaviour in order to understand the cause of contradictory results. Nevertheless it raises the question of relevant factors influencing the use of automation. If trust does not always serve as predictor of operators' behaviour, it may be due to other factors which have a similar strong, or sometimes even stronger, relevance for decisions about whether or not to use automation.

In order to extract such potential relevant factors which could influence operators' behaviour, we conducted a narrative analysis. The analysis was done in two steps with five studies regarding human-automation interaction with the focus on reliability. Based on the results of this analysis, we developed a theoretical framework more suitable for the empirical investigation of factors influencing operators' use of automation.

Narrative analysis

The studies used for the analysis were all conducted in our laboratory. For the first four studies, the PC-based simulation environment M-TOPS (Multi-Task Operator Simulation, see Domeinski, Wagner, Schoeble, Manzey, 2007) was used. With M-TOPS participants have to fulfil tasks that simulate the cognitive demands required by tasks in a control room of a chemical plant. One task is supported by automation (i.e., alarm system). The number of secondary tasks varies from one to two. The last study was conducted using the simulation environment AuotoCAMS 2.0 (Manzey et al., 2008) which simulates a live support system of a space craft. In this environment participants have to detect failures in the underlying system. Subjects were supported by an automated decision aid and in addition they had to fulfil two secondary tasks. The two conditions of this study have been analyzed separately; therefore six "cases" are shown in Table 1 and Table 2. In the first four studies, the reliability of the alarm system was varied between groups. The reliability ranged from 0.1 to 0.9. In the last study, the subjectively perceived reliability of decision aids was manipulated.

Except for one study trust was not assessed in a systematic way or it was assessed after the experimental block which did not allow calculating mediator analyses because causality could not be proven. Lacking these systematic trust ratings, we based the first step of our analysis on the assumed linear relationship of system characteristics; namely, reliability, and the resulting behaviour in terms of frequencies of compliance, reliance, and cross-checking behaviour. It was assumed that trust changes in a linear way if reliability changed. Moreover, the different levels of reliability should influence participants' use of automation in a linear way as well. Based on these assumptions, we formulated expectations regarding the outcome behaviour for every experiment. We did a comparison of expected outcomes and the behaviour shown by participants. Table 1 shows the results of this comparison.

Table 1. First step of the narrative analysis: comparison of expected and real behaviour of participants in experimental studies regarding automation use

Empirical Studies	Automation reliability	Expected Behaviour	Behaviour
1) Gérard & Manzey (2010)	Variation	Uncertainty reduction[1]/ Probability Matching[2]	Over-checking[3]
2) Gérard & Schmuntzsch*	Variation	Uncertainty reduction/ Probability Matching	Uncertainty reduction
3) Gérard & Zorn*	Variation	Uncertainty reduction/ Probability Matching	Over-checking
4) Wiczorek (2009)	Variation	Probability Matching	Extreme responding[4]
5) Manzey, Reichenbach & Onnasch (2008)	High	Incomplete checking	Complete checking
	High	Incomplete checking	Incomplete checking

*Gérard (2011) Ph.D. dissertation
[1] to cross-check more if uncertainty is high, e.g. medium level of reliability
[2] to match response frequency to system reliability
[3] to cross-check almost every advice
[4] to respond to all or no advices depending of the reliability

This comparison revealed differences in expected and real behaviour in four out of the six cases. For the remaining two cases, participants' variation in response frequencies corresponded in a linear way to the variation of reliability. It seems that in these two studies the reliability had the main effect on behaviour, probably mediated by the trust towards the system. In the other four cases there might have been other relevant factors which had a stronger influence on automation use. These results correspond to the earlier contradictory findings regarding the mediating role of trust, as the underlying factor "reliability" did not take effect in any of these cases.

To gain greater insight in other factors which could have influenced participants' behaviour we considered other aspects, i.e. environmental parameters that had been varied in the experiments and could be relevant. These aspects are presented in Table 2. To avoid the problem of multiple single parameters, we clustered the different aspects with regard to their possible influence on participants' perception of relevant characteristics of the situation which represents the basic foundation for their decision making and behaviour.

Table2. Second step of the narrative analysis: clustering relevant environmental parameters to factors influencing automation use.

	Environmental Parameters									
	Cross check option	Access to info	Ambiguity of info	Time constrains	Context	Training	Expe-rience	Payoff	Cover story	Dynamic
1	Yes	M	No	M	IC	30 min	ES	PD	CP	M
2	Yes	M	No	H	IC	30 min	ES	PD	CP	H
3	Yes	D	M	M	IC	30 min	ES	PD	CP	M
4	No	No	No	M	IC	30 min	S	PD	CP	M
5	Yes	E	No	H	IC	4 h	ES	G	LSS	H
	Yes	D	M	H	IC	4 h	ES	G	LSS	H
	situation-related				operator-related					
	FACTOR 1: Perceived interpretability of the situation							**FACTOR 2: Perceived risk**		

M =Medium
D =Difficult
E=Easy
H=High
IC=Indoor Conditions (laboratory)

ES=Engineer Students
S=Students
PD=Performance dependent
G=Global
CP=Chemical Plant
LSS = Living Support System

It turned out that seven parameters contributed in different ways to the amount of information regarding the situation participants could obtain. As decisions under uncertainty often lead to the use of heuristics (Bliss, 2003), which do not improve performance, operators tend to reduce their uncertainty if possible. Therefore we summarized these parameters to one factor named "perceived interpretability of the situation". This factor includes the two components "situation-related" and "operator-related", which can be seen as different causes that lead to the same consequence. Whereas this factor influenced on participants *capability* of decision making, the other three parameters were more related to the consequences of the decisions. These parameters have been summarized to the factor "perceived risk".

The two influencing factors

The two extracted factors are the perceived interpretability of the situation and the perceived risk. Operators' perception of risk is based on the expected negative outcomes of erratic behaviour. Therefore, they have to integrate the value of the outcome and the probability of its occurrence, which has been described as the "expected value theory" (Meyer, 2004). When the likelihood of a negative outcome increases, the risk is perceived as higher. Likewise when the resulting consequences become more negative, the perceived risk increases. In laboratory settings, the perception of risk depends on and can be manipulated with the help of cover stories, payoffs, and the dynamics of the task itself.

The interpretability of the situation consists of two components: situation-related and operator-related aspects. The perceived interpretability refers to the accessibility of information relevant for the situation as well as to operators' own ability to understand and interpret such information. The degree of interpretability influences operators' ability to validate the automations' advices. When further information such as raw data is available, the access to such data is possible and not limited by time constraints and when operators are well trained to understand the specific nature of this information, they can use it to make a cross-check. With such a cross-check option operators have the possibility to verify advices of the automation and to base their decisions on multiple sources rather than only on the information generated by the automation. Situation-related aspects which may limit the interpretability are availability of other information, quality of access to the information, context factors such as weather conditions, workload, and other time constraints in general. Operator-related aspects such as experience, training, and self-confidence constitute the other component of that factor and are necessary for a high degree of interpretability. In laboratory settings, it is possible to vary operator-related aspects by the use of different instructions or different trainings. Situation-related aspects can be varied by changing context conditions (e.g. introducing cross-check options), ambiguity of presented information, or time constraints and workload.

Model predicting the use of automated decision aids

The model aims to predict operators' use of automation, including system characteristics, perceived risk, perceived interpretability and trust in the system as relevant factors. The model is based on the assumption that perceived risk and perceived interpretability have a stronger impact on behaviour then does the reliability. That is true only if both of these two factors reach extreme values (low or high). Under these conditions certain behavioural actions are shown independently of the reliability of the automation. Moreover the model hypothesises the influence of these two factors on operators' trust towards the automation. Contradictory findings regarding the predictor role of trust can be explained with the effect of risk and interpretability on trust. For example operators normally would trust a highly reliable system, but in a high risk environment their trust can be reduced because they also consider the potential negative consequences of erratic behaviour. In such situations the perceived trust does no longer serve as a mediator. That is because the mediation of trust is moderated by the perceived risk and the perceived interpretability of the situation. Trust serves as predictor only, if the intensity of control-tendency which is due to the degree of perceived risk and perceived interpretability corresponds to the level of trust. Figure 1 illustrates these relations between the different factors and automation use.

predicting the use of automated decision aids

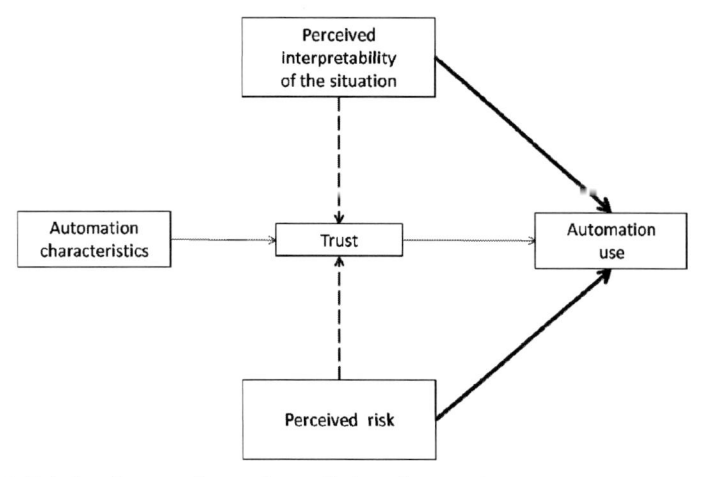

[Figure: Diagram showing "Perceived interpretability of the situation" at top, "Automation characteristics" at left, "Trust" in center, "Perceived risk" at bottom, and "Automation use" at right, with arrows connecting them.]

Figure 1. Relations between factors for prediction of automation use.

To summarize the predictions of the model, one can say that if perceived risk and perceived interpretation of the situation are high, operators tend to verify automation advices by cross-checking with other information. If on the other hand perceived risk and perceived interpretability are low, operator' diminish their cross-checking behaviour. Both happens *independent* of the reliability of the system and operators' sensation of trust in the automation. The consequence of high risk and easily available information may be an over-checking behaviour, which is often rather time consuming. In the opposite case when risk is perceived as very low and the access to information is perceived as being difficult operators will tend to rely on the system even if it is not trustworthy because of its low reliability. Likewise this combination of low risk, low interpretability and low reliability can lead to the cry wolf effect (Breznitz, 1983) when operators ignore the system's advices.

Discussion

The aim of this work was to develop a model which predicts operators' use of automated decision aids. The number of relevant aspects influencing operators' behaviour should be limited to a degree that allows an empirical validation of the model. Furthermore the model should be able to explain contradictory findings of former studies dealing with the influence of trust.

Based on a narrative analysis of five studies conducted in our laboratories, we extracted two relevant factors which influence the use of automation and moderate the relationship of trust on operators' behaviour. The two factors perceived risk and perceived interpretability of the situation which contain an operator-related and a situation-related aspect are influencing operators' tendency to cross-check further information and their level of trust in the system.

The model meets the criteria of explaining feigned contradictions of results of earlier research, because it assumes a mediation effect of trust on the influence of system characteristics on automation use and nevertheless describes situations where trust does not serve as a predictor of the behaviour. Due to the limited number of relevant factors the model allows empirical validation of its predictions. In addition, predictions of the model raise some questions with regard to the transfer of concepts from one discipline to another.

As the relation between operators and automation is often compared with operators' relation to their human co-workers (e.g. Christoffersen & Woods, 2002), it makes sense to adopt concepts derived from human-human interaction and use them to explain and analyze human-machine interaction (e.g. Lee & Moray, 1992; Muir, 1994; Lee & See, 2004). However, the parallels between these types of cooperation should not conceal the existence of some general differences between humans and machines. Lee & See (2004) mention, that automation itself never has an intention and that no reciprocity exists in the development of trust in each other because of its unidirectional nature. But the most important notation the authors make is that no joint responsibility with regard to human-automation interaction exists. Operators make the final decisions and they are being held accountable in case of emergency. With this notation in mind, it may seem rational as well as reasonable if operators do not only rely on automation but try to verify the information generated by the machine even in case of high trust. This is especially true if the perceived risk of the situation is high. Unlike human operators, machines never resent operators when they cross-check other information to reduce their uncertainty instead of relying on the - probably even highly trusted - automated co-worker.

If trust in the technical agent represents the only base for decision making and therefore guides operators' behaviour, it is likely to result in misuse and in case of low trust also in disuse of automation. Especially in the latter case, it is doubtable that design for appropriate trust is needed as often required, if appropriateness of trust is defined as "[...] the match between the trust and the capabilities of automation [...]" (Lee & See, 2004; p.72). If reliability of automation is low in terms of generating many false alarms, a low level of trust seems most appropriate but the often resulting cry wolf effect is not intended by the designers and rather dangerous. Therefore, the level of trust in the automation is probably less important than operators' actual behaviour towards the machine. Notwithstanding the relevance of the subject, intense research interest in human-machine trust should not take place at the expense of the investigation of human-machine interaction itself. The most important requirement should rather be to design for appropriate use.

One important condition for a more appropriate use of automation is to make sure that reliability of the automation does not represent the only source of available information. With the increased introduction of sophisticated technical components, a lot of information disappeared which was available for the operators in the time without automation support. In the more direct interaction with ongoing processes operators obtained a lot of information through the simple perception of noise, vibration, smell, and specific visual cues (Zuboff, 1988). The different types of information were integrated and furnished the base for operators' decisions. Nowadays information available for the operators is often reduced because it has already been integrated by the automation. If the output of the automation remains the only source of information, operators are forced to base their decisions on their trust in the machine. Those situations are often characterized by uncertainty and may result in unsafe behaviour such as ignoring automations' advices. To reduce operators' uncertainty and to enable them to work consciously and safe, it is important to offer different sources of information, if automation is not perfectly reliable, which is true in most of the cases (Macmillen & Creelman, 2005). That is especially true in safety related work environments where operators often perceive a high risk.

Thus, the most important practical implication that can be derived from the model is to rise up operators' perceived interpretability of the situation. Whereas it is not possible to reduce the risk of the situation in most of the cases, interpretability can be increased with the help of engineers and designers of technical systems. When operators are given the possibility to cross-check information using different sources in a non-time consuming manner, they tend to use that additional information to reduce their uncertainty. Furthermore, an intense education and continuous training are needed to enable operators to understand the available information. A better understanding of the available information will enhance the quality of their decision-making.

The model emphasises the relevance of the perceived risk and the perceived interpretability of the situation for operators' behaviour. The next phase will be the empirical validation of its predictions, especially with regard to the interdependence of the two factors and their respective relevance for operators' control behaviour.

References

Breznitz, S. (1983). *Cry wolf: The psychology of false alarms*. Hillsdale, N.J., USA: Lawrence Erlbaum Associates.

Bustamante, E.A. (2009). A reexamination of the mediating effect of trust among alarm systems' characteristics and human compliance and reliance. *Proceedings of the Human Factors and Ergonomics Society 53th Annual Meeting* (pp. 249-253). Santa Monica, CA: HFES.

Christoffersen, K. & Woods, D. D. (2002). How to make automated systems team players. In E. Salas (Ed.), *Advances in Human Performance and Cognitive Engineering Research. Vol. 2. Automation.* (pp. 1-12). Burlington: Elsevier.

Dixon, S.R., Wickens, C.D., & McCarley, J.S. (2007). On the independence of compliance and reliance: Are automation false alarms worse than misses? *Human Factor, 49*, 564-572.

Domeinski, J., Wagner, R., Schöbel, M., & Manzey, D. (2007). Human redundancy in automation monitoring: Effects of social loafing and social compensation. *Proceedings of the Human Factors and Ergonomics Society 51st Annual Meeting* (pp. 587-591). Santa Monica, CA: HFES.

Gérard, N. & Manzey, D. (2010). Are false alarms not as bad as supposed after all? A study investigating operators' checking behaviour in response to imperfect alarms. In D. de Waard, A. Axelsson, M. Berglund, B. Peters, and C. Weikert (Eds.) *Human Factors: A system view of human, technology and organisation* (pp. 55 - 69). Maastricht, the Netherlands: Shaker Publishing.

Gérard, N. (2011). *Verhaltenseffektivität von Alarmen: Experimentelle Untersuchungen zum Einfluss von Reliabilität und Prüfmöglichkeit auf die Anwendung von Heuristiken [Behavioural effects of alarm systems: Experimental investigation regarding the influence of reliability and cross-check options on the use of heuristics].* Unpublished Ph.D. dissertation, Berlin Institute of Technology, Germany.

Lees, M.N. & Lee, J.D. (2007). The influence of distraction and driving context on driver response to imperfect collision warning systems. *Ergonomics, 50,* 1264 - 1286

Lee, J.D. & Moray, N. (1992). Trust, control strategies and allocation of function in human-machine systems. *Ergonomics, 35,* 1243 – 1270.

Lee, J.D. & See, K.A. (2004). Trust in automation: Designing for appropriate reliance. *Human Factors, 46,* 50 - 80.

Macmillan, N. A. & Creelman, C.D. (1991). *Detection theory: A user's guide.* Cambridge: Cambridge University Press.

Madhavan, P. & Wiegmann, D.A. (2004). A new look at the dynamics of human-automation trust: Is trust in humans comparable to trust in machines? *Proceedings of the Human Factors and Ergonomics Society 48th Annual Meeting* (pp. 581-585). Santa Monica, CA: HFES.

Madhavan, P., Wiegmann, D.A., & Lacson, F.C. (2006). Automation failures on task easily performed by operators undermine trust in automated aids. *Human Factors, 48,* 241-256

Manzey D., Bleil M., Bahner-Heyne J.E., Klostermann A., Onnasch L., Reichenbach J., & Röttger, S. (2008), *AutoCAMS 2.0 Manual.* Berlin: FG AIO Psychologie.

Manzey, D., Reichenbach, J. & Onnasch, L. (2008). Performance-consequences of automated aids in supervisory control: The impact of function allocation. *Proceedings of the Human Factors and Ergonomics Society 52nd Annual Meeting* (pp. 297-301). Santa Monica, CA: HFES.

Meyer, J. (2001). Effects of warning validity and proximity on responses to Warnings. *Human Factors, 43,* 563-572.

Meyer, J. (2004). Conceptual issues in the study of dynamic hazard warnings. *Human Factors, 46,* 196-204.

Muir, B.M. (1994). Trust in automation: Part I. Theoretical issues in the study of trust and human intervention in automated systems. *Ergonomics, 37,* 1905-1922.

Parasuraman, R., & Mouloua, M. (Eds.). (1996). *Automation and human performance: Theory and applications.* Hillsdale, NJ: Erlbaum.

Parasuraman, R., & Riley, V. (1997). Humans and automation: Use, misuse, disuse, abuse. *Human Factors, 39,* 230-253.

Rice, S.R. (2009). Examining single- and multiple-process theories of trust in automation. *The Journal of General Psychology, 136,* 303-319.

Riley, V. (1996). Operator reliance on automation: Theory and data. In R. Parasuraman and M. Mouloua (Eds.), *Automation theory and applications* (pp. 19–35). Mahwah, NJ: Erlbaum.

Sheridan, T.B. (1997). Supervisory control. In G. Salvendy (Ed.), *Handbook of Human Factors & Ergonomics.* (pp. 1295-1327). New York: Wiley.

Wiczorek, R. (2009). *Alarmmanagement im Mensch-Maschine-System: Eine experimentelle Untersuchung zum Einfluss der Basisrate kritischer Ereignisse auf den Umgang mit Alarmen. [Alarm management in the human-machine system: an experimental study about the influence of the base rate of critical events on the handling of alarms]*.Unpublished M.Sc. Thesis, Berlin Institute of Technology, Germany.

Wiczorek, R. & Manzey, D. (2010). Is operator's compliance with alarm systems a product of rational consideration? *Proceedings of the human Factors and Ergonomics Sociaty 54th Annual Meeting,* (pp. 1722 – 1726). Santa Monica, CA: HFES.

Zuboff, S. (1988). *In the age of smart machines: The future of work technology and power.* New York: Basic Books.

The effects of preliminary information about adaptive cruise control on trust and the mental model of the system: a matched-sample longitudinal driving simulator study

Matthias Beggiato & Josef F. Krems
Chemnitz University of Technology
Chemnitz, Germany

Abstract

Advanced driver assistance systems, such as adaptive cruise control (ACC), aim to support the driver by automating driving subtasks, for example, speed and distance control. In order to adequately make use of such systems in a safe manner, a correct mental model of the system's functionality is required. The present study investigated the effects of preliminary information about an ACC on the mental model and trust in the system over time. A matched sample of 51 participants drawn from 396 applicants was assigned to three experimental conditions. Every group received one of three different descriptions of an ACC, realistic, idealistic and wrong. The realistic scenario informed participants of all potential system failures; the idealistic one contained no information about possible failures; and the wrong scenario gave additional information about potential failures that, however, did not occur. All participants drove the same 56-km track of highway in a driving simulator three times within 6 weeks. Results of the sampling process, effects of preliminary information on the mental model as well as changes in trust over time are presented.

Introduction

From the perspective of cognitive psychology, driving a car is a very complex task. A driver has to perceive and understand all of the relevant characteristics of a situation to make appropriate decisions about taking the best course of action. Advanced driver assistance systems (ADAS), such as adaptive cruise control (ACC), aim to support drivers by automating driving subtasks such that comfort and safety are enhanced. ACC partially automates speed and distance control by maintaining a driver-set vehicle speed. Additionally, the system adjusts vehicle speed in relation to a preceding vehicle, in order to maintain a pre-selected time headway. However, ACC detection problems may occur in some situations, such as in navigating narrow bends, in bad weather conditions or when small vehicles precede the driving vehicle. In these cases driver intervention is required. The positive effects of the automated system may be furthermore diminished or even inverted by unintentional changes in user behaviour. Behavioural adaptation, which is induced by the use of automated ADAS, is a well-known phenomenon. However, results are heterogeneous regarding

In D. de Waard, N. Merat, A.H. Jamson, Y. Barnard, and O.M.J. Carsten (Eds.) (2012). *Human Factors of Systems and Technology* (pp. 63 - 74). Maastricht, the Netherlands: Shaker Publishing.

conditions, influencing factors and direction of changes and, therefore, further research is required to understand the factors underlying this adaptation (Cacciabue & Saad, 2008). A lack of knowledge is apparent in the relation between the appropriateness of mental models, system trust and behavioural changes over time (Saad, 2007).

Several studies address the connection between mental models, trust and changes in driving behaviour in response to ACC use (Cahour & Forzy, 2009; Kazi et al., 2007; Rajaonah et al., 2008; Rudin-Brown & Parker, 2004). However, knowledge of learning processes and influences of different forms of preliminary information on the development of the mental model, trust and driving behaviour, is incomplete. Empirical evidence shows that many ACC users are not aware of all the possible situations where an ACC could fail: Jenness et al. (2008) report that 72% of ACC users are not aware of any system limitations. Similar tendencies can be observed regarding frequency and intensity, on reading the owner's manual for automotive vehicles. Mehlenbacher et al. (2002) report that approximately 41% of respondents declared not having read the manual at all and the rest read on average 50% of the owner's manual. Therefore, it cannot be assumed that all users acquire an adequate mental model about an ACC system's structure and functionality before using it. However, the initial mental model formed by drivers constitutes the basis for building system trust and supports interactions with the system. Consequently, this initial mental model influences driving behaviour. Different driving behaviour, in turn, leads to different experiences and affects system trust and further shapes the mental model (Schömig et al., 2011).

The construction and continuous updating of a mental model according to experience is theoretically based on the concept of "situation awareness". This concept originates from the field of aviation and describes human behaviour in complex and dynamic environments (Endsley, 2004). Recent approaches define situation awareness as a comprehension process, analogous to theories of reading comprehension (Durso et al., 2007; Krems & Baumann, 2009). According to this view, the perception of new elements in a situation (bottom-up processes) activates associated knowledge structures in long-term memory (top-down processes). This process forms the actual "mental situation model", which triggers the activation of coherent actions, which in turn changes the situation, and subsequently demands updating the situation model. As the amount of available information exceeds the capacity of working memory, an additional mechanism is postulated: long-term working memory stores the excess information, but provides it immediately when it becomes relevant. The schemata stored in long-term memory encodes experiences of different driving situations and therefore also knowledge about the functions of assistance systems (Endsley & Garland, 2000). The role that the ADAS mental model plays is important in the construction and maintenance of an adequate situation model, as well as in selecting appropriate actions (Cotter & Mogilka, 2007; Seppelt & Lee, 2007). Analogous to findings in reading comprehension research (Zwaan et al., 1995), an inappropriate mental representation hinders and delays the comprehension process. Moreover, experiences that demand reorganisation of the mental model due to the addition of new information should result in higher cognitive load (processing-load hypothesis: Zwaan et al., 1998) and a stronger negative impact on trust due to the mismatch between expected ADAS functionality and real system behaviour (Lee & See, 2004).

To organise relevant factors influencing behavioural changes in response to ADAS use, the conceptual driver appropriation model developed in the HUMANIST project is proposed (Cotter & Mogilka, 2007). The model integrates and extends previous theories (e.g. Rudin-Brown & Noy, 2002; Weller & Schlag, 2004) and is mainly used to control relevant confounding variables in this study. Based on a broader definition of "driver appropriation", the model includes cognitive, regulatory and motivational processes underlying observable behaviour, as well as temporal interactions. Characteristics of the system, the situation and the driver are taken into account. The most important driver factors are age, gender, driving experience, experience with the system, attitudes towards the system, driving style, cognitive abilities (primarily perceptual speed) and personality characteristics including locus of control and sensation seeking. Locus of control as individual assumptions regarding the responsibility for the outcome of events is considered important for the usage of assistance systems whereas sensation seeking is correlated with risky driving (Rudin-Brown & Parker, 2004).

The present study aims to assess the effect of different initial mental models of ACC on system trust, and the development of trust over time. Preliminary information with a different number of potential system problems stated is provided to induce the initial mental model. It is assumed that:

1. More information about potential system failures in the preliminary description should lead to lower initial trust in the ACC.
2. After gaining experience with the system, trust should steadily increase for the group presented with the appropriate system description, due to correspondence of their mental model and experience. The same result is expected when drivers are presented with information about potential failures that do not, however, actually occur during driving. Initial trust should be lower due to the higher number of expected failures. However, due to the assumed weaker impact of expected but not-experienced potential failures, trust should reach almost the same level over time as the group with the appropriate mental model.
3. An inverse relationship of trust development is predicted in cases of missing information about potential failures that really do occur during driving. Trust is expected to decrease steadily over time because of the stronger impact of unanticipated, but experienced, failures.

Method and materials

Sampling and participants

The simulator study was designed as a matched-sample study to control the effects of relevant confounding driver variables, based on the conceptual model of Cotter and Mogilka (2007). The model factors "system" and "situation" were controlled for by providing the same ACC functionality and the same simulator track for all participants. Sampling was conducted using a multi-stage procedure. In the first stage, an invitation to complete an online questionnaire was sent to 31 public mailing lists of Chemnitz University. In order to participate in this study, applicants had to be younger than 30 years of age and in possession of a valid driver's licence. The sample age limit (i.e., young drivers) was set in order to ensure a constant age range and to eliminate this variable as a confounding factor. Survey questions focussed on knowledge and experience using different ADAS, cumulative driving experience (i.e., km driven since obtaining a driver's licence), demographic information (e.g., gender, age and occupational status) and finally contact details for further correspondence.

In total, 396 prospective participants applied for the study. A first filtering was done by removing participants with ACC experience or extensive knowledge about the system, to avoid different initial conditions. In the next selection step, 3 groups of 17 participants each (51 in total) were defined and assigned to group membership such that no significant statistical differences in gender, age and driving experience resulted. The final sample consisted of 25 male and 26 female participants with a mean age of 23.8 years (SD 2.37) and an average total driving experience of 52,000 km (SD 47,000). They filled in a second online questionnaire including the following factors:

- Personality: 10-item version of the Big Five Inventory (BFI-10, Rammstedt & John, 2007)
- Locus of control (LOC): driving internality-externality scale (Montag & Comrey, 1987)
- Sensation seeking (SS): brief sensation seeking scale (BSSS, Hoyle et al., 2002)
- Driving style: 11 bipolar items (Popken, 2009) resulting in two scales "carefulness" and "routine"

Based on results of both online questionnaires, final group membership was assigned before starting the driving simulator trials.

Research design and procedure

A two-way (3×3) repeated measures mixed design was used with the three system descriptions as between-subjects factor (realistic, idealistic, wrong) and the three consecutive trials as within-subjects factor. Figure 1 shows the experimental design and procedure. Participants were paid for participating and were allowed to freely choose from among three possible appointments on an internet-based calendar. The average time interval between trials was 13 days with a minimum of 7 days and a maximum of 24 days. As a cover story, test subjects were told that the study aimed to test a driver assistance system, namely, DriveFree. The terms "adaptive cruise control" and "ACC" were never used in order to prevent information search between simulator trials.

Upon arrival at the driving simulator, all participants were informed about study aims and procedures, signed a consent form and completed a paper-and-pencil cognitive ability test to assess perceptual speed (ZVT "Zahlenverbindungstest"; Oswald & Roth, 1986). Subsequently, they read a one-page ACC description and orally answered seven questions about the system's functionality to ensure that information was well understood. Afterwards, all participants completed questionnaires related to mental model, acceptance and trust. Before driving on the simulated highway track, accompanied participants drove 5-km on a city route to practise using the driving simulator, the ACC and performing secondary tasks. Then, participants drove the 56-km simulated highway track. They were instructed to use the ACC in such a way that they could complete as many secondary tasks as safely as possible. After each drive, a semi-structured qualitative interview on their experiences was conducted and all participants filled in questionnaires related to mental model, acceptance and trust.

In the second and third trials, the mental model questionnaire was also applied before the simulator run to check for possible changes between trials. However, all other procedures remained the same as in the first trial, including instructions, highway track, interview and questionnaires applied after each trial. The system description was not shown again between trials.

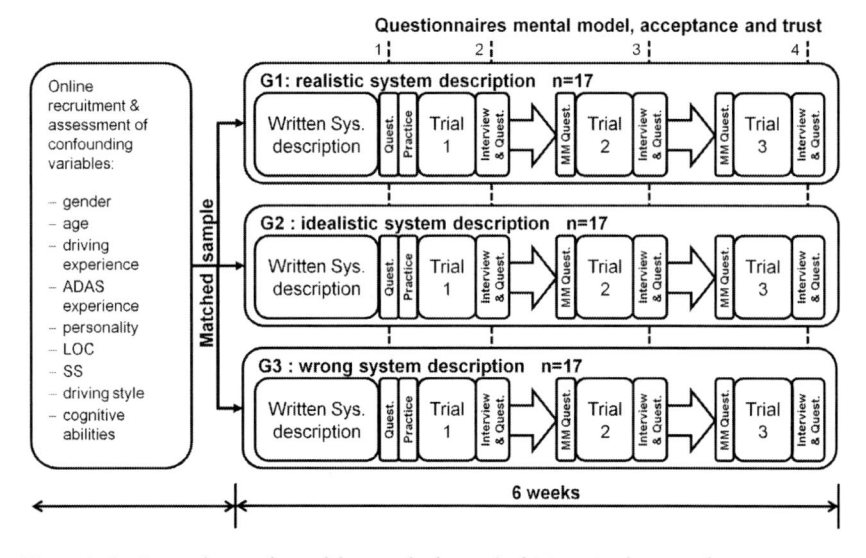

Figure 1. Design and procedure of the matched-sample driving simulator study

Facilities and apparatus

At the Chemnitz University of Technology, a fixed-base driving simulator (STISIMDrive 100w) composed of a BMW 350i with automatic transmission was used as driving cab. The simulated projection provided a 135° horizontal field of view.

The simulator route was constructed as a 56-km long two-lane highway with an average driving time of 35 min and 32 s. With the exception of a 100-km/h speed limit in construction zones, simulator speed limit was set to 120 km/h. The route consisted of five consecutive base modules of 11 km each and 1 km of straight road at the end. Every base module included:

- two left bends and one right bend
- a cut-in situation from the left after 4.7 km
- a construction zone from 8.5 to 11 km, where only the right lane could be used and a lead car with a constant speed of 80 km/h set driving pace
- an approach situation involving queuing at the beginning of the construction zone

However, the five base modules differed in various aspects:

- weather conditions (good weather / light fog / heavy fog)
- cut-in vehicle (normal car / motorbike / white car)
- last car in the queue (normal car / white truck)
- lead car in the construction zone (normal car / white truck)
- tracking lost (during heavy fog / in narrow bends)

The first base module represented a baseline for good weather conditions and no system failure (Figure 2). The second module was identical but included light fog conditions. During the third module, a motorbike was used for the cut-in situation, resulting in failure of the ACC to react and requiring the driver to intervene. Moreover, the last car in the queue was a white truck. During the fourth module a white car was used for the cut-in situation and tracking was lost in the final narrow left bend. In the final part, heavy fog conditions resulted in lost tracking on two occasions.

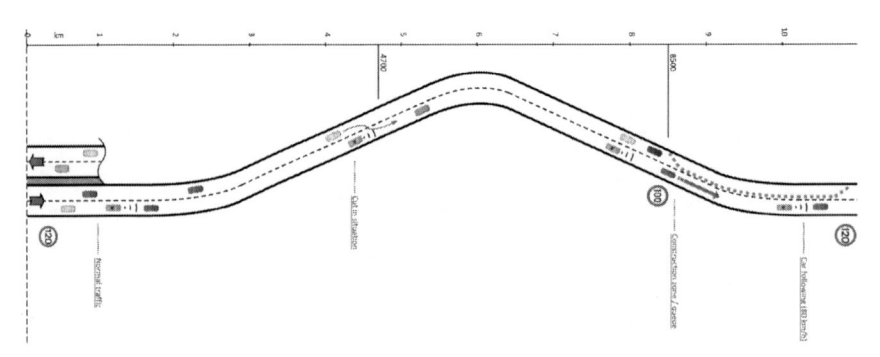

Figure 2. Schematic simulator layout of the first 11 km (base module) driven

The ACC was integrated in the simulation software and allowed specific manipulation of functionality in defined driving situations. In general, the ACC worked at every driving speed including stop-and-go functionality. During the entire simulation, ACC headway time was set to 2 s and speed to 100 km/h. To maintain constant conditions, participants were not allowed to change these settings during driving or turn off the system. Pressing the accelerator or brake pedal disabled the ACC control temporarily; when the pedals were released, ACC functions were immediately reactivated. No ACC alarms were implemented to simulate critical situations without prior warning. The system was programmed to recognise all leading vehicles except motorbikes. In three defined situations (i.e., two narrow bends, one episode of heavy fog), recognition of the lead vehicle was disabled temporarily to simulate system failures.

As a secondary task, participants had to continuously solve the surrogate reference task (Mattes & Hallén, 2009). The visual search task was used as an online measure for situation awareness, based on the work of Schömig et al. (2011). Fifty randomly arranged circles were presented on a 10-inch touch screen on the central console. Test subjects had to identify and select the one circle with a 30% greater radius.

Independent variable: system description

To create different initial mental models of the ACC, a one-page system description was designed. The first section dealt with general system functionality and was the same for all groups. This included an explanation of longitudinal control automation, speed and time headway settings, temporary deactivation due to braking or accelerating, and no system reaction upon encounter with traffic signs, pedestrians, oncoming traffic and traffic lights. The second part of the description differed between the three experimental groups (Table 1):

Table 1. Initial ACC information in the three experimental groups

Initial information about the ACC	Group 1 realistic information	Group 2 idealistic information	Group 3 wrong information	Events occurred during driving
General information on system functions	X	X	X	
No reaction for traffic signs, pedestrians, oncoming traffic and traffic lights	X	X	X	
Potential problems at narrow bends	X		X	Yes
Potential problems in bad weather conditions	X		X	Yes
Potential problems with small vehicles, such as motorbikes	X		X	Yes
Potential problems with big vehicles, such as buses or trucks			X	No
Potential problems with white or silver cars due to reflection			X	No

Group 1 (realistic) was informed about potential system failures at narrow bends, in bad weather conditions (heavy rain, fog or snow) and with small vehicles, such as motorbikes. These problems really occurred during the simulator run. Group 2 (idealistic) did not receive information about the potential failures mentioned in group 1. Group 3 (wrong) had the same realistic information as group 1, but two wrong potential failures were additionally indicated: The ACC could have detection problems with big vehicles (busses, trucks) and with white or silver cars due to reflection. These problems, however, did not occur during the driving simulation.

Dependent variables

The dependent variable, mental model of the ACC, was assessed by a standardized questionnaire consisting of 35 randomised items. All items dealt with specific aspects of ACC functionality and were answered on a 6-point Likert scale ranging from 1 (*totally disagree*) to 6 (*totally agree*). Five items focussed on the difference between the second group (idealistic) and both of the other groups, for example, "The system works on curvy roads". Another four items focussed on the specific differences between the third group (wrong) and the other groups, for example, "The system reacts to trucks driving ahead in the same lane". The remaining 26 questions were distractor items designed to not show any difference between the groups, for example, "The system reacts to pedestrians in the traffic lane".

The 12-item scale of trust in automated systems (Jian et al., 2000) was used to assess the second dependent variable, trust in the ACC. Reliability analysis showed good-to-excellent results at all four points of measurement with a Cronbach's alpha ranging from .854 to .911.

Results

Sample matching

The sampling process aimed at creating three experimental groups with no significant differences in gender, age, driving experience, ADAS experience, driving style, cognitive abilities, locus of control, sensation seeking and the five BFI factors extraversion, agreeableness, conscientiousness, neuroticism and openness. The between-group comparison showed no statistical significant differences between the three groups (p ranging from .634 to .951) in all confounding variables. Therefore, the sample can be considered as matched in regard to these factors.

Initial mental model of the system

Participants were presented with differing initial ACC descriptions to create a distinctive initial mental model of system functionality. Figure 3 shows the mental model profiles for every group and the nine relevant questionnaire items applied immediately after having read the system description. The first five items focussed on differences between the idealistic and both of the other groups. Posthoc ANOVA results showed statistically significant differences between the idealistic group and both of the other groups ($p < .001$ for all five items), whereas the realistic and wrong group did not differ significantly ($p > .650$ for all five items). Items six to nine concentrated on differences between the wrong group and both of the other groups. Posthoc ANOVA group comparisons showed statistically significant differences between the wrong group and both of the other groups ($p < .001$ for all four items) and no differences between the idealistic and realistic groups ($p > .403$ for all four items). Therefore, it can be concluded that providing different preliminary system descriptions led to the desired distinctive mental model in all groups.

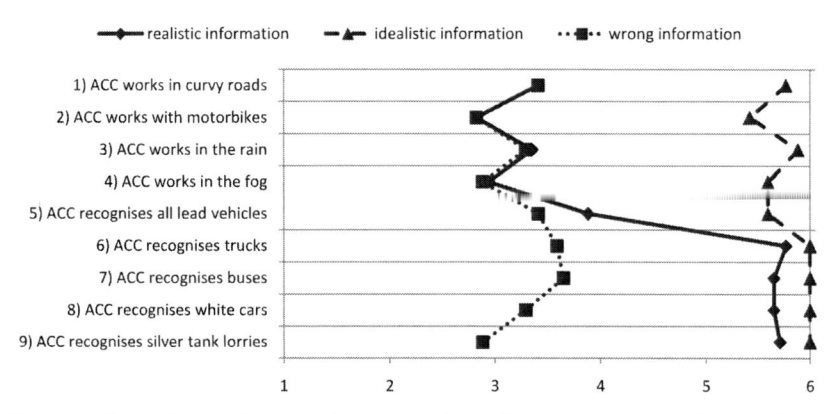

Figure 3. Mean values of the nine relevant mental model questionnaire items by type of initial ACC information after having read the system description (1 = totally disagree, 6 = totally agree)

Trust in the system

As hypothesised, more information about potential system failures in the preliminary description led to lower initial trust in the ACC. Trust scores differed significantly, $F(2, 48) = 4.47$, $p = .017$, $\eta_p^2 = .157$, immediately after having read the system description (Fig. 4, first measurement). The group with the most information about potential failures (wrong group) showed a statistically significant lower trust level than both of the other groups ($p = .004$). As predicted, idealistic information (idealistic group) led to the highest initial trust score, however, only slightly higher than the realistic group. Development of trust over time showed statistically significant differences between groups, $F(4.14, 97.26) = 3.41$, $p = .011$, $\eta_p^2 = .127$. In the realistic group, a tendency was observed for trust to increase with experience. Idealistic information led to a steady decrease in trust without recovery, from the highest to lowest level by the third trial. The group presented with wrong additional information showed a rapid growth in trust at the end of the first trial. Subsequently, trust grew steadily and almost reached the same level as the realistic group.

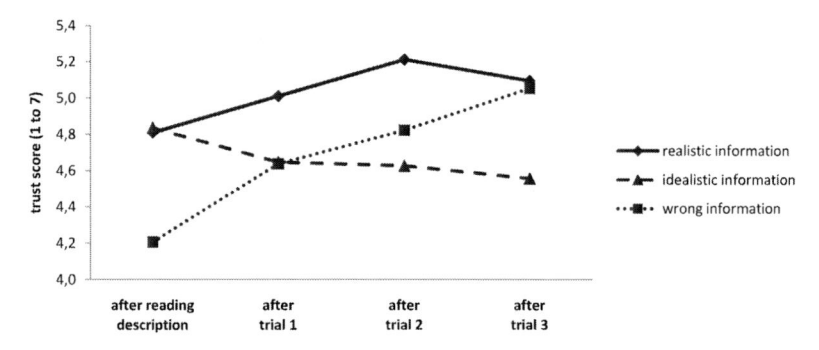

Figure 4. Mean trust level at the four times of measurement by type of initial information about the ACC. Higher trust scores represent greater trust in the system

Discussion and conclusions

This study aimed to assess the effects of different initial mental models of ACC on system trust as well as the development of trust over time. In order to induce the initial mental model, three different written system descriptions were designed. The realistic description includes all potential system failures occurring in the driving simulation, the idealistic information omits these critical aspects and the wrong description lists additional potential failures, which, however, do not occur.

Based on the conceptual model of driver appropriation from Cotter and Mogilka (2007), a matched sample controls for possible confounding factors. Results of the two-step sampling process show no statistically significant differences between the three experimental groups in gender, age, driving experience, ADAS experience, driving style, cognitive abilities, locus of control, sensation seeking and the five BFI factors extraversion, agreeableness, conscientiousness, neuroticism and openness. Therefore it can be concluded that sample matching has been successful and there has been no intervening bias on results due to these variables. However, the study focuses on young drivers under the age of 30 with no extreme values in personality variables.

As a manipulation check, the initial mental model was assessed by questionnaire immediately after reading the system description. The results show that the system description is able to induce a different understanding of ACC functionality. In the experimental conditions with realistic and wrong information, users are aware of potential failures in curvy roads, with motorbikes and in bad weather conditions, such as rain or fog. In contrast, these limitations are not part of the mental model in the idealistic group. Moreover, the initial mental model of the group with additional wrong information about potential problems with large white/silver cars includes these aspects, whereas for the realistic and idealistic groups, these aspects do not represent any potential problem.

The amount of information about potential failures influences initial trust in the ACC, regardless of gender, age, driving style and experience, cognitive abilities and personality aspects. The more potential critical situations that are stated, the more sceptical users are. However, when real system experience matches the initial mental model, trust increases steadily. System failures do not affect development of trust, if they are expected at the outset. Moreover, the provision of "over-information" about not-occurring potential failures lowers trust in the beginning only. Expected, but not-experienced, failures do not affect trust in the long run. An inverse development of trust is observed when potential problems are omitted in the initial description. The experience of unexpected failures lowers trust significantly, and repeated practice with the same system in the same situations does not restore trust. However, the study is limited to three trials in a simulated environment: Further research is needed to investigate the long-term development, both under real conditions and in different age groups.

It can be concluded that preliminary information about an ACC system, which forms a driver's initial mental model, has an important and enduring effect on trust in the system. Simple trial and error, without provision of customized information, is insufficient for building trust in an ACC system. Providing critical information only lowers trust for a short time at the beginning of use. Thus, it is recommended to avoid hiding potential problems in initial system descriptions: the more information, the better.

Acknowledgements

This research was funded by the European Commission's 7th Framework Programme (EU FP7) through the Marie Curie Initial Training Network ADAPTATION.

References

Cacciabue, P.C., & Saad, F. (2008). Behavioural adaptations to driver support systems: a modelling and road safety perspective. *Cognition, Technology & Work, 10,* 31-39.

Cahour, B., & Forzy, J.F. (2009). Does projection into use improve trust and exploration? An example with a cruise control system. *Safety Science, 47,* 1260–1270.

Cotter, S., & Mogilka, A. (2007). *Methodologies for the assessment of ITS in terms of driver appropriation processes over time.* HUMANIST Project Deliverable 6 of Task Force E. Retrieved from www.noehumanist.org.

Durso, F.T., Rawson, K.A., & Girotto, S. (2007). Comprehension and situation awareness. In F.T. Durso, R.S. Nickerson, S.T. Dumais, S. Lewandowsky, and T.J. Perfect (Eds.), *Handbook of applied cognition* (2nd ed.) (pp. 163–193). Chichester, UK: Wiley.

Endsley, M.R. (2004). Situation awareness: Progress and directions. In S. Banbury and S. Tremblay (Eds.), *A cognitive approach to situation awareness: Theory, measurement and application* (pp. 317-341). Aldershot, UK: Ashgate Publishing.

Endsley, M.R., & Garland, D.J. (Eds.). (2000). *Situation Awareness Analysis and Measurement.* Mahwah, NJ: Lawrence Erlbaum Associates.

Hoyle, R.H., Stephenson, M.T., Palmgreen, P., Lorch, E.P., & Donohew, R.L. (2002). Reliability and validity of a brief measure of sensation seeking. *Personality and individual Differences, 32,* 401-414.

Jenness, J.W., Lerner N.D., Mazor, S., Osberg J.S., & Tefft, B.C. (2008). *Use of Advanced In-vehicle Technology by Young and Older Early Adopters. Survey Results on Adaptive Cruise Control Systems.* Report No. DOT HS 810 917. National Highway Traffic Safety Administration.

Jian, J.Y., Bisantz, A.M., & Drury, C.G. (2000). Foundations for an empirically determined scale of trust in automated systems. *International Journal of Cognitive Ergonomics, 1,* 53-71.

Kazi T.A., Stanton N.A., Walker G.H., & Young M.S. (2007). Designer driving: drivers' conceptual models and level of trust in adaptive cruise control. *International Journal of Vehicle Design, 45,* 339-360.

Krems, J.F., & Baumann, M. (2009). Driving and Situation Awareness: A Cognitive Model of Memory-Update Processes. In M.W. Greenlee (Ed.), *New Issues in Experimental and Applied Psychology* (pp. 56-75). Lengerich, Germany: Pabst.

Lee, J.D., & See, K.A. (2004). Trust in automation: Designing for appropriate reliance. *Human Factors, 46,* 50-80.

Mattes, S., & Hallén, A. (2009). Surrogate Distraction Measurement Techniques: The Lane Change Test. In M. Regan, J.D. Lee, and K.L. Young (Eds.), *Driver Distraction: Theory, Effects and Mitigation* (pp. 107-122). Boca Raton: CRC Press.

Mehlenbacher, B., Wogalter, M.S., & Laughery, K.R. (2002). On the reading of product owner's manuals: perceptions and product complexity. *Proceedings of the Human Factors and Ergonomics Society 46th Annual Meeting* (pp. 730-734), Santa Monica, CA: Human Factors and Ergonomics Society.

Montag, I., & Comrey, A.L. (1987). Internality and Externality as Correlates of Involvement in Fatal Driving Accidents. *Journal of Applied Psychology, 72,* 339-343.

Oswald, W.D., & Roth, E. (1986). *Der Zahlen-Verbindungs-Test.* Göttingen, Germany: Hogrefe Verlag für Psychologie.

Popken, A. (2009). *Drivers' reliance on lane keeping assistance systems as a function of the level of assistance.* Doctoral dissertation, Chemnitz University of Technology. Retrieved from http://archiv.tu-chemnitz.de/pub/2010/0048/index.html.

Rajaonah, B., Tricot, N., Anceaux, F., & Millot, P. (2008). The role of intervening variables in driver–ACC cooperation. *International Journal of Human-Computer Studies, 66,* 185-197.

Rammstedt, B., & John, O.P. (2007). Measuring personality in one minute or less: A a 10-item short version of the Big Five Inventory in English and German. *Journal of Research in Personality, 41,* 203-212.

Rudin-Brown, C.M., & Parker, H.A. (2004). Behavioural adaptation to adaptive cruise control (ACC): implications for preventive strategies, *Transportation Research Part F: Traffic Psychology and Behaviour, 7,* 59-76.

Rudin-Brown, C.M., & Noy, Y.I. (2002). Investigation of behavioral adaptation to lane departure warnings. *Transportation Research Record (1803),* 30-37.

Saad, F. (2007). Dealing with Behavioural Adaptations to Advanced Driver Support Systems. In P.C. Cacciabue (Ed.), *Modelling Driver Behaviour in Automotive Environments* (pp. 147-162). London: Springer.

Schömig, N., Metz, B., & Krüger, H.-P. (2011). Anticipatory and control processes in the interaction with secondary tasks while driving. *Transportation Research Part F: Traffic Psychology and Behaviour, 14,* 525-538.

Seppelt, B.D., & Lee, J.D. (2007). Making adaptive cruise control (ACC) limits visible. *International Journal of Human-Computer Studies, 65,* 192-205.

Weller, G., & Schlag, B. (2004). Verhaltensadaptation nach Einführung von Fahrerassistenzsystemen. In B. Schlag (Ed.), *Verkehrspsychologie. Mobilität – Verkehrssicherheit – Fahrerassistenz* (pp. 351-370). Lengerich, Germany: Pabst Science Publishers.

Zwaan, R.A., Magliano, J.P., & Graesser, A.C. (1995). Dimensions of situation model construction in narrative comprehension. *Journal of Experimental Psychology: Learning, Memory, and Cognition, 21,* 386-397.

Zwaan, R.A., Radvansky, G.A., Hilliard, A.E., & Curiel, J.M. (1998). Constructing multidimensional situation models during reading. *Scientific Studies of Reading, 2,* 199–220.

The impact of a large-screen projection of the technical process on shared mental models and team performance in a furnace control room

Vera Hagemann, Annette Kluge, & Björn Badura
University of Duisburg-Essen
Germany

Abstract

Large screen projections (LSP) are applied in control rooms in order to facilitate a shared overview of running processes for shift staff. So far, little is known about the teamwork-related impact of LSP. It was assumed that LSP affect a) the congruency of mental models of interdependently working teammates and b) team performance. Congruency is a prerequisite for effective teamwork due to its impact on team-coordination processes. Shared mental models (SMM) regarding task and team interaction affect mutual expectations regarding the teammates' behaviours. They facilitate tacit coordination, mutual performance monitoring, and proactive offering of support. In a pre-post-test design, 21 operator teams of 3 engineering students each (N = 63), divided into two groups (LSP on/off), were investigated regarding their SMM acquisition within a furnace simulator. Following 45 minutes of training, task- and team-SMM were measured. Afterwards, the groups worked at the simulator "SteelSim" either with or without LSP to produce raw iron for 45 minutes. This was followed by the post-test SMM. Team performance was measured as the amount and quality of produced raw iron. It was found that although LSP did not significantly affect task and team-interaction SMM, LSP increased team performance significantly.

Introduction

Given the increasing complexity of organisations and task fulfillment, teamwork is deemed essential for success in meeting constantly changing requirements and in reacting flexibly to turbulent business environments (Cannon-Bowers & Bowers, 2011; Morgan et al., 1993; Salas et al., 2005). However, teamwork is not always successful. Often, it is also afflicted with communication or coordination problems (Hofinger, 2005), and how these problems are handled will have an impact on the team's performance (Baker et al., 2006; Stout et al., 1997). Among other things— e.g. team training strategies (see Cannon-Bowers & Bowers, 2011; Smith-Jentsch et al., 2008) such as cross training (team members switch their roles) or guided team self-correction (teams are instructed to observe teammates' behaviour and provide and accept performance-enhancing feedback)—technical resources play a crucial role in dealing with such kinds of teamwork affordances. One of these technical resources is a "large screen projection" (LSP). An LSP generates an overview of diverse sub-processes of a task for interdependently working team members and is used to present and integrate information and its connections and to detect current system states such as temperatures in normal range. In control rooms—which function as an operations centre where a facility can be monitored or controlled—LSPs are used in order to illustrate whole (production) processes and also to indicate incidents and the need to react immediately to emergencies. Thus, LSPs should improve safety and efficiency in control rooms (Veland & Eikas, 2007) and are often used in High Responsibility Teams (cf. Hagemann et al., 2011), working, for example, in nuclear power plants or on offshore oil rigs in order to enhance effective teamwork while reducing workload and human error (Veland & Eikas, 2007). The LSP's illustration of all ongoing processes fosters a team's situation awareness (SA) (Endsley, 1995; Salmon et al., 2009). The team members receive relevant information about the system states from the LSP (SA level 1). Then, it is their task to interpret and integrate this information into their current knowledge about the system (SA level 2) and to anticipate future system states and possible malfunctions or severe states (SA level 3; cf. Endsley, 1995). It is generally assumed that the overall demonstration of relevant information of an entire technical process on an LSP supports the mental model of the individual operator and enhances the shared mental model (SMM) of the team (Heidepriem, 2004; Härefors, 2008; Veland & Eikas, 2007). The SMM, in turn, as a congruency of the individual mental models of the team members, influences the team performance positively (Cannon-Bowers et al., 1993). The LSP—by supporting SMMs—is able to reduce possible negative effects of isolated work stations; the operators are able to freely contribute to other ongoing sub-tasks of the overall production process. Furthermore, the generated social interaction due to the LSP might decrease the risk of undetected events in the system by the operators (Härefors, 2008).

The aim of the present study is thus to explore systematically and empirically the assumed positive influence of the technical resource—the LSP—on the teams' SMM and performance.

Shared Mental Models

In teamwork research, SMMs are a fundamental construct in order to explain the functioning of successful teams. "SMMs are defined here as knowledge structures held by members of a team that enable them to form accurate explanations and expectations for the task, and, in turn, to coordinate their actions and adapt their behaviour to demands of the task and other team members" (Cannon-Bowers et al., 1993, p. 228). SMMs support team members to optimise coordination and communication processes, to anticipate and explain the information needs and actions of other team members, to search for information more effectively, to jointly interpret information from the environment, and to solve problems more effectively (Johnson & Lee, 2008; Smith-Jentsch et al., 2008). Cannon-Bowers et al. (1993) divide SMMs into four sections. 1) the equipment mental model – describes the knowledge about the materials and tools required for task fulfillment, 2) the task mental model – refers to the detailed knowledge about procedures and strategies and about requirements in the environment for performing the task, 3) the team interaction mental model – describes the knowledge about the roles, responsibilities, and interdependencies as well as communication channels between the team members, and 4) the team mental model – refers to the knowledge about the competencies, knowledge, and preferences of the other team members.

Studies conducted so far have already shown that both a shared task and a team mental model affects safety and team work efficiency (Smith-Jentsch et al., 2005) as well as performance in general (Peterson et al., 2000), or that only the team interaction mental model has an impact on performance but not the task mental model (Mathieu et al., 2000). For the present study, the *team interaction* and *task mental model* are particularly relevant.

Hypotheses

Based on the idea of the commonly applied input-process-output (IPO) model in teamwork research (cf. West, 2004), it is investigated whether an LSP supports the development of congruent SMMs, whether the SMM congruency in turn affects the team performance, and also whether the LSP directly influences the team performance. The research hypotheses are as follows (see also Figure 1):

Hypothesis 1: Teams with an LSP will have a higher degree of "sharedness" within their mental models (task mental model & team interaction mental model) than teams without an LSP.
Hypothesis 2: Teams with an LSP will perform better than teams without an LSP.
Hypothesis 3: Teams with a higher degree of "sharedness" within their mental models (task mental model & team interaction mental model) will show a better team performance than teams with less shared mental models.

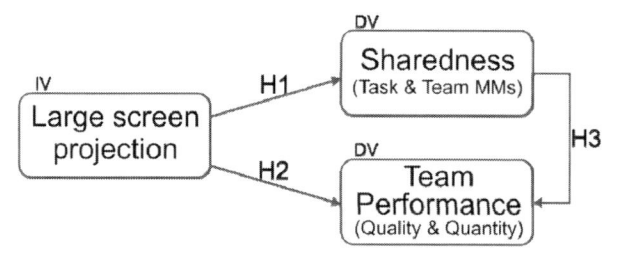

Figure 1. Illustration of the three hypotheses (IV = Independent variable, DV = Dependent variable, MMs = mental models)

Method

From January to March 2011, 63 students (aged 19 to 34 years) from the Faculty for Engineering participated in the study. Participants were randomly[*] assigned to teams of three. The teams had to operate the simulator "SteelSim" in order to produce raw iron. The experimental design was a between-subject design with two groups; teams were randomly assigned to the control group (CG, $n = 10$ teams/ 30 participants,

---

[*] Due to the random assignment of participants, there were significantly more women than men in the control group. The experimental group was balanced. All other characteristics of participants were also balanced between the groups (see the "Results" section)

large screen projection "off") or the experimental group (EG, $n = 11$ teams/ 33 participants, large screen projection "on"). Only the EGs were shown an LSP of the entire technical process of the raw iron production. The mean age of the participants was 22.74 years ($SD = 3.05$). There were no significant differences ($F_{(1,62)} = .276$, $p > .05$) between the EG ($M = 22.53$; $SD = 2.89$) and the CG ($M = 22.94$; $SD = 3.22$). However, 62% of the participants were female, which led to a significant difference in terms of sex ($\chi^2 = 7.95$, $p = .01$). As can be seen in Table 1, there were much more women than men in the CG.

The simulator "SteelSim"

The "SteelSim" simulates a blast furnace control room with three operator workstations (see Figure 2). The operators have to work as a team by coordinating the sub-processes for which the individual operator is responsible in order to produce raw iron within a smelter. The control room teams have to monitor, control, and coordinate sub-processes of the whole raw iron production process. The main task is to produce raw iron through the reduction of iron ore. The teams have to achieve three objectives simultaneously: to maximise a) the quantity and b) quality of the produced raw iron, and c) to optimise the trouble-shooting.

Each operator's position is equipped with a desk and a desktop PC. Because of the diagonal positioning of the desks, the operators are not able to look at the screens of their co-workers.

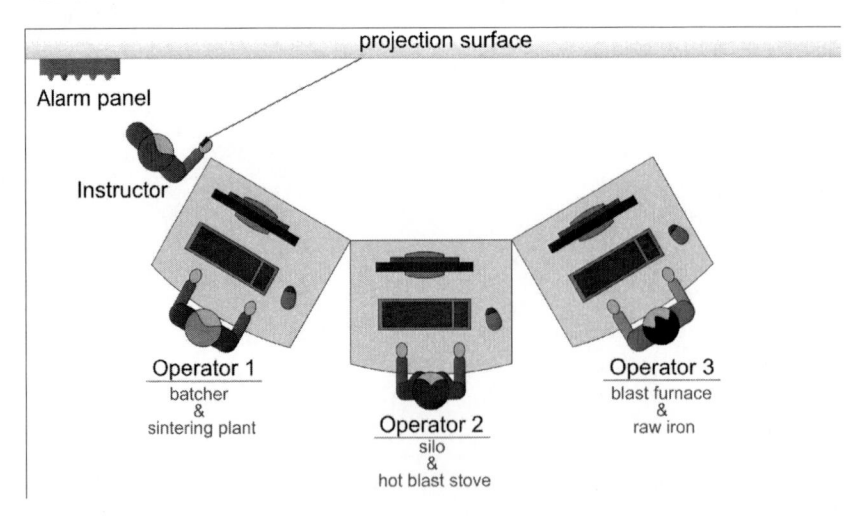

Figure 2. Layout of the control room

In addition to the three workstations, a position for the instructor exists. The instructor starts, monitors, and shuts down the simulation for the experiments and has the authority to program 16 different malfunctions (e.g. breakdown of a torpedo car or temperature in the hot blast stove is too low) to which the team members have to react. The instructor is also able to display a large screen projection of the entire technical process of the raw iron production (see Figure 3) on a projection surface (see Figure 2). Furthermore, there is an alarm box in the room which indicates malfunctions through light and sound. The "SteelSim" was the primary instrument in order to manipulate the independent (variation of LSP) and to measure the dependent (SMM and team performance) variables.

The operators' interdependent tasks
The raw iron production process consists of three interdependent sub-tasks. These sub-tasks are a result of the stringent division of the responsibilities of the three operators. Each operator has only limited access to the sub-processes of the "SteelSim". Two sub-processes are within the responsibilities of each operator (Operator 1: batcher & sintering plant; Operator 2: silo & hot blast stove; Operator 3: blast furnace & raw iron). The operators do not have direct access to the sub-processes of one of the other operators. Team work tasks are deliberately designed as interdependent. For example, operator 1 has to produce sinter and carry it into the silo; otherwise, operator 2 is unable to dose the materials (sinter) into the torpedo car. A strong interdependence also exists between operators 2 and 3: Operator 2 blends the resources and doses them into the torpedo car, and as soon as the torpedo car is loaded, operator 3 has to move it to the blast furnace top. Hence, the production of raw iron managed by operator 3 also depends on the amount of the produced sinter by operator 1. In order to coordinate these processes, the operators have to communicate with each other or they have to collect the relevant information from the LSP if possible (EG).

Each operator is able to retrieve the required feedback data and information concerning quality and quantity of his/her sub-process from his/her individual computer screen. Thus, he/she receives all relevant information and cues concerning the sub-process for which he/she is responsible, but whether the operators are able to watch all sub-processes depends on the availability of the LSP (independent variable).

Manipulation of the independent variable

The large screen projection: In addition to the information displayed on the individual screens, all sub-processes necessary to produce raw iron are integrated and displayed on the LSP. That means that each operator, in addition to his/her own sub-process information, also receives feedback about the quality and quantity of the sub-processes managed by the other two operators. The sub-processes batcher and sintering plant (1 + 2), silo and hot blast stove (3 + 5), and blast furnace and raw iron (4 + 6) are displayed in Figure 3. For example, the filling level of the torpedo car ((6) top left) is visible for all operators, even though only operator 3 is in charge of this sub-process.

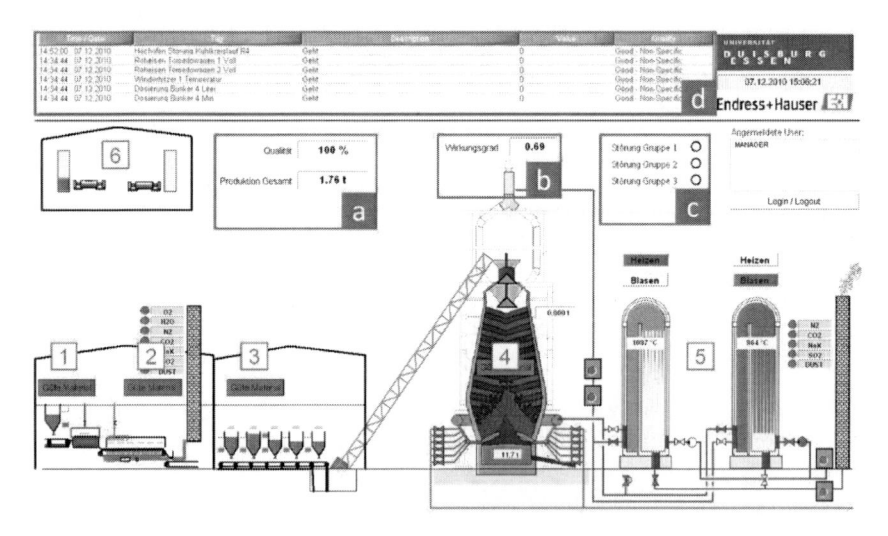

Figure 3. The technical sub-processes of the raw iron production (the LSP)

Furthermore, the operators receive feedback about the quality and quantity of the produced raw iron ((a) centre-top; quality varies between 50% and 100%) and the degree of efficiency of the blast furnace ((b) centre-top; varies between 0.63 and 0.69). The degree of efficiency quantifies the amount of raw iron smelted in the blast furnace as a function of heating temperature and oxygen transfer. These indicators are important for team performance because they are the central criteria for meeting the predefined goals. If malfunctions—programmed by the instructor—occur, everyone can see which sub-processes are affected ((c) top right). An overview of the alarms with detailed information about the malfunctions is also visible for each operator ((d) top).

Person-related variables have proven to be important predictors for team performance in several studies. General mental ability (GMA) significantly influences the teams' performance with regard to troubleshooting performance, system control, and system knowledge (Burkolter et al., 2009; Morris & Rouse, 1985; Schmidt & Hunter, 1998). Thus GMA was measured by applying a subtest of the LPS 4. Moreover, prior knowledge about the raw iron production process was also evaluated in order to set the baseline for assessing knowledge improvement and to control for threats to internal validation. The degree of friendship between the participants was recorded as well. It was assumed that friends would have a higher willingness to communicate with each other whether there was an LSP or not. The quality of their relationship might influence problem solving behaviour positively. These variables were also measured in order to control for possible "unhappy randomization" (Mohr, 1995).

Measurement of the dependent variables

The evaluation of the teams' SMMs and of the team performance (quality and quantity of produced raw iron) is described in the following.

Task mental model and team interaction mental model
Two aspects of the SMM were measured: 1) *task mental model* (knowledge about the procedures and strategies for task fulfillment) and 2) the *team interaction mental model* (knowledge about the roles, responsibilities and interdependencies of all operators as well as communication channels). In order to evaluate the *sharedness of the teams' mental models* due to the LSP, a paper-and-pencil questionnaire—containing 60 items—was developed and applied in a pre- and posttest. The pretest of the SMM was administered after the initial training and before the production process, and the posttest measure was administered after the production process (see Figure 4). Detailed information about the questionnaire, exemplary items and the answer format are displayed in Table 1. The *team interaction mental model* was further divided into two parts, namely "interaction awareness" and "team awareness". *Interaction awareness* referred to the operators' awareness of the interdependencies between all operators in order to manage the task successfully. *Team awareness* referred to the operators' awareness of the actions within the team and their consequences.

Table 1. Overview of the subparts of the Shared Mental Model questionnaire

	Task Mental Model	Interaction Awareness	Team Awareness
Number of Items	30	15	15
Exemplary Item	The raw iron is taken from the blast furnace by pouring at regular intervals.	Operator 2 has to fill the torpedo car completely, and then operator 3 can move it.	How does the outlet belt of the torpedo car by operator 3 influence the overall productivity?
Exemplary Item	If the hot blast stove reaches 1000°C, it is almost cooled down so that it has to be heated up again.	Operator 3 is in charge of the additional delivery of resources for operator 2.	How will the quality of the raw iron produced by operator 3 be affected if the two hot blast stoves controlled by operator 2 cool down?
Response format	True, false, don't know*	True, false, don't know*	True, false, don't know*

Note: *A correct answer was coded with "1" and an incorrect answer with "0"

The SMM sharedness-Index
Based on the questionnaires introduced above, an index was developed to operationalise the congruence between the operators' mental models in one team. For each item, the congruence of the operators' team answers was defined in terms of inter-operator agreement, e.g. if the following answers were given in a team: operator 1: 1 (correct answer), operator 2: 0 (incorrect answer), operator 3: 1 (correct answer), the congruence reached a value of 66. The value of 66 was the result for each operator in the team. Then, the congruence of all items of the task and of the team interaction mental model was defined by calculating the average. The resulting values ranged between 0 and 100 (see also Table 3).

As the SMM was measured before and after the simulation phase (production process), four values resulted with regard to the sharedness, namely task mental model before (t_1) and after (t_2) as well as team interaction mental model before (t_1) and after (t_2). Furthermore, an overall index was developed for the SMM at t_1 and t_2 which included the task and the team interaction mental model.

Measurement of the team performance

The team performance was measured on the basis of the quality and quantity of the produced raw iron. Quality is defined as the amount of the right composition of the different materials to produce raw iron. It is calculated in percentage of 50% to 100%. Quantity refers to the overall production in 45 minutes and is measured in tonnes (see Figure 3). The processes hot blast stove, blast furnace and raw iron determine the quantity. The degree of efficiency of the blast furnace—which varies between 0.63 and 0.69—influences the quantity, because it is determined by the hot blast stove. The two indicators (quality and quantity) were introduced to the operators at the beginning of the experimental study as the primary goals. The participants were trained on how to manipulate these factors (see also section "The training phase"). The total amount of produced raw iron as well as the quality is read out the log files of the "SteelSim" software. The quality of the produced raw iron was tracked every minute in the log files during the production process and averaged over time for each team. Both values (quantity and quality) were computed into one team performance index by multiplying the produced amount of raw iron (tonnes) by the quality value. For example, if a team produced 21 tonnes of raw iron with a quality of 74%, then the team performance index was 21 x 0.74 = 15.54 (see also Table 3).

Test procedure

The experimental study (see Figure 4) took around 2.5 hours per team. After an introduction phase, all participants were requested to complete the questionnaires in order to measure the control variables (prior knowledge test, LPS4, degree of friendship) and also to name their sex, age, and study course. Then, the training phase started, in which team members were instructed about the responsibilities and duties of each operator and how to operate the subtasks. The interdependencies between operator tasks were also addressed.

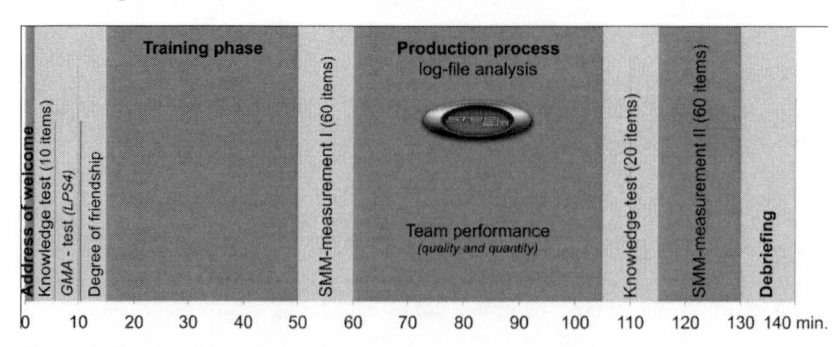

Figure 4. The experimental procedure

The training phase
The training started with an introduction of the simulator environment "SteelSim". Initially, an overview of the operating mode of the blast furnace was given. Subsequently, the user interface of the simulator, with all its functions and elements and their interdependencies, was explained in detail. It was stressed that the team had to produce as much as possible with a high quality, which should be considered throughout the production process. The explicitly given goals were: a) to maximise the produced raw iron quality, b) to maximise the produced raw iron quantity, and c) to optimise the trouble-shooting.

In order to develop the overall comprehension of the technical and complex raw iron production process, a computer animation was used. The relevance of the particular system components was not specifically explained to the participants, meaning that they were not explicitly told the priorities for enhancing the quality and quantity of the raw iron. The animation demonstrated the steps of the production process and also the division of the interdependent sub-processes. The respective sub-processes and corresponding responsibilities and sub-tasks were explained to the participants. The trouble-shooting (detection of malfunctions and deactivation of alarms) was also part of the training. Additionally, the participants received checklists for the trouble-shooting.

The production process
After the training and the t_1 SMM measurement, the 45-minute production process started. During the production of raw iron, malfunctions occurred as programmed by the instructor. All teams had to handle the same malfunctions in the same order. The malfunctions were also equally distributed to the three operators, so that each operator experienced the same workload. During this production process, only a small number of questions were answered by the instructor, e.g. questions due to misunderstandings during the training phase. After the production process (t_2), the SMM test was applied again.

Results

As displayed in Table 2, the CG did not differ significantly from the EG with regard to the control variables. A multivariate analysis of variance was conducted – Levene tests were not significant, meaning that the assumption of homogeneity of variance was confirmed. There were no significant differences between the CG and the EG with respect to age, GMA, prior knowledge and friendship index. Thus, there was no "unhappy randomization" (Mohr, 1995). Only for sex did the chi-square value reach significance ($\chi^2 = 7.95, p = .01$). There were significantly more women than men in the CG, while this was not the case for the EG.

In order to test hypotheses 1 and 2, a multivariate analysis of variance was conducted – Levene tests were not significant, meaning that the assumption of homogeneity of variance was confirmed. Hypothesis 1 focused on the *task and team interaction mental model* and stated that the teams with the LSP on during the production process would have a higher degree of sharedness than teams with the LSP off. Hypothesis 2 focused on the *team performance* instead, but also stated that the teams with the LSP on would show a better team performance than teams with the LSP off.

Table 2. Descriptive statistics regarding control variables (N = 63; SD in brackets)

	Control group	Experimental group	Significance
Age (19-34)	22.53 (2.89)	22.94 (3.22)	$F_{(1,62)} = 0.276, p = .60$
Sex	24 (♀) : 6 (♂)	15 (♀) : 18 (♂)	$\chi^2 = 7.95, p = .01$
General mental ability (0-40)	30.63 (3.61)	30.70 (4.47)	$F_{(1,62)} = 0.004, p = .95$
Prior Knowledge (t_1) (0-100)	73.30 (10.00)	71.50 (12.00)	$F_{(1,62)} = 0.382, p = .54$
Friendship index (0-2)	1.30 (0.85)	1.23 (0.78)	$F_{(1,62)} = 0.115, p = .74$

Note: Values in the brackets (left column) indicate the range of the response format

Means, standard deviations and results are shown in Table 3. There were no significant differences between the CG and EG regarding the variables SMM before (t1) the production process. It was important to control for this to enable significant results after the production process (t2) regarding the SMM to be traced back to the treatment. However, after the production process (t2), there were also no significant differences between the CG and EG regarding the dependent variables SMM. Both groups showed similar results. Therefore, hypothesis 1 was rejected. Teams with the LSP on during the production process did not have a higher degree of sharedness within their mental models (task mental model & team interaction mental model) than teams with the LSP off.

Table 3. Results regarding hypotheses 1 and 2 (N = 21; SD in brackets)

	Control group	Experimental group	Significance
SMM Team t_1 (0-100)	75.11 (7.25)	77.16 (5.59)	$F_{(1,62)} = 0.533, p = .47$
SMM Task t_1 (0-100)	73.67 (4.21)	77.76 (7.37)	$F_{(1,62)} = 2.380, p = .14$
SMM total t_1 (0-100)	74.39 (5.10)	77.46 (5.89)	$F_{(1,62)} = 1.617, p = .22$
SMM Team t_2 (0-100)	73.70 (6.98)	76.79 (7.74)	$F_{(1,19)} = 0.913, p = .35$, $\eta^2_p = 0.05$
SMM Task t_2 (0-100)	76.69 (6.54)	77.57 (7.52)	$F_{(1,19)} = 0.081, p = .78$, $\eta^2_p = 0.00$
SMM total t_2 (0-100)	75.19 (5.73)	77.18 (6.60)	$F_{(1,19)} = 0.537, p = .47$, $\eta^2_p = 0.03$
Team performance	15.80 (2.74)	18.30 (2.43)	$F_{(1,19)} = 4.909, p = .04$, $\eta^2_p = 0.21$

Note: t_1 = after training/ before the production process; t_2 = after the production process; SMM = Shared Mental Model

With respect to the second dependent variable—team performance—the analysis of variance reached significance and the effect size was satisfactory ($F_{(1,19)} = 4.909, p =$

.04, η^2_p = 0.21). The quantity (produced tones) ranged from 11.68t to 23.57t; the quality ranged from 62.43% to 93.59%. The EG (LSP on; M = 18.30; SD = 2.43) showed a significantly higher team performance than the CG (LSP off; M = 15.80; SD = 2.74). Thus, hypothesis 2 was supported – teams with an LSP performed better than teams without an LSP (Figure 5).

In addition, one-sample t-tests with repeated measures were conducted to compare the SMMs from t_1 to t_2. None of the results reached significance, either for the EG or for the CG regarding task mental model (EG: p = .92; CG: p = .17), team interaction mental model (EG: p = .88; CG: p = .68) or SMM total (EG: p = .88; CG: p = .74). In other words, the sharedness of the mental models did not increase over time (45 minutes) in any of the groups. Thus, the production process did not enhance the teams' SMMs.

In order to test hypothesis 3, partial correlation analyses were conducted. Hypothesis 3 focused on the relationship between the degree of *sharedness of mental models* and the *team performance*. It was stated that teams with a higher degree of sharedness of their mental models (task mental model & team interaction mental model) show a higher team performance than teams with less shared mental models.

The partial correlation between team performance and shared *task* mental model did not reach significance ($r_{\text{performance taskSMMT2.taskSMMT1}}$ = .20, p = .40). Moreover, the partial correlation between team performance and shared *team interaction* mental model also did not reach significance ($r_{\text{performance teamSMMT2.teamSMMT1}}$ = .38, p = .10). Thus, unsurprisingly, the partial correlation between team performance and *SMM total* also failed to reach significance ($r_{\text{performance totalSMMT2.totalSMMT1}}$ = .35, p = .13). In other words, hypothesis 3 could not be supported – teams with a higher degree of sharedness within their mental models (task mental model & team interaction mental model) did not show a higher team performance than teams with less shared mental models (Figure 5).

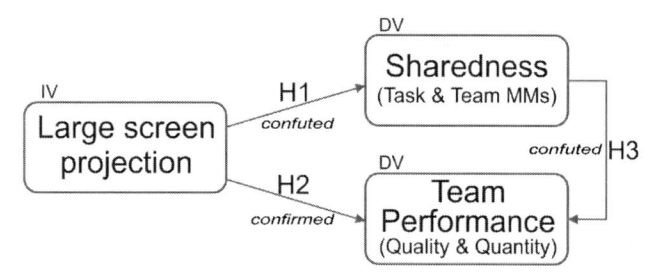

Figure 5. Summary of the hypothesis-testing (IV = Independent variable, DV = Dependent variable, MMs = mental models)

Discussion

The goal of the present study was to demonstrate the positive influence of an LSP of a technical production process on, firstly, the congruence of SMMs in operator teams and secondly, the team performance with regard to the produced raw iron. The LSP had a positive effect on the team performance, with the EG (LSP on) producing more raw iron with higher quality. Thus, it is concluded that an LSP in control rooms—e.g. in nuclear power plants or on offshore oil rigs—enhances the performance of the interdependently working team members. This, in turn, is assumed to support more reliable teamwork, an important issue for High Responsibility Teams (Hageman, 2011). A positive effect of the LSP on the team's SMM could not be demonstrated in the present study. There were no significant differences between the two groups, and the teams' SMMs did not increase in sharedness over time. The third assumption referred to the relationship between the teams' SMMs and the team performance. This relationship could not be found, contrary to other studies that supported the positive influence of SMMs on team performance (e.g. Johnson & Lee, 2008; Peterson et al., 2000; Rasker et al., 2000; Smith-Jentsch et al., 2005) due to more effective communication and (explicit and implicit) coordination processes.

The reasons for these non significant results are diverse. With regard to *methodological reasons*, the questionnaire developed for evaluating the SMM might have not measured validly that which it was supposed to measure. Due to the fact that this study was the first to use SteelSim, there was no previous validation study of this instrument. If the questionnaire did not measure the SMM—perhaps it evaluated the participants' general knowledge of the production processes—this could have threatened the internal validity of the present study. On the other hand, assuming that the SMM measures were valid, the internal validity could have been threatened by the design of the independent variable. The results of the present study seem to illustrate that it is not the SMM that supports the team performance influenced by the LSP; perhaps the LSP supports other cognitive processes such as an increase in the situation awareness (SA) of the team members (cf. Endsley, 1995; see also introduction). This means that the LSP supports a better perception of information from a dynamic system (SA level 1), such as the "SteelSim". Moreover, the information could be interpreted properly (SA level 2) and future system states could be anticipated correctly (SA level 3). For example, the achievement of the torpedo car's maximum capacity can be anticipated by all team members and included in the current decisions, meaning that precautions could be taken, which would in turn improve team performance. Thus, tests for evaluating the team members' SA should be developed in future studies.

With regard to *study design, t*he production process lasted for 45 minutes only. It is possible that the time span was too short to increase the sharedness of the SMMs and to additionally get used to all processes and sub-tasks of the simulation. The time span should be varied and extended to 90 minutes or more, for example, and the effects should be investigated systematically. Moreover, the training phase (before the production process started) could also have had an impact on the SMMs. Perhaps this training phase was too informative and in-depth, rendering a further increase in the degree of the SMMs impossible with a production process of only 45 minutes. It should be considered whether an SMM test should also be applied at the beginning of the study.

With regard to *group composition,* the non-balanced assignment of participants with regard to sex might have had an impact on the SMM-building process. Although this is highly speculative, the higher number of women in the CG might have led to more communication or rather information exchange between the operators. This, again, might have supported the teams' SMM acquisition within the CG. This influence of sex on the SMM acquisition should be examined under controlled conditions in a further experiment. If this effect does exist, teams of three women only should have higher degrees of SMM than teams of three men only, also when controlling for LSP off and LSP on.

Last but not least, possible effects of team diversity on team performance should have been taken into consideration more carefully. Jackson and Joshi (2011) stated in their review that work team diversity "is likely to impede frequent and effective communication among team members" (p. 661) and has diverse—positive as well as negative—influences on team performance (p. 666). Nevertheless, they differentiate between various types of diversity, which could either be within relationship-oriented attributes and readily detected attributes (e.g. gender, age, nationality); within relationship-oriented attributes and underlying attributes (e.g. personality, attitudes, values); within task-oriented attributes and readily detected attributes (e.g. educational level, organizational tenure, unit membership); or within task-oriented attributes and underlying attributes (e.g. task knowledge, cognitive abilities, mental models). The first and last types of this taxonomy of diversity were evaluated and controlled in the experiment (see results), but the other two types were not. It would therefore be interesting to measure the other types (relationship-oriented attributes and underlying attributes & task-oriented attributes and readily detected attributes) as well and to ascertain whether they influence communication processes or team performance in interdependently working teams. This might be especially relevant for the types of teams examined in the present study, as Jackson and Joshi (2011) report that studies have shown that (relations-oriented) diversity had the most negative effects in moderately interdependent teams.

Outlook

The fact that the team performance is significantly better in the EG (LSP on) than in the CG indicates that the LSP must have an influence – not on the SMMs, but perhaps on other cognitive processes. So far, the mode of action of the LSP on teamwork and the factors that influence this relationship are not fully understood. In a further study, it should be examined whether the LSP supports the team members' SA and whether this, in turn, leads to a better team performance. Moreover, the respective parts of the LSP should also be taken into consideration, and it should be examined in detail what information the teams should be exposed to. This aspect might also be important with regard to a distinction between experts and novices, as Wickens and Holland (2000) showed that their cognitive patterns in their work domain and detection of anomalies differ from each other. Furthermore, the duration of the production process and the degree of detail of the training phase should be varied and examined. As mentioned above (section "The training phase"), the relevance of the particular system components for meeting the predefined goals was not particularly explained to the participants in the training. Thus, the three operators were not explicitly told the priorities for enhancing the quality and quantity of the produced raw iron. In a further study, the participants should receive more detailed information about influencing the goals effectively. Finally, it can be concluded that research regarding the influence of LSPs on team performance might be fruitful in providing a better understanding of the underlying effects and causes of the cognitive processes that are supposed to be supported.

References

Baker, D. P., Day, R., & Salas, E. (2006). Teamwork as an Essential Component of High-Reliability Organizations. *Health Services Research, 41*, 1576-1598.

Burkolter, D., Kluge, A., Sauer, J., & Ritzmann, S. (2009). Predictive qualities of operator characteristics for process control performance: The influence of personality and cognitive variables. *Ergonomics, 52*, 302-311.

Cannon-Bowers, J.A. & Bowers, C.A. (2011). Team Development and Functioning. In S. Zehdeck (Ed.), *APA Handbook of Industrial and Organizational Psychology Building and Developing the Organization* (Vol. 1, pp. 597-650). Washington: American Psychological Association.

Cannon-Bowers, J. A., Salas, E., & Converse, S. (1993). Shared mental models in expert team decision making. In N.J. Castellan (Ed.), *Individual and Group Decision Making* (pp. 221-246). Hillsdale: Lawrence Erlbaum Associates.

Endsley, M. R. (1995). Toward a theory of situation awareness in dynamic systems. *Human Factors, 37*, 32-64.

Hagemann, V. (2011). *Trainingsentwicklung für High Responsibility Teams [Training development for High Responsibility Teams]*. Lengerich: Pabst Science Publishers.

Hagemann, V., Kluge, A. & Ritzmann, S. (2011). High Responsibility Teams - Eine systematische Analyse von Teamarbeitskontexten für einen effektiven Kompetenzerwerb [A systematic analysis of teamwork contexts for effective competence acquisition]. *Psychologie des Alltagshandelns [Psychology of everyday activity], 4* (1), 22-42.

Härefors, E. (2008). *Use of large screen displays in nuclear control room.* Master thesis. ISSN: 1650-8319, UPTEC STS08 024.

Heidepriem, J. (2004). *Prozeßinformatik 2: Prozessrechentechnik und Automationssysteme [Process informatics 2: process computer science and automation systems]*. München: Oldenbourg Industrieverlag GmbH.

Hofinger, G. (2005). *Kommunikation in kritischen Situationen [Communication in crucial situations]*. Frankfurt: Verlag für Polizeiwissenschaft.

Jackson, S.E. & Joshi, A. (2011). Work Team Diversity. In S. Zehdeck (Ed.), *APA Handbook of Industrial and Organizational Psychology Building and Developing the Organization* (Vol. 1, pp. 651-686). Washington: American Psychological Association.

Johnson, T.E. & Lee, Y. (2008). The Relationship Between Shared Mental Models and Task Performance in an Online Team-Based Learning Environment. *Performance Improvement Quarterly, 21*, 97-112.

Mathieu, J.E., Heffner, T S., Goodwin, G.F., Salas, E. & Cannon-Bowers, J.A. (2000). The Influence of Shared Mental Models on Team Process and Performance. *Journal of Applied Psychology, 85*, 273-283.

Mohr, L.B. (1995). *Impact analysis for program evaluation.* Thousand Oaks: SAGE.

Morgan, B. B., Salas, E., & Glickman, A.S. (1993). An analysis of team evolution and maturation. *The Journal of General Psychology, 120*, 277-291.

Morris, N.M. & Rouse, W.B. (1985). Review and evaluation of empirical research in troubleshooting. *Human Factors, 27*, 503-530.

Peterson, E., Mitchell, T.R., Thompson, L., & Burr, R. (2000). Collective Efficacy and Aspects of Shared Mental Models as Predictors of Performance over Time in Work Groups. *Group Process and Intergroup Relations, 3*, 296-316.

Rasker, P.C., Post, W.M., & Schraagen, J.M.C. (2000). Effects of two types of intra-team feedback on developing a shared mental model in Command and Control teams, *Ergonomics, 43*, 1167-1189.

Salas, E., Sims, D., & Burke, S. (2005). Is there a "Big Five" in Teamwork? *Small Group Research, 36*, 555-599.

Salmon, P.M., Stanton, N.A., Walker, G.H., & Jenkins, D.P. (2009). *Distributed Situation Awareness. Theory, Measurement and Application to Teamwork.* Farnham, Burlington: Ashgate.

Schmidt, F.L. & Hunter, J.E. (1998). The validity and utility of selection methods in personnel psychology: Practical and theoretical implications of 85 years of research findings. *Psychological Bulletin, 124*, 953-967.

Smith-Jentsch, K.A., Mathieu, J.E., & Kraiger, K. (2005). Investigating Linear and Interactive Effects of Shared Mental Models on Safety and Efficiency in a Field Setting. *Journal of Applied Psychology, 90*, 523-535.

Smith-Jentsch, K., Cannon-Bowers, J.A., Tannenbaum, S.I. & Salas, E. (2008). Guided Team Self-Correction. *Small Group Research, 39*, 303-327.

Stout, R.J., Salas, E., & Fowlkes, J.E. (1997). Enhancing Teamwork in Complex Environments Through Team Training, *Group Dynamics: Theory, Research and Practice, 1*, 169-182.

Veland, Ø. & Eikås, M. (2007). A Novel Design for an Ultra-Large Screen Display for Industrial Process Control. In M.J. Dainoff (Ed.), *EHAWC 2007 - Ergonomics and Health Aspects of Work with Computers* (pp. 349-358). Beijing, China

West, M.A. (2004). *Effective teamwork: practical lessons from organizational research.* BPS Blackwell.

Wickens, C.D. & Holland, J.G. (2000). *Engineering psychology and human performance.* New Jersey: Prentice Hall.

Exploring the acceptance of mobile technologies using walking interviews

Frances Hodgson[1], Yvonne Barnard[1], Mike Bradley[2], & Ashley D. Lloyd[3]
[1]*Institute for Transport Studies, University of Leeds*
[2]*Middlesex University*
[3]*The University of Edinburgh*
UK

Abstract

To investigate aspects of acceptance and use of digital technologies an ethnographic method was developed to elicit responses and opinions of older persons on new mobile technologies. Large numbers of older people in Europe never use the internet or a computer, and miss out on communication and travelling benefits new mobile devices could bring. A scenario was designed in which people over 65 years old could experience the use of a tablet computer. In the scenario participants and researchers walked around the university campus, using a tablet computer with Google maps and other relevant applications. Firstly, a route was previewed on Google maps and Google Earth, and then it was walked. Activities in the scenario were: navigating by using map and a GPS signal, using landmarks, searching for information about nearby services including bus stops, using Skype for a video chat, taking photographs, and using a travel-planner for a bus journey home. During the 1.5 hour interview, questions were asked about the participants' experiences. Walking and talking stimulated naturalistic and informal conversations, and was an excellent method for (a) getting a deeper understanding of the impact of technology on older people's daily life; (b) their concerns and problems in using technology and mobile technologies; and (c) the importance of the social context for positive benefits.

Introduction

Throughout the world the largest group of people who are not, or only to a limited degree, engaged with new technologies are the oldest age group of 65 and over. Large numbers of older people in Europe never use the internet or a computer. In 2010, 60% of people over 65 years old in the UK reported never to have used the internet, see Table 1 (ONS, 2010). In the groups that have never used the internet, females, those on lower incomes and with lower levels of education are over-represented. 68% of UK widows never used the internet; they are a group where the different demographic characteristics of the several groups who do not use the internet come together.

Table 1. Internet users and non-users, 2010, UK

		Internet use Ever used	Never used
		Per cent	
Age	16-24	99	1
	25-44	96	4
	45-54	89	11
	55-64	78	20
	65+	40	60
Sex	Male	84	16
	Female	79	21
Marital status	Single	92	8
	Married	81	19
	Widowed	32	68
	Divorced	75	25
Occupation	Managerial and professional	91	9
	Intermediate	84	16
	Small employers and own account workers	80	20
	Lower supervisory and technical	77	23
	Semi-routine and routine	67	33
Gross Income	<£10,399	69	31
	£10,400 – £20,799	83	17
	£20,800 -£ 31,199	95	5
	£31,200 - £41,599	95	5
	£41,600>	98	2

Table adapted from ONS, (2010) 'Table 5 Internet users and non-users, 2010', *Statistical Bulletin: Internet Access 2010*, last accessed online 1/09/10, http://www.statistics.gov.uk/pdfdir/iahi0810.pdf

New mobile devices and applications are currently coming on the market, opening up a wide range of opportunities to support mobilities and independent life-styles. For the older population, who may be more vulnerable and isolated, this may give new possibilities and benefits, contributing to independent living. However, there may be many reasons for people from this group not to engage with these new technologies, such as financial reasons, lack of access, difficulties using new systems, or negative attitudes (see also Hanson, 2010, and Renaud & van Biljon, 2008).

The study presented in this paper was performed within the BRIDGE project: Building Relationships with the Invisible in the Digital (Global) Economy Project (Bradley et al., 2010). The novelty of this study is the use of qualitative mobile methods to investigate the experiences, acceptance and uses of digital technologies among the older adults community.

The aim of this study was to look at acceptance and use of mobile technologies. The specific research questions included: What are the skills and competencies involved in accessing the internet and using mobile technologies for walking? How could the use of these technologies improve the quality of life of older people? What is the significance of the social environment?

To answer these questions the study was informed by theories of technology acceptance including the Technology Acceptance Models such as TAM and Unified Theory of Acceptance and Use of Technology (UTAUT) (Venkatesh & Davis, 2000; Venkatesh et al., 2003). Research based on these models often uses questionnaires, in which (potential) users are asked about their intention to use a new system, and their expectance of how useful the system would be for them, the effort they have to make to learn and use the system, and the influences of their social environment. Studies using these theories and approaches often indicate that key determinants of acceptance are those associated with social context, for example, normative beliefs and the competencies, skills or resources required for behavioural adoption. Attitude, expectance and intention, however, only provide a part of the picture. In the study reported here, a different approach was taken in order to study technology acceptance of older people. The study aimed to create a more ecologically valid context, in which people could experience and use mobile technologies, with the aim of developing better informed research and insights about the role these technologies play in everyday lives. At the same time, it was expected that being in a user context would make it easier for the participants to express ideas, fears, difficulties and needs, as well as reflections on their personal history with technologies, and easier for the researchers to observe social and behavioural practices in the use of the technology, and the comparative ease or difficulty of use of the technology in situ.

In this paper the focus is on the method of performing "walking interviews" (Evans & Jones, 2011). Firstly an explanation is given of the method and scenarios used to perform these interviews. To illustrate the kinds of findings from these interviews, the preliminary findings are discussed. Finally, the usefulness of the method is discussed in the conclusions.

The qualitative interviews

The interviews generally lasted for 90 minutes in total. The walking interviews consisted of three elements: (1) general questions about the participant and his or her history with regard to computing and other technology, (2) preparing and conducting the walking interview, using both a laptop and a handheld device with a variety of applications, during which questions were posed and issues discussed as they arose from experiences during the walk, (3) a debriefing and a short exercise in using the tablet. A Samsung Galaxy tablet was used, as well as a PC laptop. The walking interviews were performed by two experienced researchers. The participants and one of the researchers wore a recording device.

Interview questions and issues

The overall goal of the interviews was to study acceptance and potential use of mobile technologies through an exploration of existing patterns of use and experiences of learning of technologies, and by discussing attitudes towards technology, as well as needs for the future. Throughout the walk, questions were asked, using non-technical language, and issues discussed, not only about the use of the tablet computer and its applications, but also about previous experiences with different technologies. During the interview, a checklist of questions was used. Some questions were posed explicitly by the researchers; other issues were raised spontaneously by the participants. The issues and questions to be raised in the interviews were derived from existing work with a basis in Psychology and Sociology, including factors in the Unified Theory of Acceptance and Use of Technology (UTAUT, Venkatesh et al., 2003) concerning, intention to use, performance expectancy (e.g. how well will it work for you?), effort expectancy (e.g. how hard is it to learn?) and social influences (e.g. how will your children react if you use this technology?) and factors from the STAM (Senior Technology Acceptance and Adoption Model, Renaud & Van Biljon, 2008, Van Biljon & Kotze, 2008), in which the experimentation and exploration phase is emphasised as precursory for acceptance or rejection. In addition sociological explorations of the competencies and skills in walking and networking (Hodgson, 2012) were used to complement the individualised perspective of the psychology-based models. In Table 2 a list of the questions and topics used in the interview is given. Note that this is only a checklist, not a structured interview schedule, to make sure that all the issues were addressed at some point during the interview.

Table 2. Interview topic guide

Briefing:	General questions	What other kinds of technologies do you use?
		How do you feel about new technologies in general?
After the preview of the walk:		Have you ever used anything like this before?
		What is your first impression, what do you like/dislike?
During the walk, when using different features	Perceived usefulness	When would this feature be useful, in what situations (walking purpose, (un)familiar environment)?
		For what kind of people would this be useful?
		What are the advantages and disadvantages of using this feature?
	Perceived ease of use	What is difficult about using this feature? And what is easy?
		Would it be easy to learn how to use this feature? What would make it easier?
Debriefing (if not addressed during the walk	Intention to use	If you would get the features for free, would you want to have them?
		Would you buy this system? If so for how much?
	Trust	Would you rely on the information?
		Would you be concerned about incorrect information?
		Are you concerned about privacy issues?
	Perceived social acceptability	What would family and friends think if you used these kinds of technology?
		Would you encourage friends and family to use these kinds of technologies?
	Encouraging adoption	What would encourage you to use this kind of technology?
		How would you like to learn how to use the features?

The walking scenario

The walking scenario was developed to explore mobility skills and competencies of navigation, communication and information searching by bringing participants in an ecologically valid situation. The scenario consisted of six different activities including, 1. Preview of the route to be taken; 2. Walking the previewed route; 3. Comparing the electronic map with a static map; 4. Walking freely, heading towards a building; 5. Sitting in the hall of the building, performing several activities; and finally, 6. Walking back, and taking a picture. Each of these activities is described in more detail below.

1. Preview of the route to be taken
Google maps were used to preview a route from the office in which the briefing took place to a main square on campus. Map, satellite and street views of the same route were presented to the participants on a laptop and subsequently on the tablet. Google Earth was used to present a fly-over of the route. In streetview, several landmarks, such as a red postbox, were brought to the participant's attention. In addition, an interactive map of Leeds University campus was shown as this was the area in which the walk took place. In Figure 1, the map and the previewed route is shown.

Figure 1. Previewed route, shown on Google maps, satellite view

2. Walking the previewed route
During the walk, the GPS signal on the map was shown to the participant to show where we were on the rehearsed route, as well as directions given in the Google maps views. Pictures of some landmarks were shown, to verify the walk was on the right track, and to compare and discuss what was seen in the virtual preview and in reality.

3. Comparing the electronic map with a static map
At the destination square, a static map is posted. Participants were asked to find a certain building on this map (Figure 2 gives some photographs near the static map

made during the walking interview.), and the differences between that map and the electronic maps were discussed, to evoke opinions about electronic maps and explore differences in the resources available to those using and not using internet enabled electronic media and walking.

Figure 2. Photographs near the static map made during the walking interview

4. Walking freely, heading towards a building

At this point researchers indicated the large tower of the main university building, and it was suggested to go and visit that building. This was a short walk of around 5 minutes during which time discussion continued around experiences using technology, but also other topics that interested the participants were discussed, stimulated by the topics and by the events and surroundings and included grandchildren, Christmas presents, holidays, dresses of students graduating, encountered during the walk, to give a few examples. This contributed to a friendly, interested and informal atmosphere.

5. Sitting in the hall of the building, performing several activities

The hall of the main University building provided room to sit at a table and to use both the tablet and the laptop. The following activities were performed:

a) Using the satellite view of Google maps to locate the building.
b) Finding Wikipedia information about the building, starting from the Wikipedia icon on Google maps.
c) Using Google maps to find a bookshop, going to their website and showing how to phone them directly with an enquiry about a book. Other shops and coffee shops were also explored.
d) Finding the nearest bus stop on Google maps, looking at the stop with Streetview. From the bus stop icon, the website of the bus company was accessed, and the travel-planner used to view the options for the journey home.
e) Accessing the website of the travel-planner directly by using voice commands with the internet browser of the tablet.
f) Looking at a map to locate other persons. This was not a real-time application; photos of colleagues were placed on the map. Using the laptop, Skype was used to establish a video connection with a colleague, who was invited to join us for a cup of coffee. The colleague always declined but engaged in a conversation with the participant, asking about how the walk was going and their impressions.

6. Walking back, and taking a picture.
During the walk back to the office, the GPS signal was regularly consulted to keep track of the way back. At an historic building, the participant was asked if they wanted to use the tablet to take a photograph of one of the researchers.

Participants

Thirteen participants took part in the study: six male and seven female. Their age was between 65-79 years old, the mean age being 68. None of the participants had a university background. None used mobile internet technologies (e.g. iPhone). Five participants never used a computer, described below as non-users; five used a computer regularly and could be classified as experienced users, three participants used a computer from time to time, but only for a limited set of functionalities (such as email), and could be described as intermediate users. Most of them were married, having children and grandchildren, three of them lived alone.

Findings

All interviews were recorded and subsequently transcribed. Quotes and sections in the discussions were classified using a priori codes derived from a review of existing knowledge including the use of the models described earlier. The analytical technique used here was to create a series of responses to each of the issues and questions in the interview guide (Table 2). These were related to the research questions, and then additional observations from the walking interviews were added. The situation of the responses from participants was also noted so it was possible to tell if the discussion had taken place in particular circumstances, surroundings and sequences.

The scenario and the method of walking interviews allowed addressing the research questions of the study. The analysis is ongoing but preliminary results are given below of how the study may provide answers to three research questions.

The first research question is: *What are the skills and competencies involved in accessing the internet and using mobile technologies for walking?*

The qualitative interview and the walking element of the time spent with participants significantly enabled observation of the interaction with the technology, and key skills in walking. Walking may often be considered as an unskilled activity, yet complex sets of skills are involved, such as skills of negotiation and navigation (Hodgson 2012a and b). The preview allowed exploring skills of navigation and rehearsal practices. The interview questions asking about past experiences and how the participants envisaged using the technology identified issues around exploring a new environment, anxieties about unfamiliarity, personal security and the fear of assault, and fear about getting lost. The route previewed and walked was very short and not very challenging, but participants could easily imagine how such a preview could be re-assuring.

> *"If I am going to a meeting somewhere I have not been before I would go the night before, so that I know exactly where I am going. With something like this you would not have to do that"*, female, 67, non-user.

Some of the computer users had already seen something of Google maps and streetview, and used it to view destinations, and the interview questions prompted a recounting of their previous experiences and their assessments of usefulness. In addition the previews and questions prompted participants to recall their experiences with maps and with Google maps. For some who had travelled as part of their professional lives prior to retirement maps were a tool and technique that they were very familiar with and trusted more than the electronic versions.

During the walking interview skills of landmarking and navigation were observed and discussed. Walking around meant accompanying the participants as they were placed in real-life environments, and in one instance this meant observing and discussing the use of the rehearsed route maps and the actual electronic map to overcome a missing streetname sign. It was observed how the participants triangulated with other landmarks and remembered the rehearsed journey. The handheld dynamic map also acted as a reassurance that they were on the right route. Again, the route walked was limited but participants compared and extrapolated the understanding they had of the way the technology was working to other situations and scenarios such as visiting foreign cities and unfamiliar destinations.

The second research question is: *How could use of these technologies improve the quality of life of older people?*

As part of the scenario, the destination of the rehearsed walk was a static map. Participants were given the task of locating a building which was actually next to them on the static map. The static map was difficult to use, involving matching of the buildings displayed on the map with a list of numbers giving the names and functions of the university buildings. While looking at this map, several times other pedestrians were trying to find their way, and participants entered into discussions with them. The advantages of interactive maps became clear.

Using the tablet in a building, such as in a coffee shop, was experienced while sitting in the hall of the main university building, a rather grandiose hall in which there are some tables to sit at, among students. Participants were interested to visit this building and to search for historic information about it. Looking for shops and bus information in order to plan the rest of the journey was seen as a useful activity, making, for example, shopping trips more enjoyable. The travel-planner evoked conversations about the use of public transport. Contacting a friend using Skype was welcomed by most of the participants, who had never used Skype themselves, although several participants knew people who used it to contact distant relatives. Participants told the colleague contacted that they enjoyed the walk and quite liked the device, freely speaking via the video link.

> *"I would feel very safe. And it's saving you time. For example, if I were in City Square and my husband rang and said I'm at so and so, in town and do you want to meet me? And I would say how do I get there? And if I had this, I*

would go on it and I could find out which bus to get, where it were, and how to walk, and that would save me time and effort" female, 74, non-user.

During the walk back, experiences with the applications used were further discussed. Taking the picture turned out to be surprisingly easy for most participants, especially for the ones that did not use digital cameras or had struggled with this technology in the past. The picture stimulated discussion about sharing pictures with others sending them via phones.

The third research question is: *What is the significance of the social environment?*

Our analysis addresses two environments: the social environment of the interview and the social environment of participants' everyday life. During the walk discussions about a variety of topics occurred and as the time passed the situation and the researchers became familiar to the participants, and the discussions developed a more informal and conversational style with participants initiating topics and talking fluently about their experiences and interests.

Initial results show the social environment to be important in three ways. It may provide a significant stimulus to use technologies; it is the social network to connect and to travel to; and finally, acting as a support network if things go wrong and to provide help in the learning process.

The results of our analysis show that there is often a division of skills within households, and that the social environment for the non-user of technology can be off-putting because they don't feel a sense of entitlement to use the technology, being scared of negatively impacting on the resources of the family by breaking it or because they did not know how to ask someone to teach them use it. One of our participants said:

"My wife's computer, [laughs], it's in the little bedroom: I'm not allowed to touch it... well see, the thing is, she said 'if you touch it you'll press the wrong button or something and I won't be able to get whatever back up or whatever'" male, 72, non-user

The social environment also needs a social network to connect with, to travel to (e.g. friends to email, visits, skyping children living abroad). Using the Skype application participants were often delighted with it, and of all the responses only one participant was indifferent, however, many recognised that the usefulness was influenced by the numbers and types of people they could contact using it – so the (lack of) density of use within their network was the real barrier.

"One of my friends, she has a son in Hong Kong and a daughter in Taiwan, so she speaks to them with ..... skype," female, 66, non-user

One participant told us her son had been asking her to get Skype for a number of months. Her son lives in Australia with her young grandchildren. Once she saw how it worked she said she would go home and download it as she could see why her son had told her to get it. She expressed great pleasure at the realisation that she would be able to use it to see her grandchildren.

For the participants there were a number of places they turned to when things went wrong. For most this was a member of their social network that they would call upon to help them out. What was interesting was the intergenerational exchange as the younger members of the family were called upon for downloads and to fix computers. Others called on commercial experts or retailers to fix problems. One respondent reported that their computer had been set up by a small one-man business who they had found through recommendations in their social network. Others said they had their computers set up at the shop they purchased them from for an extra fee and that they would take them back if there were problems.

> *"Now it's my daughter that comes over and helps me, sort out things, she lives close by, she pops in. She is brilliant, doing things with computer"* female, 66, experienced user

Conclusions

The tablet and the applications that were used during the walk may be useful for older people. Mobile technologies influence the range of competencies and practices associated with travelling, finding information and contacting and meeting people. Navigation facilities are useful if fitting with a lifestyle where people go to unknown places, or when they want to synchronise with others. Most participants could think of situations in which this might be useful and they liked the idea; some computer users already used Google maps. Several participants were uncomfortable with going to unknown places, especially when alone; they were afraid of losing their way and worried about safety. Prior knowledge of what to expect may reduce anxiety and make it easier for older people to go out alone, meet others and be more independent.

During the debriefing, the participants started to come up with ideas themselves, for example, wanting to see the route from the university to the city centre, to find the bus information they needed to get home, asking to see their holiday destinations, and becoming interested in the name of the tablet and the cost, and in the addresses of the websites visited. Among the non-computer users several started to think aloud about the need to take a computer course or taking up the use of computers or other digital devices. Participants with a computer at home (although sometimes only their spouses would use it) said they had learned new things and would have a look for themselves as soon as they got home.

During the visit, sometimes problems were encountered with the wireless connection. Failing technology and seeing the researchers struggle a bit with the technology did not seem to put the participants off and several recounted experiences of difficulties with using technologies.

In this study, the participants experienced the use of mobile technologies under the guidance of the researchers. Having to use a tablet on their own might be quite a different experience, with different consequences for acceptance. Further research may shed more light on the difference between exploring and experiencing technology in a supportive environment and using it independently.

Walking and talking stimulated naturalistic and informal conversation and was an excellent method for getting a deeper understanding of (1) the impact of technology on the daily life of older people, (2) their concerns and problems in using technology and mobile technologies, and (3) the importance of the social context for positive benefits.

All participants said they enjoyed the experience, and that they had learned something new. Sessions sometimes took longer than planned because the participants did not want to leave and liked to discuss things further. For the researchers it was also a very enjoyable experience and they are grateful and privileged that participants were so open about themselves and their relation with technology.

Acknowledgement

The BRIDGE project is a collaboration between three universities (Leeds, Edinburgh, and Middlesex) and financed by Research Councils United Kingdom (Grant EP/H006753/1), the study reported in this paper was completed by a team at the University of Leeds.

References

Bradley, M.D., Barnard, Y., & Lloyd, A.D. (2010). Digital inclusion: is it time to start taking an exclusion approach to interface design? In M. Anderson (Ed.) *Proceedings of the International Conference on Contemporary Ergonomics and Human Factors 2010* (pp. 549–553). London: Taylor & Francis.

Davis, F.D. (1989). Perceived usefulness, perceived ease of use, and user acceptance of information technology. *MIS Quarterly, 13*, 319-340.

Evans, J., & Jones, P. (2011). The walking interview: Methodology, mobility and place, *Applied Geography, 31*, 849-858.

Hanson, V.L. (2010). Influencing technology adoption of older adults. *Interacting with computers, 22*, 502-509.

Hodgson, F.C. (2012a). Structures of encounterability: space, place, paths and identities. In M.S. Grieco and J. Urry, (Eds.) *Mobilities: new perspectives on transport and society* (pp. 41-64). Aldershot, UK: Ashgate.

Hodgson, F.C. (2012b). Escorting economies: Networked journeys, household strategies and resistance. *Research in Transportation Economics, 34*, 3-10.

Office of National Statistics (2010). *Statistical Bulletin: Internet Access 2010*, London: SO.

Renaud, K., & Van Biljon, J. (2008). Predicting technology acceptance and adoption by the elderly: a qualitative study. In: *Proceedings of the 2008 Annual Research Conference of the South African institute of Computer Scientists and information Technologists on IT Research in Developing Countries: Riding the Wave of Technology* (pp. 210-219). SAICSIT '08, vol. 338. New York, NY: ACM.

Van Biljon, J., & Kotze, P. (2008). Cultural Factors in a Mobile Phone Adoption and Usage Model. *Journal of Universal Computer Science*, *14*, 2650-2679.

Venkatesh, V., Morris, M.G., Davis, G.B., & Davis, F.D. (2003). User acceptance of information technology: toward a unified view. *MIS Quarterly, 27*, 425-478.

Venkatesh, V., & Davis, F.D. (2000). A theoretical extension of the technology acceptance model: Four longitudinal field studies. *Management Science, 46*, 186–204.

Aviation

Controller-Pilot communication in a Multiple-Airport-Control scenario

Nora Wittbrodt & Adeline Nelsiana Chandra
Technische Universität Berlin
Germany

Abstract

In a future remote airport traffic control centre, a single air traffic controller will be responsible for the management of several small airports at the same time during times of low traffic volume. The increased complexity of such a multiple-airport-control scenario certainly requires major changes of many well established routines and procedures. An example is the controller-pilot communication via voice radio. In a first experimental study, participants were asked to manage three airports simultaneously using two different design options for the communication environment. Reaction times, error rates and subjective mental effort ratings were collected as dependent variables. Results of the study will be presented and discussed with respect to their impact on the design of the communication environment from the controllers' point of view. In addition, implications of these findings for a second experimental study considering the pilots' perspective will be addressed.

Introduction

To increase both economical efficiency and ergonomic quality in airport management, ATC experts, engineers and human-factors researchers are currently developing a controller workplace to manage several small airports from a central remote location during times of marginal traffic (e.g. Fürstenau, Schmidt, Rudoph, Möhlenbrink, Friedrich, Papenfuß & Kaltenhäuser, 2009; Van Schaik, Lindqvist & Sundberg, 2010; Woods, Francis & Lee, 2008). Such a multiple-airport-control scenario certainly requires major changes of many well established routines and procedures. An example is the voice-radio communication between ATCOs and pilots. Since the beginning of air traffic control, ATCOs use a shared radio frequency to issue instructions to pilots in order to coordinate airport traffic within their area of responsibility (Prinzo, Lieberman & Pickett, 1998). Due to the shared radio frequency, which is also known as the 'party line', pilots of each aircraft are able to overhear both their own conversations as well as those of the other pilots (Midkiff & Hansman, 1992). This method offers both pros and cons. On the positive side, overhearing messages from nearby aircraft provides pilots with a lot of supplementary information about the current traffic situation (Midkiff & Hansman, 1992). As Pritchett and Hansman (1995) ascertained through a survey among pilots of different operational groups, the importance of such additional information was rated highest in the busier phases of flight near the airport. They also found, that regardless of the phase of flight, General Aviation pilots stated a higher necessity for 'party line' information (PLI) as opposed to pilots from other operational groups. On the negative side, taking advantage of these 'party line' benefits requires pilots to continuously shadow the shared radio frequency in order to filter out relevant information. This places a high strain on working memory and may distract attention from the primary task. For instance, Hodgetts, Farmer, Joose, Paramentier, Schaefer and Hoogeboom (2004) found a significant increase in subjective workload ratings when pilots were exposed to the 'party line' while at the same time performance of flight task mental activities such as checklist completion decreased ("irrelevant sound effect").

Despite these disadvantages, voice radio communication will continue to play a central role in ATC especially in the General Aviation sector. However, with regard to a future remote airport traffic control centre, both controllers and pilots will have to adapt to certain alterations in the communication environment in order to deal with the increased complexity of the multiple-airport-control situation. So far, two basic design options have been considered for the organisation of controller-pilot communication: (a) the retention of the existing local 'party line' frequencies or (b) the implementation of a single global 'party line' frequency across several airports (Figure 1).

Figure 1. Two possibilities for the organisation of controller-pilot communication in a future remote airport traffic control centre

Both options offer advantages and disadvantages likewise (cf. Wittbrodt, Gross & Thüring, 2010). On the one hand, maintaining separated local 'party line' frequencies still enables pilots to enhance their situation awareness by receiving additional PLI. On the other hand, since only information concerning their own traffic situation is conveyed, pilots have a smaller chance to fully assess the ATCOs' current workload. Thus, the likelihood of overlapping messages from different airports increases which is detrimental to the ATCOs. In addition, ATCOs need to constantly switch between radio frequencies when addressing pilots at different airports, which constitutes an additional source of error. Nevertheless the retention of the local 'party lines' may also support ATCOs in assigning incoming messages to the corresponding traffic situation. A prerequisite is, however, that radio messages from different airports are presented via separated loudspeakers located next to the display unit of the corresponding airport (as shown in Figure 1). In that case, ATCOs may use the spatial direction of an acoustic signal as a valuable cue to quickly draw their attention to the corresponding traffic situation without having to analyse message content first (cross modal priming, see e.g. Schröger, Kaernbach & Schönwiesner, 2002). It is assumed that this helps to reduce switching costs and error risks. In contrast, the implementation of a single global 'party line' frequency requires the ATCOs to decide from content and context to which airport a radio message belongs given that no additional audio cue is available. It is expected that the information processing required for this decision is more time consuming and error-prone since the ATCOs are obliged to store and recall the whereabouts of each aircraft in order to assign them to the correct traffic situation. However, from the ATCOs' perspective, using a global 'party line' frequency eliminates the need to switch between radio frequencies when talking to pilots at different airports. Furthermore, it prevents an overlapping of radio messages since pilots are able to fully assess the ATCOs' current workload. On the downside, this requires pilots to listen to irrelevant information from other airports which involves more information filtering on their behalf and increases mental workload. As a consequence, the 'irrelevant sound effect' mentioned by Hodgetts, et al. (2004) may even be reinforced compared to the present situation. In addition, receiving PLI from other airports may impede the pilots' ability to develop an accurate situational picture and may lead to confusions, for instance, when two aircraft at different airports are at the same position expecting similar instructions.

As the discussion of the pros and cons shows, any decision for one of the proposed design options is most likely to involve drawbacks either from the controllers' perspective or the pilots' point of view. In order to satisfy both groups effectively, it is important to determine the severity of the different consequences with regard to the overall system performance. For that purpose, a study focussing on the controller's viewpoint was conducted. The paper describes the experimental design and methodological approach of the study. Results are presented and discussed with respect to their implications on the organisation of controller-pilot voice radio communication in a multiple-airport-control scenario.

Method

Participants

Thirty-two participants between the age of 19 and 33 ($M = 23.9$), twenty-one males and eleven females were recruited, at both Technische Universität Berlin and Technische Hochschule Wildau. Participants were either current students or alumni of the mentioned Universities. All participants had a background in engineering or applied sciences. 52% of the students had attended courses with 1st degree exposure to air traffic control. Additionally, two qualified air traffic controllers of DFS Deutsche Flugsicherung[*] participated in the study to allow an estimation of differences in performance between novices and professional controllers. All participants were rewarded a remuneration of 30€ to take part in the study.

---

[*] German air traffic control service provider

Apparatus and materials

Three standard PCs with Nvidia graphic boards were used for simulation purposes. The setup for each airport included a 24" monitor and a set of loudspeakers. In the global 'party line' condition, individual loudspeakers were interconnected using a mixing console to simulate the acoustic properties of a single radio frequency. Microphones and compact keyboards were assembled according to the design option under investigation (e.g. Figure 2). PCs were interconnected to enable data exchange.

Figure 2. Experimental setup in the local 'party line' option

Fourteen traffic scenarios were embedded in the system using the Arise V4.2.4 software[*] of which two scenarios were used for training. During each scenario, simulated pilots requested a total of 16 clearances containing four landing clearances, four taxi clearances to the parking position, four taxi clearances to the runway holding point and four take-off clearances. Participants needed to reject one out of four clearance requests in compliance to preassigned rules. In addition to the clearance requests, each traffic scenario also contained 16 pilot questions concerning the current or past traffic situation (information requests). To minimise sequence effects, information requests, clearance requests and their rejections as well as aircraft call signs were randomised across scenarios. Both, clearance requests and information requests were formulated in a standardised phraseology based on professional controller-pilot communications in German.

Data collection was conducted both manually and electronically. Participants' demographics and subjective mental effort ratings were recorded using paper and

---

[*] Developed by ATRiCS Advanced Traffic Solutions

pencil. Likewise, protocols were scripted by the experimenter on the basis of participants' verbal utterances during each experimental run. Electronic audio files and time stamped keystrokes were recorded automatically.

Task

The participants' main task consisted in the coordination of various aircraft movements through the correct handling of simulated pilots' clearance and information requests during a particular traffic scenario. Participants were asked to either approve or reject clearance requests in compliance with a given rule and to answer information requests as completely and correctly as possible. All responses had to be verbally executed in accordance with the following communication guidelines:

- All requests should be answered promptly,
- Clearance requests should always be answered first since they have higher priority than information requests,
- All responses should include the call sign of the addressed aircraft and the name of the associated airport,
- All responses should be indicated by pressing the spacebar to determine reaction times.

The task was designed to resemble professional controllers' activities as closely as possible while at the same time being easy to learn for novices.

Experimental design

The study was conducted as a mixed between-within subject design. Participants were allocated in two groups representing the different design options but had to perform the same 12 traffic scenarios in a random order.

Independent variables. Two independent variables were examined. The first variable represented the two design options (local 'party lines' and global 'party line', Figure 1). The second variable described the time interval between two consecutive pilot requests. This variable was broken down into the four levels: full overlap, partial overlap, no interval, and with interval (Figure 3).

As illustrated in Figure 3, a pair of two consecutive pilot messages always consisted of one clearance request and one information request. In case of a partial overlap, the second message started halfway through the first message. In condition type D: with interval, messages followed each other with a given time interval of 12 seconds. To make situations less predictable, the order of clearance requests and information requests was randomised within the pairs.

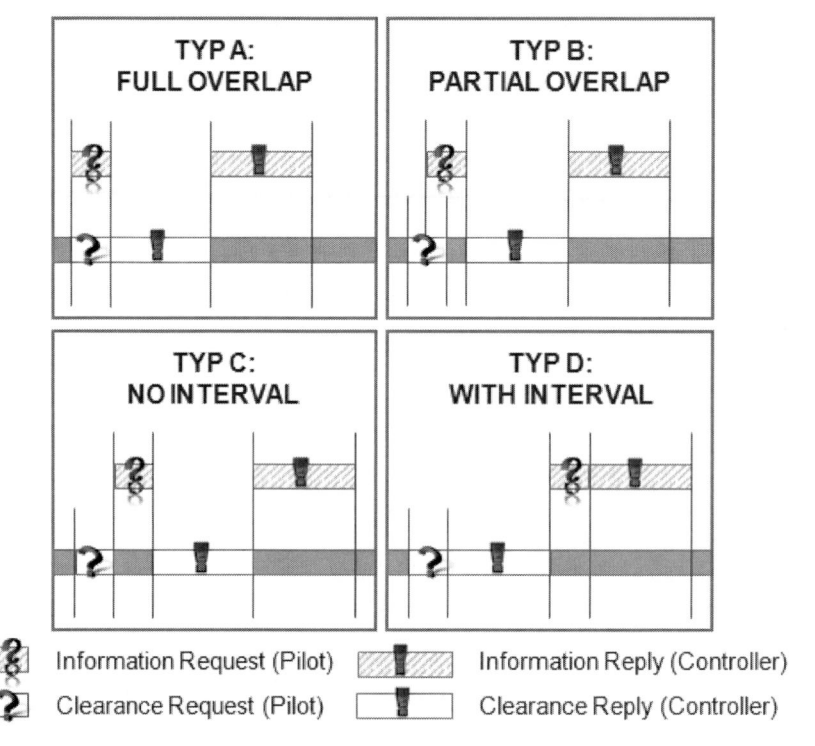

Figure 3. The four different types of time interval between pilots' information request and clearance request. Question marks denote the duration of pilots' requests whereas exclamation marks indicate the available participants' response time

Dependent variables. Three dependent variables including response errors and reaction times to pilots' requests as well as the subjective perception of mental effort during scenario execution were investigated.

Failure or incorrect denomination of airports and aircraft callsign, incorrect approval or refusal of clearances and incorrect or incomplete responses toward information requests were regarded as response errors. Reaction times were measured as the time between the end of a pilot request and a keystroke indicating the beginning of the participant's response. Perceived mental effort ratings were collected manually using the SEA scale* introduced by Eilers, Nachreiner and Hänecke (1986).

Hypotheses. According to the points mentioned in the introduction, it was hypothesised that participants would show a better performance in the local 'party line' condition compared to the global 'party line' condition leading to shorter reaction times, fewer response errors and lower subjective mental effort ratings. With respect to the temporal sequence and overlap of messages, it was assumed that the shorter the stimulus onset time between two radio messages the worse the

---

* The SEA scale (Skala Subjektiv Empfundener Anstrengung) is an adapted German version of the Rating Scale Mental Effort RSME developed by Zijlstra and Van Doorn (1985)

participants' performance would be. Especially in situations where two radio messages overlap, error rates and reaction times were expected to increase significantly. This assumption was based on findings by Cherry (1953) who discovered great deficiencies in participants' ability to separate and understand two simultaneously spoken messages. However, it was assumed that due to the spatial separation of the loudspeakers participants in the local 'party lines' condition would show a better performance when listening to overlapping messages (cf. Bronkhorst, 2000).

Procedure

The total duration of the experiment was around three hours, consisting of a 40 minutes introduction phase and a 140 minutes data collection phase. After receiving detailed instructions, participants were given time to familiarise with the experimental setup. They were then asked to perform two 5-minute practice trials to ensure that all relevant instructions were understood correctly. During the data collection phase, participants had to execute 12 traffic scenarios with an average duration of about 10 minutes each. Between scenarios, participants were requested to rate their mental effort with respect to the completed traffic scenario using the SEA scale.

Results

The differences between the levels of the independent variables design option and time interval were examined through the participants' reaction times, error rates as well as the subjective mental effort ratings[*]. Since participants were instructed to pay more attention to the processing of clearance requests, both error rates and reaction times were analysed separately for clearance requests and information requests.

Mental effort ratings

A one-way independent measures ANOVA was conducted to analyse the differences between participants' mental effort ratings in a given scenario when using one of the two design options (local 'party lines' or global 'party line'). As shown in Figure 4, a significant difference between both design options was observed, $F(1, 31) = 7.09$, $p < .05$, $\eta_p^2 = .196$. Participants in the local 'party lines' group felt less mentally challenged ($M = 77.94$, $SD = 33.12$) compared to participants in the global 'party line' group ($M = 114.47$, $SD = 40.44$).

Error rates

A 2x4 mixed-model ANOVA was conducted to analyse the differences between the two design options (local 'party lines' and global 'party line') with respect to participants' response rate to pilots' requests in the four different types of time interval (type A: full overlap, type B: partial overlap, type C: no interval, and type D: with interval).

---

[*] Data collected of the two professional air traffic controllers were excluded from statistical analysis

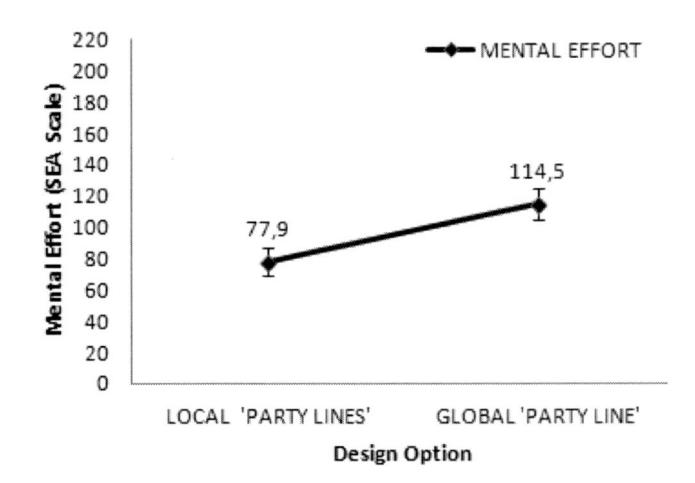

[Figure 4 chart: Mental Effort (SEA Scale) vs Design Option. LOCAL 'PARTY LINES' = 77.9; GLOBAL 'PARTY LINE' = 114.5]

Figure 4. Subjective mental effort ratings

Clearance requests. The main effect of the between subjects variable design option was significant, $F(1, 29) = 12.53$, $p < .05$, $\eta_p^2 = .30$. Participants were more likely to make errors using the global 'party line' option ($M=7.14\%$) compared to the local 'party lines' option ($M=4.2\%$).

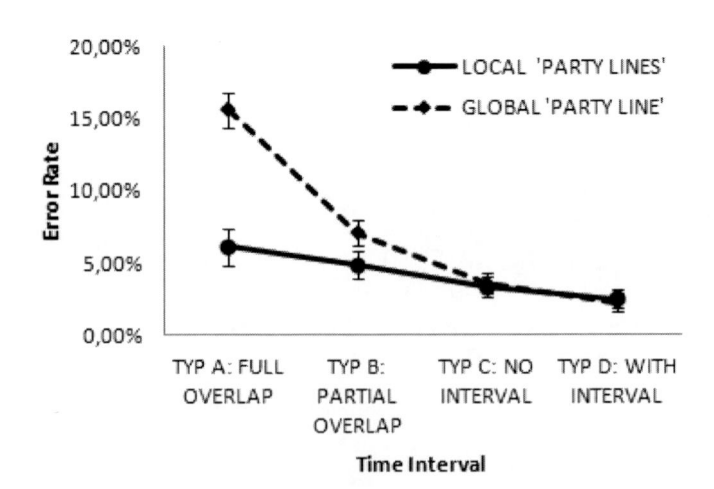

[Figure 5 chart: Error Rate vs Time Interval for LOCAL 'PARTY LINES' and GLOBAL 'PARTY LINE' across TYP A: FULL OVERLAP, TYP B: PARTIAL OVERLAP, TYP C: NO INTERVAL, TYP D: WITH INTERVAL]

Figure 5. Error rates for responses to clearance requests

Furthermore, the main effect of the within subjects variable time interval was significant, $F(2.10, 61.01) = 47.25$, $p < .001$, $\eta_p^2 = .62$. Error rates increased from 2.37% (time interval type D) to 10.85% (time interval type A) the shorter the stimulus onset time between messages.

In addition, data analysis showed a significant interaction between the independent variables design option and time interval, $F(2.10, 61.01) = 16.75$, $p < .001$, $\eta_p^2 = .37$. As is shown in figure 5, error rates were significantly higher in the global 'party line' condition when messages overlapped. However, results obtained through a pairwise comparison indicated that despite of the overall main effect of time interval, no significant difference between both design options was found for time intervals of type C and D.

Information requests. The main effect of the between subjects variable design option was significant, $F(1, 30) = 5.07$, $p < .05$, $\eta_p^2 = .15$. As for the clearance requests, results indicate that participants were more likely to make errors in the global 'party line' condition ($M=30.97\%$) compared to the local 'party lines' condition ($M=23.9\%$).

Furthermore, the main effect of the within subjects variable time interval was significant, $F(2.02, 60.49) = 157.73$, $p < .001$, $\eta_p^2 = .84$. Table 1 shows the error rates in participants' responses to information requests as a function of the time interval between messages. Participants were more likely to produce errors the shorter the time interval between two messages.

Table 1. Means and standard deviations of error rates for information responses according to time interval between requests

time interval	mean M	standard deviation SD
type A: full overlap	45.65%	18.27%
type B: partial overlap	29.26%	12.55%
type C: no interval	22.91%	9.79%
type D: with interval	11.88%	3.03%

In addition, there was a significant interaction between design option and time interval, $F(2.02, 60.49) = 31.02$, $p < .001$, $\eta_p^2 = .51$. Again, results showed a significant difference between design options when messages overlapped but no difference in case of time intervals type C and D.

Reaction times

A 2x4 mixed-model ANOVA was conducted to analyse the differences between the two design options (local 'party lines' and global 'party line') on participants' reaction times in the four conditions of time interval between messages (type A: full overlap, type B: partial overlap, type C: no interval, and type D: with interval).

Clearance requests. The main effect of the between subjects variable design option was insignificant, thus participants showed similar reaction times irrespective of the design option.

However, the main effect of the within subjects variable time interval was significant, $F(1.94, 56.20) = 46.38$, $p < .001$, $\eta_p^2 = .62$. Fastest reactions were recorded for responses to messages in time interval type D ($M=2.07$s), followed by messages in time interval type A ($M=3.27$s). Slowest reactions were registered for responses to messages in time interval type C (*M*=4.4s) and type B (*M*=4.13s). In both cases participants usually waited until the second radio message was fully transmitted before they started to respond, hence the increase in reaction time.

Furthermore, there was a significant interaction between design option and time interval, $F(1.94, 56.20) = 5.48$, $p < .05$, $\eta_p^2 = .16$. As is shown in figure 6, reaction times were significantly longer in the global 'party line' condition for time intervals type B and C. However, results obtained through a pairwise comparison indicated that despite the overall main effect of time interval, no significant difference between both design options was found for time intervals of type A and D.

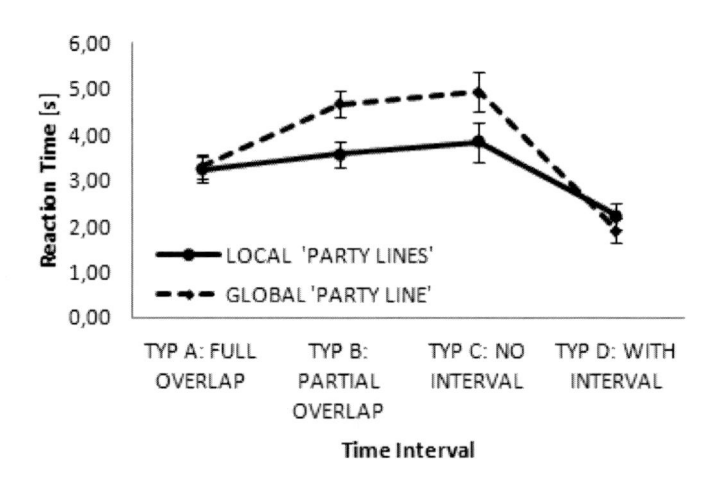

Figure 6. Reaction times for responses to clearance requests

Information request. As for responses to clearance requests, the main effect of the between subjects variable design option was insignificant. However, the main effect of the within subjects variable time interval was significant, $F(2.16, 62.70) = 295.28$, $p < .001$, $\eta_p^2 = .87$. Table 2 shows the reaction times in participants' responses to information requests as a function of the time interval between messages. Participants were more likely to take longer in providing their responses to information requests when they had to answer a clearance request first (time intervals type A, B, and C). There was no significant interaction between the independent variables design option and time interval.

Table 2. *Means and standard deviations of reaction times for information responses according to time interval between requests*

time interval	mean *M*	standard deviation *SD*
type A: full overlap	15.64 s	4.15 s
type B: partial overlap	14.95 s	4.22 s
type C: no interval	15.75 s	4.56 s
type D: with interval	2.88 s	1.45 s

Discussion

It was expected that participants would show a better performance in the local 'party lines' condition compared to the global 'party line' due to the separated loudspeakers and the associated cross modal priming effect. This hypothesis was partially supported. Overall, participants perceived a lower mental effort through the use of several local 'party lines' for controller-pilot communication. Moreover, in the global 'party line' condition significantly more response errors to both clearance and information requests were recorded. However, against the hypothesis, reaction times showed no significant differences between the two design options.

With respect to the time interval between requests, participants' performance was expected to decrease the shorter the stimulus onset time between two consecutive messages. The obtained data affirms this assumption. Participants' responses to both, clearance and information requests were more error-prone the closer two messages followed each other. In addition, participants showed longer reaction times when they had to comprehend two messages at once (time intervals type A – type C) compared to situations where they were required to listen to only one request at the same time (time interval type D).

Regarding the interaction between the two independent variables, it was expected that due to the spatial separation of the loudspeakers participants in the local 'party lines' condition would show a better performance when listening to overlapping messages. This hypothesis is only supported with respect to the error rates. Especially in the condition type A: full overlap, participants committed considerably more errors in the global 'party line' group compared to those in the local 'party lines' group. However, no difference in error rates between the two design options was found for time intervals type C: no interval and type D: with interval in both clearance and information requests. Furthermore, the separation of loudspeakers in the local party lines' condition had only a slightly positive effect on reaction times, which was unexpected.

In general, participants demonstrated a better performance in responding to clearance requests which accordingly resulted in lower error rates and faster reactions compared to results obtained for information requests. This is partially due to the fact that participants were instructed to pay more attention to the processing of clearance requests. It is also arguable that responses to clearance requests only required a binary decision (clearance approved or clearance denied) whereas responses to information requests had to be chosen from a larger pool of possible answers, thus required a higher mental effort leading to more errors and slower reactions.

The comparison between novices and professional controllers showed that despite of being better trained and more experienced, professional controllers surprisingly performed slightly worse when responding to pilot requests compared to the corresponding group means.

Conclusion

In a future remote airport traffic control centre, both ATCOs and pilots will be exposed to substantial changes in every aspect of the flight conduct. One important issue to be considered revolves around the shared voice radio frequency, the 'party line'. In a first empirical study, two design options were examined, with which the participants were supposed to provide accurate responses toward simulated clearance and information requests in a simultaneous control of traffic at three different airports. Further manipulation included the different time intervals between two consecutive pilot messages (full overlap, partial overlap, no interval, with interval).

Results revealed that in general, participants achieved a better performance using several local 'party lines' for controller-pilot communication resulting in lower mental effort ratings, lower error rates and slightly shorter reaction times. However, other than expected the separation of the loudspeakers in the local 'party lines' condition did not enhance participants' performance equally across all time intervals. By analysing the significant interaction effects, it turned out that the associated benefits of the local 'party lines' were only fully exploited when messages overlapped. No relevant difference between both design options was found for responses to non-overlapping messages. Since similar observations were made for professional controllers, it can be concluded that the perceived difficulties in understanding concurrent audio messages indeed seem to result from a general bottleneck in human information processing rather than a lack of training or experience.

However, even with an additional audio cue results showed an intolerable increase in error rates (especially for responses to information requests) in case of overlapping messages. Obviously, these situations constitute a major problem for the ATCO regardless of the design option and should be avoided in any case. Although a single global 'party line' is considered beneficial in overcoming this problem, this design might well expose pilots to very serious consequences as discussed in the introduction. To address this issue, a second experiment focusing on the pilots' perspective should be conducted to assess the impact of 'party line' information on pilots' situation awareness in a multiple-airport-control situation. To start with, it seems reasonable to focus on General Aviation pilots since they show a higher need for additional 'party line' information (c.f. Pritchett & Hansman, 1995). In a real-time full mission simulation of a remote airport traffic control centre, participants' responses to critical 'party line' information could be recorded and evaluated as proposed by Midkiff and Hansman (1992). In addition, participants' situation awareness of specific 'party line' information could be assessed using a freeze-and-query technique in order to detect possible confusion errors caused by the perception of information from several airports via a global 'party line'. Results could be used to further improve the design of the communication environment and communication concept for a future remote airport traffic control centre.

Acknowledgements

The study was funded by the German Research Foundation as part of the research training group prometei and supported by Technische Universität Berlin, HFC Human-Factors-Consult GmbH and Deutsche Flugsicherung DFS. Technical assistance and support was provided by ATRiCS advanced traffic solutions. The authors would like to thank all participants for their time and collaboration.

References

Bronkhorst, A.W. (2000). The Cocktail Party Phenomenon: A Review of Research on Speech Intelligibility in Multiple-Talker Conditions. *ACUSTICA united with acta acustica, 86,* 117-128.

Cherry, E.C. (1953). Some Experiments on the Recognition of Speech, with One and with Two Ears. *Journal of the Acoustical Society of America, 25,* 975-979.

Eilers, K., Nachreiner, F., & Hänecke, K. (1986). Entwicklung und Überprüfung einer Skala zur Erfassung subjektiv erlebter Anstrengung. *Zeitschrift für Arbeitswissenschaft, 40,* 215-224.

Fürstenau, N., Schmidt, M., Rudoph, M., Möhlenbrink, C., Friedrich, M., Papenfuß, A., & Kaltenhäuser, S. (2009). Steps Towards the Virtual Tower: Remote Airport Traffic Control Center (RAiCE). In *Proceedings of the ENRI International Workshop on ATM/CNS 2009* (pp.67-76). Tokyo: Japan.

Hodgetts, H., Farmer, E., Joose, M., Paramentier, F., Schaefer, D., Hoogeboom, P., Van Gool, & Jones (2004). The effects of party line communication on flight task performance. In D. de Waard, K.A. Brookhuis, R. van Egmond, and T. Boersema (Eds.), *Human Factors in Design, Safety, and Management* (pp. 327–338). Maastricht: Shaker Publishing.

Midkiff, A. & Hansman, R.J. (1992). *Identification of Important „Party Line" Informational Elements and the Implications for Situational Awareness in the Datalink Environment.* (Report ASL-92-2). Cambridge, MA, USA: Aeronautical Systems Laboratory, Massachusetts Institute of Technology.

Prinzo, O.V., Lieberman, P. & Pickett, E. (1998). *An Acoustic Analysis of ATC Communication* (Report DOC/FAA/AM 98/20). Oklahoma City, OK, USA: FAA Federal Aviation Administration.

Pritchett, A. & Hansman, R.J. (1995). Variations on 'Party Line' Information Importance between Pilots of Different Characteristics. In *Proceedings of the 8th International Symposium on Aviation Psychology*. Columbus, OH, USA: Ohio State University.

Schröger, E., Kaernbach, C., & Schönwiesner, M. (2002). Auditive Wahrnehmung und multisensorische Verarbeitung. In J. Müsseler and W. Prinz, (Eds.), *Allgemeine Psychologie (*pp. 66-117), Heidelberg: Spektrum Akademischer Verlag.

Wittbrodt, N. & Thüring, M. (2009). Neue Konzepte in der Flugplatzverkehrskontrolle: Anthropotechnische Herausforderungen. In M. Grandt and A. Bauch (Eds.) *Kooperative Arbeitsprozesse DGLR Report 2009/02* (pp. 93-106). Bonn: Germany.

Wittbrodt, N., Gross, A. & Thüring, M. (2010). Challenges for the Communication Environment and Communication Concept for Remote Airport Traffic Control Centres. In *Proceedings of the 11th IFAC/IFIP/IFORS/IEA Symposium on Analysis, Design, and Evaluation of Human-Machine Systems, 1(1)*. Valencienne: France.

Woods, S., Francis, M. & Lee, J. (2008). Tower Information Display System (TIDS): the System Architecture. In *Proceedings of the 8th Integrated Communications, Navigation, and Surveillance (ICNS) Conference* (pp. 1-9). Bethesda, MD: USA.

Van Schaik, F.J., Lindqvist, G., & Sundberg, M. (2010): *ART Validation Report* (Report WP3.4 Deliverable D3.4.1 Rev 01.00 2010-03-17). Växjö, Sweden: SAAB Security Systems AB.

Zijlstra, F.R.H. & Van Doorn, L. (1985). *The construction of a scale to measure perceived effort* (Report). Delft, The Netherlands: Department of Philosophy and Social Sciences, Delft University of Technology.

Air Traffic Controller assistance systems for attention direction: comparing visual, acoustical, and tactile feedback

Maik Friedrich[1], Bernhard Weber[1], Simon Schätzle[1], Hendrik Oberheid[3], Carsten Preusche[1], & Barbara Deml[2]
[1]German Aerospace Centre
[2]University of Magdeburg
[3]Aeronautical Information Service Centre
Germany

Abstract

Modern aircraft have the capability to perform fuel-efficient and noise-reduced continuous descent approaches (CDA). Using these capabilities for approaching today's airports would mean that CDA-traffic and conventional non CDA-traffic would have to be synchronized by the air traffic controller (ATCo). ATCos can be supported when merging both aircraft streams by providing assistance systems that direct attention towards relevant events in merging operations. By actively directing the ATCos' attention potential, complacency effects might be diminished. In the present experiment, a visual, an acoustical and a tactile assistance system were tested in a micro world simulation with 55 participants. Participants had to merge CDA and non CDA-traffic shortly before landing. Non CDA-traffic was controlled via radio communication, while CDA traffic had to be monitored at the same time. Whenever a CDA airplane passed a critical waypoint, visual, acoustical, or tactile feedback was given to maintain a sufficient level of situation awareness. Visual feedback was presented via aircraft label blinking; acoustical feedback was given via radio and tactile feedback was given via a vibrotactile feedback device attached to the wrist. Indeed, results indicate that additional feedback about the CDA traffic has a positive effect on visual attention and situation awareness. Practical implications are discussed.

Introduction

How to manage modern technology is an important challenge that every society has to face. There is a need to use and manage existing technologies in a safe and efficient way; and then to make progress by introducing new technologies. Introducing new technologies is usually expected to provide benefit through improved efficiency, more safety or handling a growing demand for a limited resource. New technology in the area of air traffic management combines these three expected benefits at the cost of higher system complexity and new features of the user interface that an operator has to understand and manage. According to predicted growth of worldwide aviation traffic (Eurocontrol, 2008) the resource airspace seems to become increasingly scarce every year. Thus, there is a serious need for technology to improve the efficiency of today's air traffic management, especially at the bottleneck airport. New approach procedures that promise increased throughput and the reduction of pollution are realized through modern flight management systems (FMS). FMS provide new functionalities for managing and executing highly optimized trajectories for aircraft movements with a high degree of time precision. So-called 4D-trajectory planning (including time as the fourth dimension in aircraft trajectories) constitutes a core concept element of the single European sky ATM research (SESAR) program.

Aircraft equipped with a FMS-4D (henceforth labelled as equipped aircraft) are able to fly an ideally fuel- and noise-optimized continuous decent approach (CDA) and reach a certain location at a previously planned point in time. Figure 1 shows the difference between equipped aircraft and conventional aircraft without FMS-4D equipment (unequipped aircraft). Within the terminal manoeuvring area, the equipped aircraft have the advantage of a more direct approach without being controlled actively by the ATCo in contrast to the conventional approach where the ATCo guides aircraft along the trombone pattern. Consequently, an ATCo will have to monitor/ guide equipped and unequipped traffic (i.e. mixed traffic) at the same time and therefore build one mental approach sequence including both aircraft streams.

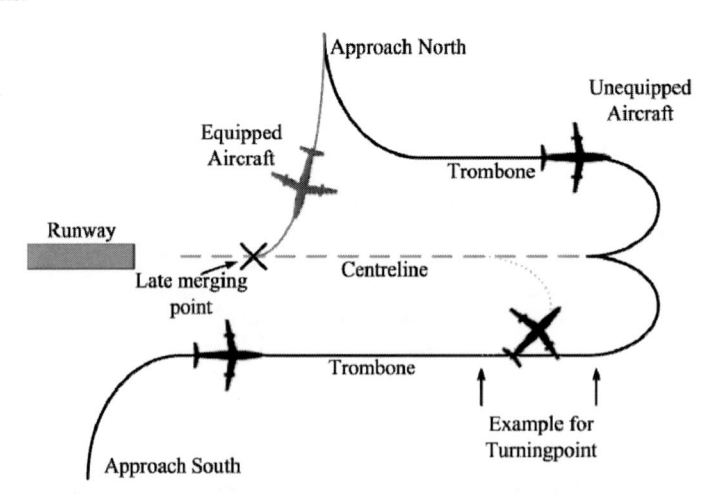

Figure 1. Approach path for equipped and unequipped aircraft (Weber et al., 2010)

A possible solution for supporting the mental sequencing for the ATCo is the use of ghosts (MacWilliams & Porter, 2007; Mundra et al., 2003) as projections of the equipped aircraft. In case of CDA, an aircraft label (ghost) is displayed on the centreline (Figure 2) as a placeholder for the equipped aircraft (Figure 2). In contrast to the (real) equipped aircraft, the ghost has a constant speed (e.g. 230 knots) and merges with the equipped aircraft at the late merging point. Oberheid et al. (2009 a, b) provided evidence that the application of visual assistance systems could increase efficiency and safety when performing late merging operations. Using ghosts has influences on ATCo's situation awareness (SA) in terms of additional visual information and resultant cognitive load. SA in the context of air traffic control is the ATCo's ability to perceive the current situation with respect to time and aircraft position (level 1), the comprehension of their meaning (Level 2), and the projection of aircraft positions into the future (level 3) (Endsley, 1995). Weber and Oberheid (2011) reported that ghosts might draw attention from the equipped aircraft and hence cause losses of SA. For analysing the attention eye tracking data is an indicator for the main fixation areas. Therefore eye data will be captured to support the results of SA.

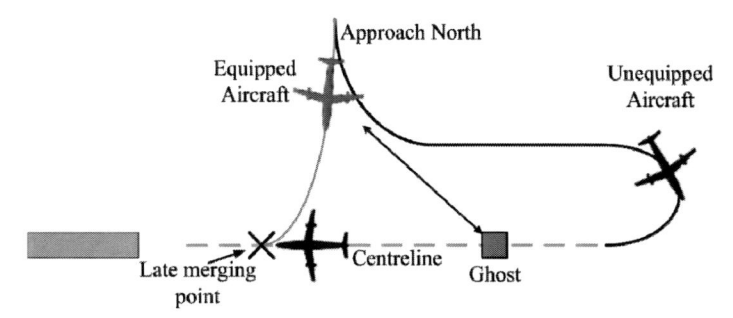

Figure 2. Projection of a Ghost on the centreline as assistance system (Weber et al., 2010)

To avoid that ATCos' visual attention is completely drawn from the equipped aircraft; active attention (re-)direction seems to be a promising approach. In this experiment, additional cues were introduced, indicating that an equipped aircraft reaches relevant waypoints along their approach paths. Yet, to avoid information cluttering, the additional waypoint feedback should not interfere with information processing when performing late merging operations.

Usually, a high density of visual and acoustical information characterizes the ATCo task en-vironment. Therefore, we assume that proving haptic assistance system seems to be most promising, to preserve valuable visual and auditory resources (cf. Wickens, 2002). Indeed, there is empirical evidence from aerospace research that attention direction with tactile feedback is effective and does not impair or interfere with visual information processing (Hameed et al., 2007; Hopp et al., 2005; Sklar & Sarter, 1999). Compared to acoustical or visual feedback, working with additional haptic feedback should also lead to improved SA for equipped aircraft, but workload should be lower , because it does not interfere with the visual and auditory task demands of an ATCo (monitoring, radio communication). In the following experiment, visual, acoustical, and haptic assistance systems were implemented in a microworld simulation to compare the effects on performance, attention allocation, SA, and workload.

Method

Prior to the main experiment, a cross modality matching experiment (Stevens, 1959) with $N = 10$ participants (employees of the DLR) was conducted, to determine visual, acoustical and tactile stimuli that are perceived as equally intensive. The visual stimulus is displayed by blinking of aircraft labels and can be adjusted via the blinking frequency. The acoustical stimulus is presented by a synthetic voice and can be adjusted via volume. A "VibroTac" bracelet developed at the German Aerospace Centre (Schätzle, Ende, Wüsthoff, & Preusche, 2010) presented the tactile stimuli. Attached to the human wrist, the VibroTac bracelet provides vibrotactile feedback via six equidistant vibration segments.

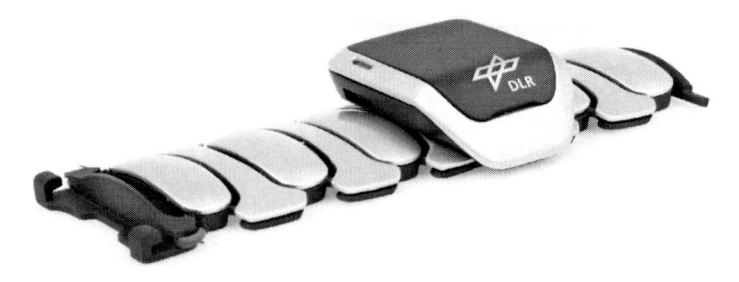

Figure 3. The VibroTac bracelet (Schätzle, Ende, Wüsthoff, & Preusche, 2010)

In a first step, participants were asked to adjust stimuli in each modality to a level they perceived as sufficiently intense for drawing attention, but not too aversive during prolonged working sessions. Seven standard stimuli for every modality were generated that were distributed symmetrically around the determined values (visual = 2.4 Hz, acoustical = 64 dB and tactile = 74 Hz). Standard stimuli were presented randomly and participants adjusted the stimulus intensities of the other modalities, until the intensities were perceived as being equal. The result from the cross modality matching experiment, using the standard stimuli, showed that a visual feedback with a frequency of 3 Hz equates to an acoustical feedback with a volume of 64 dB and a tactile feedback with 73 Hz. These values were chosen for the additional waypoint feedback in the main experiment.

Participants

Fifty-five (21 female, 34 male) German undergraduate students (different fields of study) took part in the experiment. Participants' age ranged between 19 and 34 years ($M = 23.8$; $Mdn = 23$; $SD = 2.8$) and had large variety in fields of study. As an incentive for participation, each participant was paid € 25.

Apparatus

A simplified version of an ATCo radar display (Figure 4) was used as microworld simulation (Oberheid et al., 2009a). Situation awareness and workload were measured with the SAGAT method (measuring SA level 1 to 3; Endsley, 1995) and the NASA-TLX (Hart & Staveland, 1988).

The focus of the experimental setting is the investigation of additional feedback for equipped aircraft reaching critical waypoints while performing a late merging operation. Since our sample consisted of non-experts, task complexity was reduced by using a simple route structure and by standardizing the flight characteristics of equipped and unequipped aircraft (flight profiles, weight categories, etc.). The longitudinal separation for equipped and unequipped aircraft was three miles independent form the weight categories or other flight rules. The radio communication was also reduced to heading and speed clearances, e.g. heading "Aircraft XXX change heading to XXX degrees" and speed change "Aircraft XXX change speed to YYY knots". While the equipped aircraft approached the late merging point autonomously, all unequipped aircraft had to be controlled actively via a North- or South trombone (Figure 4). Participants issued direction and speed clearances for the unequipped aircraft via a simulated radio contact. These clearances were entered in the simulation system by the experimenter on a different workstation and were read back by a synthetic voice to standardize communication and avoid potential effects of open bidirectional communication. Every controller had a timeline (up right Figure 4) that showed the traffic for the next nine aircraft in an optimized order. No other controller assistance systems were available and participants had no secondary tasks.

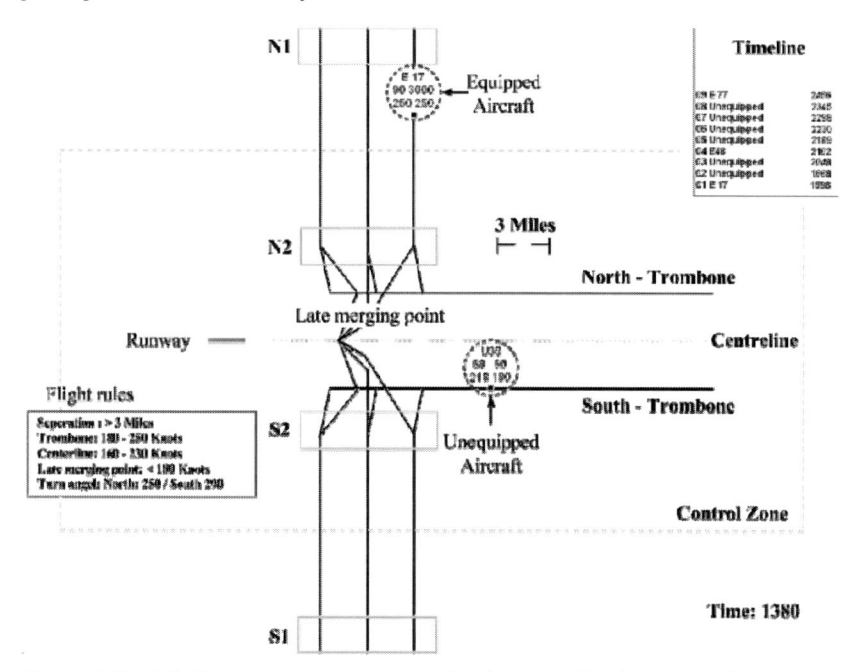

Figure 4. Simplified routing structure combined with equipped and unequipped aircraft used as simulation environment (Weber et al., 2010)

Whenever an equipped aircraft passed one of four pre-defined waypoints (N1, N2, S1 or S2, also see Figure 4) feedback was triggered. Visual feedback was presented by an aircraft label blinking of 3 Hz. Acoustical feedback was given by a synthetic voice (e.g. "Aircraft at North 1") with a volume of 64 dB. In case of two simultaneously triggered messages (e.g. clearance read back and waypoint feedback), the second message was delayed. The VibroTac bracelet applied the vibrotactile feedback with a frequency of 73 Hz.

Procedure

A single factor between subjects design was utilized in the experiment. The four experimental groups with the independent variables visual feedback (V), acoustical feedback (A), tactile feedback (VT),and no feedback (CG). Table 1 shows the dependent variables. To control the impact of task-related capabilities, attention performances and working memory capacities were measured prior to the main experiment with the d2 attention performance test (Brickenkamp & Zillmer, 2002) and the KAI-N working memory test (Lehrl & Blaha, 2004). Based on their test results, individuals were assigned to one of the experimental groups to guarantee similar capability distributions in each experimental condition.

Also prior to the experimental session, the eye tracking system (iViewX, with a 200 Hz sampling rate) was calibrated. The head position was tracked with the A.R.T. optical head tracking system. In a 20 minutes training session, participants were familiarized with the experimental task, the waypoint feedback and the radio communication. Every participant had to merge an average of 15 aircraft (6 equipped and 9 unequipped). Issuing clearances (e.g. headings) was practiced stepwise and in a standardized procedure. Participants were instructed that avoiding separation losses should be their primary objective. Yet, they should also try to maximize throughput as secondary objective. After 10 min training session, the simulation was interrupted and participants had to complete the SAGAT query (measuring SA level 1 to 3; Endsley, 1995) and the NASA-TLX (Hart & Staveland, 1988) for workload measurement.

After the training session, the 40 min main session was started. Every participant had to merge an average of 30 aircraft (15 equipped and 15 unequipped) in every condition. In the main session, the SAGAT method and the NASA-TLX questionnaire was administered at four randomly chosen points of time.

Results

A one-way ANOVA was performed to test whether task-relevant characteristics vary significantly between the different experimental groups. As intended, no significant differences between the experimental conditions (= independent variable) occurred; neither for attention performance (d2) ($F\,(3,\,51) = .06$) nor for working memory capacity (KAI-N) ($F\,(3,\,51) = .80$).

Since the number of aircraft was identical in all experimental groups, the communication and the coordination demands were also similar. In the following section, the results for the performance data, visual attention, situation awareness, and workload are presented.

Performance data

The main performance indicator was the number of longitudinal separation losses (< 3 NM) between equipped and unequipped aircraft when merging on the centreline. Table 1 presents the average number of merging conflicts in each experimental condition. Performing a one-way ANOVA on the number of conflicts showed a marginally significant difference between the experimental groups ($F (3, 51) = 2.8$; $p = .05$). Post-hoc contrast analysis revealed significantly fewer conflicts in the conditions with vibrotactile and acoustical feedback as compared to the control group (both $ts (51) > 2.2$; $ps < .05$). There was no performance gain when working with visual feedback.

Table 1. Summary of the important results

Experimental Conditions / Dependent variable		Visual Feedback (V) n = 15	Acoustical Feedback (A) n = 15	Tactile Feedback (VT) n = 15	Control Group (CG) n = 10
Performance	longitudinal separation losses	1.7 (1.2)	0.9 (1.0)	1.0 (0.7)	2.1 (2.0)
Fixations on the equipped a/c		25.6 (12.9)	21.7 (8.0)	20.3 (8.3)	16.4 (7.9)
Situation Awareness	Level 1 (equipped a/c)	2.0 (0.8)	1.9 (0.8)	2.1 (1.0)	1.9 (1.1)
	Level 2 (merging conflicts)	0.1 (0.4)	0.0 (0.0)	0.1 (0.3)	0.0 (0.0)
	Level 3 (future merging conflicts)	0.6 (0.7)	0.8 (0.9)	0.5 (0.7)	1.1 (1.3)
Workload	NASA-TLX (Scale from 0 to 20)	9.9 (2.4)	9.8 (2.2)	10.4 (2.0)	10.8 (2.5)

Visual attention

For the examination of visual attention, the eye data recorded during the main session were analysed. The number of fixations on the equipped aircraft was used as an indicator for visual attention (Rayner, 1977). The minimum duration for fixations was 100ms, which is a usual threshold for visual tasks involving reading (Duchowski, 2007). The average number of fixations (standard deviations in parentheses) on equipped aircraft per experimental conditions is summarized in Table 1. A one-way ANOVA on the number of fixations on equipped aircraft failed to reach the conventional level of significance ($F (3, 49) = 1.8$; *ns.*). Yet, individuals devoted more attention to the equipped aircraft when there was visual feedback compared to no feedback at all in the control condition ($ts (49) = 2.2$; $ps < .05$).

The results indicate a trend that different types of feedback have influence on the distribution of visual attention on the equipped aircraft. Obviously, acoustical and tactile feedback can be processed without explicit visual response i.e., participants might have gathered relevant information without fixating the equipped aircraft. Thus, the next step of analysis is the investigation whether the differences in visual attention also affected situation awareness and workload.

Situation awareness

Level 1 to 3 of SA was measured with the SAGAT questionnaire. As indicator for level 1 SA, individuals had to reproduce equipped, unequipped aircraft and ghost positions within the simulation environment. For level 2 SA the number of undetected merging conflicts and for level 3 SA the number of undetected future merging conflicts was counted. The results for SAGAT level 1, 2 and 3 are presented in Table 1. An ANOVA showed no significant (F (3, 51) = 0.17; $ns.$) difference between the Level 1 SA results. Similar results were obtained for Level 2 SA (F (3,51) = 1.08, $ns.$) and Level 3 SA (F (3,51) = 0.95, $ns.$).

Workload

The NASA-TLX ratings were also averaged for each experimental group (last row in Table 1). The results of ANOVA indicate (F (3, 51) = 0.57, $ns.$) no significant between different types of feedback. Yet, the highest average value was found in the CG. This indicates that the use of feedback for equipped aircraft at least did not increase workload, although additional information had to be processed by the controller.

Altogether, results indicate that additional feedback for equipped aircraft has a positive influence. The results for the control group have the lowest values in performance, SA (except Level 2 SA) and workload. A difference between the tactile feedback, the acoustical and visual feedback was found on a nominal level comparing the merging conflicts and the number of fixations on the equipped aircraft. The differences between CG, A, and VT were significant for the merging conflicts. This indicates that feedback has positively influenced the performances of participants and that this differs between the different types of feedback.

Discussion

The experimental setting of our microworld study was based on a number of simplifications compared to the real working environment of ATCos. However, key aspects of the ATCo work environment were depicted in facets, such as traffic control, coordination, and communication. Thus, the general challenges of late merging operations could be analysed with a high degree of experimental control. The experiment demonstrates the usefulness of the simulation environment for future experiments that might have similar research objectives.

Yet, it would be interesting to explore the impact of feedback under more realistic conditions, with equipped aircraft deviating from their optimal trajectories, cancellations of autonomous approaches, for instance. In the current study, equipped aircraft behaviour was fully deterministic and the displayed ghosts were reliable. The SAGAT and the NASA-TLX are both subject assessment tools and therefore the measurement dependent on a variety of personal properties.

As an additional limitation of our study, the non-expert sample , of course, is not representative for expert controllers Thus, the experimental findings should be validated with an expert sample in future research.

References

Brickenkamp, R. & Zillmer, E. (2002). Test d2, Aufmerksamkeits-Belastungs-Test. Göttingen, Germany: Hogrefe.

Duchowski, A.T. (2007). *Eye Tracking Methodology: Theory and Practice*. New York: Springer.

Endsley, M.R. (1995). Measurement of Situation Awareness in Dynamic Systems. *Human Factors, 37*, 65-84.

Eurocontrol. (2008). STATFOR Forecasts. Retrieved August 30, 2011, from http://www.eurocontrol.int/statfor/public/standard_page/forecast_methodology.html

Hart, S.G., & Staveland, L.E. (1988). Development of NASA-TLX (Task Load Index): Results of empirical and theoretical research. *Human mental workload, 1* (pp. 139-183). Amsterdam: Elsevier.

Hameed, S., Jayaraman, S., Ballard, M., & Sarter, N.B. (2007). Guiding visual attention by exploiting crossmodal spatial links: An application in air traffic control. *In Proceedings of the Human Factors and Ergonomics Society 51st Annual Meeting* (pp. 220–224). Santa Monica, CA: HFES.

Hopp, P.J., Smith, C.A.P., Clegg, B.A., & Heggestad, E.D. (2005). Interruption Management: The Use of Attention-Directing Tactile Cues. *Human Factors, 47*, 1-11.

Lehrl, S. & Blaha, L. (2004). *Messung des Arbeitsgedächtnisses*. Ebersberg, Germany: *KAI-N*.

MacWilliams, P., & Porter, D. (2007). An assessment of a controller aid for merging and sequencing traffic on performance-based arrival routes. *MITRE Technical Paper*. McLean, VA, USA: MITRE Corp.

Mundra, A.D., Bodoh, C.L.P., Domino, D.A., El-Sahragty, A., Helleberg, J.R., & Smith, A.P. (2003). Capacity enhancements in IMC for airports with converging configurations with knowledge of aircraft's expected final approach speeds: A case study. *USA-Europe ATM Seminar*, Budapest, Hungary.

Oberheid, H., Weber, B., & Rudolph, M. (2009). Visual Controller Aids to Support Late Merging Operations for Fuel Efficient and Noise Reduced Approach Procedures. *Human Factors and Ergonomics Society Annual Meeting Proceedings, 53*, (pp. 11-15).). Santa Monica, CA: HFES.

Oberheid, H., Weber, B., Temme, M.M., & Kuenz, A. (2009). Visual Assistance to Support Late Merging Operations in 4D Trajectory-Based Arrival Management. In *Proceedings of 28th Digital Avionics Systems Conference*, Orlando, FL. Orlando, FL, USA: IEEE

Rayner, K. (1977). Visual attention in reading: Eye movements reflect cognitive processes. *Memory & Cognition, 5*, 443-448.

Schätzle, S., Ende, T., Wusthoff, T., & Preusche, C. (2010). VibroTac: An ergonomic and versatile usable vibrotactile feedback device. *19th International Symposium in Robot and Human Interactive Communication* (pp. 670-675). Viareggio, Italy: IEEE.

Sklar, A.E., & Sarter, N.B. (1999). Good vibrations: Tactile feedback in support of attention allocation and human automation coordination in event-driven domains. *Human Factors, 41*, 543-552.

Stevens, S.S. (1959). Cross-modality validation of subjective scales for loudness, vibration, and electric shock. *Journal of Experimental Psychology, 57,* 45-51.

Weber, B. & Oberheid, H. (2011, Mai). Visual Assistance for Merging Operations: Effects on Controllers' Performance, Situation Awareness, and Mental Workload. *presentation at the 15th conference of the European Association of Work and Organizational Psychology,* Maastricht, Netherlands

Wickens, C. (2002). Multiple resources and performance prediction. *Theoretical Issues in Ergonomics Science, 3,* 159–177.

The Eurofighter Typhoon cockpit assessment process

Suzanne Broadbent
BAE Systems, Preston
UK

Abstract

Since the early days of the Eurofighter Typhoon, Human Factors (HF) has had a major part to play in its development. The early cockpit layout and philosophy were informed by initial mission analyses, an ergonomics handbook and principles of cockpit design. Today's current processes still rely on these methods, alongside HF specialists working together with engineers and the end users, in order to provide a platform that is both highly usable and operationally effective. As the aircraft continues to be developed, Cockpit Assessments remain an integral part of the design cycle and are carried out regularly as a de-risking exercise that allows aircrew to review and comment on the proposed design, whilst also allowing Eurofighter Cockpit Group the opportunity to collect data prior to design freeze. A variety of assessments are carried out by Cockpit Group to evaluate such issues as layout, lighting and the moding of displays and controls. This paper provides an overview of the initial design process, discusses the steps involved in the development of the assessment methodology and presents the various tools and techniques used throughout the assessment to ensure acceptability of the intended product.

Introduction

The Eurofighter Typhoon is a multi-role combat aircraft, built by the Eurofighter Consortium of BAE Systems, Cassidian and Alenia Aeronautica and is currently in service with the air forces of the UK, Spain, Italy, Germany, Austria and the Kingdom of Saudi Arabia. Available in a single seat or twin seat training variant, and in development since the late 1980s, the Eurofighter Cockpit is treated as a subsystem in its own right (rather than different aspects being owned by differing functions, e.g. radar, fuel etc.) and as such has a dedicated team with design authority. The multi-disciplinary team was originally made up of designers, engineers, HF specialists and aircrew. This team worked together to inform the initial layout, geometry, lighting and moding based on HF and operational requirements for single seat operation, alongside knowledge of the capabilities of the systems and equipment. This work is continued today by the current Eurofighter Cockpit Group at BAE Systems, Warton.

Cockpit design requirements

As a swing role combat aircraft, responsible for carrying out both Air-to-Air and Ground Attacks, it is imperative that Human Factors are considered within the cockpit and that the right data, in the right format, at the right time are available to the pilot to aid them carry out these tasks. As a single seat aircraft, it is also necessary to ensure acceptable pilot workload by making use of automation, spreading the tasks across the pilot's attentional resources and employing advanced cockpit technologies such as Direct Voice Input (DVI) and the Head Equipment Assembly (HEA).

Initial Human Factors work

A great deal of work was carried out initially in order to define the cockpit philosophies and ultimately satisfy the design requirements. A number of documents were produced which outlined the requirements from the users' perspective and set out "style guides" to help define the cockpit controls and displays. The main documents are the Mission Analysis (Wilkinson, 1998), the Principles of Cockpit Moding (Wykes 1988) and the Ergonomics Handbook (Wykes 1988).

Mission analysis

The mission analysis was carried out in order to inform the initial cockpit design. It involves a number of "forcing missions" (Wilkinson, 1992) being broken down into flight segments and operational modes, which then allow the specification of function, information and control requirements within the cockpit.

Principles of cockpit moding

This document outlines how and when information from the aircraft systems is presented to the pilot; whether by the multi-function displays, the head-up display (HUD), the Helmet displays, or the Get-u-Home (GUH) instruments, as well as how the pilot interacts with these subsystems; either via DVI, the Manual Data Entry Facility (MDEF) or the Hands on Throttle and Stick (HOTAS) functions.

Ergonomics handbook

It was recognised that the cockpit design would be conducted by a multidisciplinary team with varying levels of experience and therefore this handbook attempts to disseminate some practical specialist advice relating to control, display and layout design and ensure a consistent style is applied throughout the cockpit. Therefore the Ergonomics Handbook contains information from a number of Military Standards, Defence Standards, Standardisation Agreements (STANAGS), Ergonomics textbooks and practical cockpit design experience.

Current design process

Cockpit Group remains a multidisciplinary team, with Systems Engineers and HF specialists working together, along with support from the aircrew community. As such, the Mission Analysis, Principles of Cockpit Moding and the Ergonomics Handbook are still being used within Cockpit Group to influence the design and also updated where appropriate to reflect the latest advice and information available from

HF best practice and from the customer aircrew on how the aircraft is being used within service.

A formalised user-centred design process has been devised in order to ensure that the initial requirements from the customer are evaluated in terms of implications for the pilot together with the more traditional equipment considerations. Figure 1 below captures an overview of this process and the various "gates" that the design proposals must pass through before making it onto the aircraft. Each of these gate reviews has differing inputs and outputs that are necessary for progression into the next phase of design, e.g. documentation updates, completion of task analysis, agreement of proposals from aircrew etc.

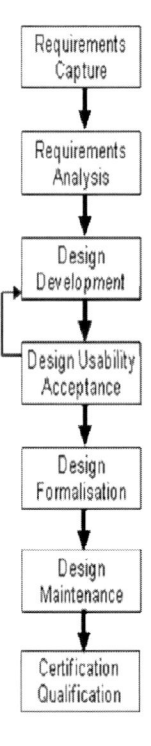

Figure 1. Cockpit design process

Assessment types

The stage of Design Usability Acceptance (above) is generally passed by means of a Cockpit assessment. Cockpit assessments can take on a variety of forms, focusing on layout, lighting or moding (each discussed below). All assessments are carried out in accordance with the Cockpit Group Human Factors Assessment Guidelines (Fairburn & Ek 2008) and governed by the Ethics Guidelines for Human Factors Assessments (Fairburn 2008).

Layout

The cockpit alterations over the last few years have predominantly involved small additions to the original cockpit (e.g. extra stowage requirements or extra control switches required for new functionality) and therefore assessments can be carried out in either the Mission Development or Lighting Cockpit (MDLC) or on real aircraft rather than prototypes. The MDLC is a standalone production standard cockpit that utilises flight standard (or appropriate prototype) displays, controls, structure and seat. Assessments can cover tasks such as ingress and egress, clearance, reach and whether the proposed layout will support task performance. A mixture of engineering and aircrew subjects are often used in order to cover the anthropometric requirements imposed upon the cockpit and the participants are required to don the various aircrew clothing and equipment.

Lighting

Lighting assessments generally occur more frequently than layout assessments, due to regular improvements in display hardware and new information being introduced onto the displays. These changes require assessment under different lighting conditions to ensure readability throughout the varying ambient levels. The visibility of symbology, colour perception, glare and reflections are all considered within these assessments.

The MDLC is used to carry out these assessments in the Ambient Lighting Facility (ALF) dome at BAE Systems Warton. This facility uses a roof-mounted low ambient 'starlight' generator, periphery-mounted arc lamps and moveable 'sun guns' to generate ambient conditions from overcast starlight to un-obscured sunlight at an altitude of 30000ft. The facility also includes a separate Lighting Laboratory for the provision of objective data such as luminance, chromaticity and reflectivity

Moding

The majority of cockpit assessments carried out within Cockpit Group are Moding assessments. Due to regular updates of new functionality and capability, there is a steady influx of design changes or additions required in the software, such as adding extra symbology onto the displays or further controls onto the softkeys and HOTAS. These designs are rapidly prototyped on the Active Cockpit Rig (ACR) at Warton and can then be assessed in terms of suitability and usability, as detailed further in the sections below.

Moding usability assessment process

Establish need/requirement/scope

It is now expected practice within Cockpit Group to carry out moding assessments for each major new software package, in order to de-risk the design intent upfront in an operational context and allow industry and customer pilots to have input into the proposed alterations. Depending on the point in the product lifecycle and the amount of change required to the previous standards, the scope of these assessments can vary through concept design, design development, risk reduction and ultimately qualification and certification.

Establish objectives, method and plans

From an HF perspective, the ultimate objective of the assessments can vary at different stages of product development. Early assessments may focus on HF issues at a functional level ("Can you distinguish that symbol from the others?", "Is that control accessible?") and can be used to inform further rapid prototyping of other options if required. Later assessments may assess the new upgrade as a whole coherent package and focus more on interoperability of new functions in an operational scenario and on gathering data for a workload baseline.

In order to formulate and formalise the required methods for assessment a Human Factors Proposal is generated reflecting the considerations shown in Figure 2. The proposed new design elements are considered for assessment in terms of how they are expected to be used (contextual use) and the likelihood of them resulting in any particular HF concerns, in terms of workload (WL), situation awareness (SA) or human error. Taking into account any constraints there may be in terms of the demonstration (e.g. timescales, risk etc.), design elements are then prioritised for inclusion in the assessment and it is decided whether they should be covered in dynamic mission segments, training exercises (see "Devise Training Plan" below), or simply during training. This rationale is fully covered within the Human Factors Proposal document, allowing the HF specialist to evaluate the designs and propose the best way of gathering data in order to satisfy the assessment aims.

The length of an assessment also varies depending on the amount of change that is proposed; some assessments involve just half a day of training and half a day of missions, whilst others can last up to two weeks per slot. Separate assessment slots are provided for each nation and each of the four partner nations are requested to provide two pilots each. Therefore assessment periods can last for under a week or over two months.

Prepare and validate facilities

The ACR is based around representative cockpit geometry and houses a comprehensive suite of displays and controls. The cockpit is located in a 9m truncated dome onto which an outside world view and the HUD and HEA displays are projected. The ACR is driven by a suite of software models that emulate the operation of the real aircraft systems without the complexities associated with using real aircraft software. This allows for rapid prototyping and early demonstration and alteration of proposed cockpit HMI in advance of actual software coding work.

Prior to assessment, Cockpit Group Engineers validate this facility by ensuring that the design in the ACR is representative of that in the design documents they have produced and described in the briefing guide sent to assessing pilots. Often the designs are complex and the ACR implementation has to be reviewed to ensure that the moding works in the correct manner across a variety of different scenarios that pilots may wish to explore during the assessment.

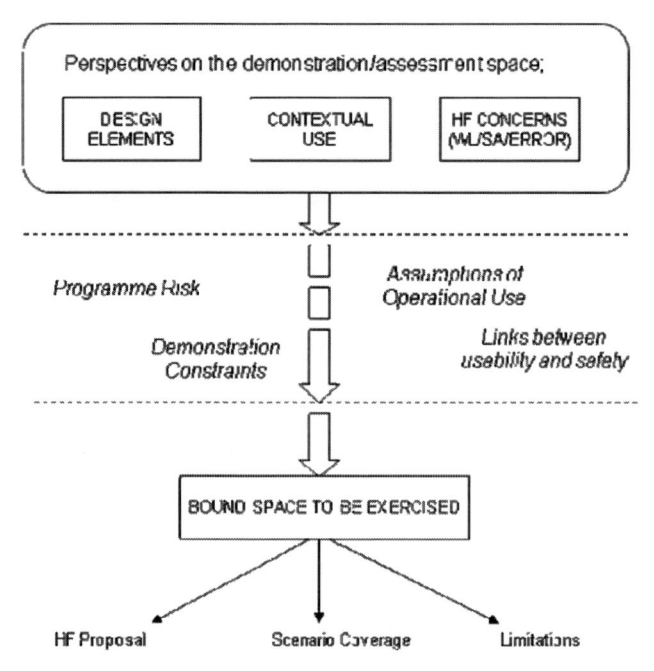

Figure 2. Considerations for the HF Proposal

Devise training plan

Again, dependent on the scale of the assessment, the level of training the participants require can alter dramatically. All participants are trained Eurofighter pilots, some operational/frontline pilots and some qualified test pilots. As a relatively new platform, only recently are there pilots who have been trained solely to fly the Eurofighter Typhoon; previously pilots have come from different platforms (e.g. Tornado F3, Jaguar, Harrier) and have either been Air-to-Air or Air-to-Surface specific The swing role of the Eurofighter allows pilots to carry out both these mission types and therefore it is often found that training needs to be tailored to differing prior experience. Similarly, pilots are invited from all the four partner nations to take part in the assessments (UK, Spain, Italy and Germany) and therefore they have differing experience on previous and alternate aircraft, which again may influence the level of training they require, and also lead to preferences when it comes to display options.

Generous amounts of training time are provided during the assessments to ensure that the pilots are able to complete the tasks and provide valid feedback. Training is generally carried out in the ACR by an experienced Cockpit Group Engineer. During this training, checklists are used by the engineer to ensure that the pilot has seen all the new aspects of design. The pilot is also invited to provide any comments on the design throughout training which will be recorded on the checklists, for later discussion and possible inclusion in the recommendations.

"Training gates" are employed to ensure that the pilot is familiar enough with the new design functionality to progress to the mission aspect of the assessment. These can also be used to fulfil the secondary purpose of exercising either elements of the design that it may be difficult to force during mission scenarios, or certain "edge cases" that may not merit an entire mission to cover them. These gates often take the form of "recovery exercises" where the pilot is "dropped" somewhere in airspace and has to use the cockpit displays to complete various tasks, relying on their interpretation of the symbology to determine what is happening and what they need to do to reach the target/engage the hostile aircraft/land the aircraft. These exercises also allow Cockpit Group to gain confidence in the usability of the symbology even when the pilot has low SA.

Devise and prepare missions

A large aspect of the cockpit assessment involves the participants carrying out representative missions or segments of missions in the ACR. A communications link allows members of Cockpit Group to act as other players during dynamic mission scenarios, such as air traffic control (ATC), Command and Control (C^2) units and wingmen or co-operators. These individuals are fully trained beforehand on best practice in each of these fields to ensure the assessment runs as realistically as possible.

The missions themselves are driven out of the assessment objectives – e.g. if there are specific HF concerns that require specific data to be collected, then the scenario can be designed to exercise this functionality. Alternatively, if a workload baseline is being carried out the missions can be developed so that they will cover off all aspects of the mission analysis to ensure a full understanding of workload across all the different tasks the pilot and aircraft are likely to be requested to perform. Specialist mission briefs are also devised to help the pilot envisage the scenario and the operational theatre, Rules of Engagement, weapon configuration and mission taskings are all covered, along with code-words and information about co-operators and contacts.

Design data assessment methods

A number of assessment methods are available for moding assessments. The ACR automatically logs all button presses and displays presented throughout the course of the mission, along with any other objective measures that are deemed appropriate (e.g. accuracy of task completion, flight path, timings etc). An eye-tracker is available to record where the pilot is looking throughout the sortie and in the ACR control room there is a suite of monitors that repeat the cockpit displays and selections from the dome itself, allowing for expert observers to comment on the pilot's tasks and make notes to discuss in debrief. A separate control panel is also available in order to select various failures within the ACR, so data can be gathered on the how the pilot would cope and react to real-life failures within the aircraft.

Further to this, subjective measures are taken during and post-assessment, with workload and situation awareness ratings often taken in flight, using simple rating scales such as the Bedford Rating Scale (Roscoe and Ellis 1990) and a modified version of the Crew Awareness Rating Scale (CARS, McGuiness & Foy 2000). Further questionnaires relating to specific concerns or more detailed workload and SA measurements are taken post-mission during a debrief session. This debrief also allows the pilot the opportunity to make further comments on the designs, having used them in a representative scenario.

Shakedown trial

Once the assessment methods have been agreed upon, the facilities validated and the missions in place then the assessment design progresses to a "Shakedown Trial" with Industry aircrew. This involves an industry test pilot from one of the partner companies, running through the proposed training and assessment in order to check that the practicalities, timings and methods are all acceptable, realistic and appropriate and can be carried forward to the customer assessment. Although Industry aircrew are thoroughly involved in the work up of the design and the assessments, this trial allows them to actively see it in practice and provide their own additional comments into the design if required.

Conduct assessment

The assessment itself follows a set procedure, which is repeated for each of the assessing nations in order to maintain consistency throughout the assessment and hopefully generate valid and comparable results. Following completion of the training and training gates, as described above, the pilots are allowed a generous amount of "free play" in the ACR in order to further familiarise themselves with the designs and explore differing scenarios. Once both the engineers/trainers and the pilots are happy, then they are ready to continue to the mission scenario aspect of the assessment.

During the missions, the pilot flies the ACR and the participating members of Cockpit Group run the mission from the control room. The HF specialists monitor the progress of the mission and the pilot actions from the control room and make numerous notes for discussion post-mission, injecting any failures into the cockpit as required. "Measures of Effectiveness" forms are completed to capture whether the pilot has carried out all the tasks requested of them by the pre-mission brief or ATC. These can be used post-mission to ensure all pilots flew similar missions and provide a prompt for data to be discarded if not comparable. The HF specialists are able to segment the ACR data, so that they can gather specific objective data for the different aspects of the mission for comparison later e.g. an Air-to-Air segment versus a Navigation segment. Data can also be "flagged" if of particular interest for further review and analysis.

Post-mission, a debrief is carried out involving the participating pilot, one of the expert observers and the HF specialist. First the pilot is requested to fill in any questionnaires that are required, not only to capture their immediate reactions to the mission, but also so they are not biased by the comments of any of the assessors that may follow. Once the questionnaires have been completed, the pilot is asked for any other comments and then talked through the mission by the expert observer. This gives the observer the opportunity to question the pilot on anything they did not understand in their behaviour throughout the mission and also allows the pilot to ask any further questions of the design or mission scenario, and even see a replay of certain events if necessary. Following completion of all mission scenarios, the pilots are requested to fill in a Design Acceptance Questionnaire (reflecting those used in flight test) to provide an overall acceptability rating on the various new design features.

Analyse data, wash-up and report

Following assessments, all data collected is analysed by the HF specialists in order to highlight initial findings, concerns and recommendations. A presentation is put together and all assessing pilots invited back to participate in a "wash up" session, where all nations can view each others comments and agree on a consolidated way forward. Following the wash up meeting, Observations and recommendations can then be assessed by Industry and a route identified for them to be progressed into the product.

Future assessments

With the Eurofighter Typhoon now being actively used on operations and with increasing export opportunities, the operators are discovering new functionality they desire and new scenarios that they need to be equipped for, so the cockpit will continue to evolve to deliver the functionality the pilots need to take part in such operations. By maintaining a close relationship with the customer nations, Cockpit Group can reflect in the cockpit assessments the scenarios customers are facing in the operational theatre and ensure that any future HMI alterations are both usable and fit for purpose. Each cockpit assessment is also seen as an opportunity to explore further HF tools and techniques. In forthcoming assessments it is intended to research further the use of freeze probe techniques to objectively measure SA and trial different measures of workload. The potential to link up the ACR with other such facilities to enhance the realism of the flight lead/wing man dynamic is also under investigation. It is fully expected that updates to the Eurofighter Typhoon will continue throughout the life of the aircraft, and that the cockpit assessment processes currently defined will also continue to be updated in light of best practice and research to ensure the platform remains both operationally effective and highly usable in the future.

References

Fairburn, C. (2008) *Ethics Guidelines for Human Factors Assessments*. (BAE-RE-W-7N3-PC-000031). Preston: BAE Systems.
Fairburn, C. & Ek, A.M. (2008) *HF Assessment Guidelines* (BAE-WSE-RP-EFA-CPT-1163). Preston: BAE Systems.
McGuiness, B. & Foy, L. (2000) *A subjective measure of SA: The Crew Awareness Rating Scale (CARS)* presented at the Human Performance, Situational Awareness and Automation Conference, Savannah, Georgia, 16-19 Oct 2000.
Roscoe, A. & Ellis, G. (1990) *A Subjective Rating Scale for Assessing Pilot Workload in Flight* (TR90019), Farnborough, UK: RAE.

Wilkinson, P.R. (1992) *The Integration of Advanced Cockpit and System Design.* (AGARD-CP-521 Paper 26). AGARD Avionics Panel Symposium, May 1992 Madrid.

Wilkinson, P.R. (1998) *Eurofighter 2000 Mission Analysis: The Complete Works.* (BAe-WSE-RP-EFA-CPT-881). Preston: BAE Systems.

Wykes, K. M. (1988) *EFA Ergonomics Handbook.* (BAe-WSE-R-EFA-CPT-077). Preston: BAE Systems.

Wykes, K.M. (1988) *Principles of Cockpit Moding.* (EFJ-R-EFA-000-106). Preston: BAE Systems.

Introduction of ramp-LOSA at KLM Ground Services

Robert J. de Boer[1], Bekir Koncak[1], Robbin Habekotté[1], & Gert-Jan van Hilten[2]
[1] Amsterdam University of Applied Sciences
[2] KLM Ground Services, Amsterdam
The Netherlands

Abstract

Airline ground operations are subject to the conflicting demands of short turn-around times and safety requirements. They involve multiple parties, but are less regulated than airborne processes. Not surprisingly, more than a quarter of all aircraft incidents occur on the ground. These incidents lead to aircraft damage and associated costs, risk of injuries, and can potentially impact in-flight safety. KLM Ground Services has targeted platform safety performance as an area for improvement. However, existing safety awareness programs have had limited effect. A direct link between safety culture surveys and safety performance has not been established, and therefore these are insufficient to give adequate feedback on interventions. Newly developed by the Texas University are the Line Operations Safety Assessments (LOSA), first targeted at cockpit operations. Variants are available since October 2010 for the platform and maintenance environments. The research group for Aviation Engineering at the Amsterdam University of Applied Sciences has used the original platform LOSA material and tailored these to the specific circumstances at KLM. Results to date show that with these modifications, platform LOSA is a useful tool to quantify safety performance and to generate trend data. The effect of safety interventions can now be monitored.

Background

Ramp Line Operations Safety Assessments (LOSA) are part of an audit system developed for airport ground operations (i.e. activities on the so-called platform or ramp) based on the cockpit LOSA system (ICAO, 2002; FAA, 2006). Cockpit LOSA has been effective in identifying areas to target to improve safety, triggering a 70% reduction of checklist errors and a 60% reduction in unstable approaches (Gunther, 2006). Both cockpit and ramp LOSA build on the threat and error management model (Helmreich et al. 1999) and adopt standard LOSA guidelines: peer to peer observations, anonymity, confidential and non-punitive data collection, voluntary participation, trusted and calibrated observers, union cooperation, systematic observations, secure data collection repository, data verification roundtables, targets for enhancement and feedback to workers. Ramp LOSA tools are available on the FAA website (FAA, 2010) in the form of a threat and error management model, threat and error codes, observation forms, software and training material.

KLM Ground Services is the platform handling department of Air France – KLM at Amsterdam Airport Schiphol. The department consists of three operational sections which are controlled by a Hub Control Centre: Passenger Services, Baggage Turnaround Services and Aircraft Services. Besides the operational departments Ground Services operates six staff departments. Operational Integrity is one of these and handles amongst others ground safety. Activities on the platform that are executed under supervision of Operational Integrity include baggage services, pushback and towing, catering and onboard supply, cleaning, aircraft refuelling, and water and toilet services.

KLM Ground Services has a relatively high number of incidents compared to other divisions within KLM, leading to lost labour time, damages and delays. In the past, KLM Ground Services utilized so-called Ground Safety Audits to proactively manage safety, but these were discontinued because they did not deliver sufficient diagnostic data. Ramp LOSA seems to suit KLM because it focuses on the whole turnaround compared to the arrival-only scope of the Ground Safety Audits, it generates useful data, and supplements current safety initiatives at KLM Ground Services.

Theoretical foundation

Line Operations Safety Assessments are based on two theoretical concepts: the safety pyramid and threat and error management. The significance of ramp LOSA for safety improvement lies in the former. The theory of threat and error management is as yet insufficiently mature to perform as a framework for ramp LOSA.

Safety pyramid or iceberg

As in many safety-critical companies, KLM classifies safety related occurrences into four different categories: substantial (safety is not ensured, enhanced protective measurements are urgently required), high (safety is not ensured, protective measurements are urgently required), medium (safety is partially guaranteed, normal protective measures are required), and small (safety is largely guaranteed). The relative frequency of these for Ground Services is approximately[*] 1 : 100 : 500 : 1000. Because occurrences are reported ex post facto, the latter two categories are underrepresented: "occurrences" without consequences are only reported in isolated cases (Hobbs & Kanki 2008).

LOSA constitutes "a principled, data-driven approach to prioritize and implement actions to enhance safety" (ICAO, 2002). Ramp LOSA is a program for the measurement of human error in ground handling. The system captures the whole ground handling process in normal operations from arrival to departure using tools which are openly available. A LOSA report will include threats and errors that do not lead to occurrences and are not usually reported elsewhere. The well-managed turn-arounds that are also sampled are the frame of reference for the interpretation of error data. According to the FAA, "LOSA provides unique data about an airline's

---

[*] Exact figures are company confidential

defenses and vulnerabilities [and] forms a unique and complementary tool to incident reporting and [existing] safety audits" (FAA, 2006). LOSA therefore identifies errors (or lack thereof) at the bottom at the iceberg. Interventions aimed at improving performance here will logically reduce occurrences throughout the iceberg.

Threat and error management

The threat and error framework that ICAO and the FAA propagate is shown in figure 1 (ICAO, 2002; FAA, 2010). A threat is defined as an event or error that occurs outside the influence of the individuals being observed, increases the operational complexity of their task, and requires their attention and management if safety margins are to be maintained. A mismanaged threat is defined as a threat that is linked to or induces an error. Errors are defined as (in)actions that lead to a deviation from the individual's or the organizational intentions or expectations. Errors in the operational context tend to reduce the margin of safety and increase the probability of adverse events. A mismanaged error is defined as an error that is linked to or induces additional error or an undesired state.

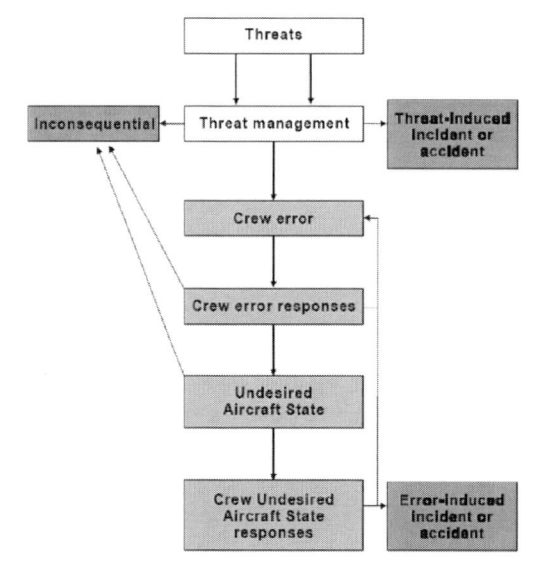

Figure 1. ICAO threat and error model

The model is based on earlier models developed at the University of Texas at Austin Human Factors Project (Helmreich et al. 1999). The schematic representation of the model and the wording in the ramp LOSA guidance material imply that errors are a result of external or internal threats, such as faulty equipment, adverse weather (external) or fatigue (internal). However, "errors can [also] be the result of a momentary slip or lapse" (Merritt & Klinect 2006) and therefore a threat need not necessarily precede an error. In fact, Klinect, Wilhelm et al. found for cockpit LOSA that less than 10% of crew errors had a threat as precedent (Klinect et al. 1999). Similar low numbers (if not even less) are expected on the platform, where tasks are often executed with less adherence to formal procedures. As a consequence, the link between threats and errors is expected to be sporadic. The identification and resolution of threats may only have a limited effect on safety, despite the desirability for those being observed to focus on external factors to justify errors. Also, the identification of threats is susceptible to hindsight bias (Dekker 2006; Woods et al., 2010).

A model reflecting a less prominent role for threats has been developed by Delta Airlines (Delta Airlines 2007) and is reproduced in figure 2.

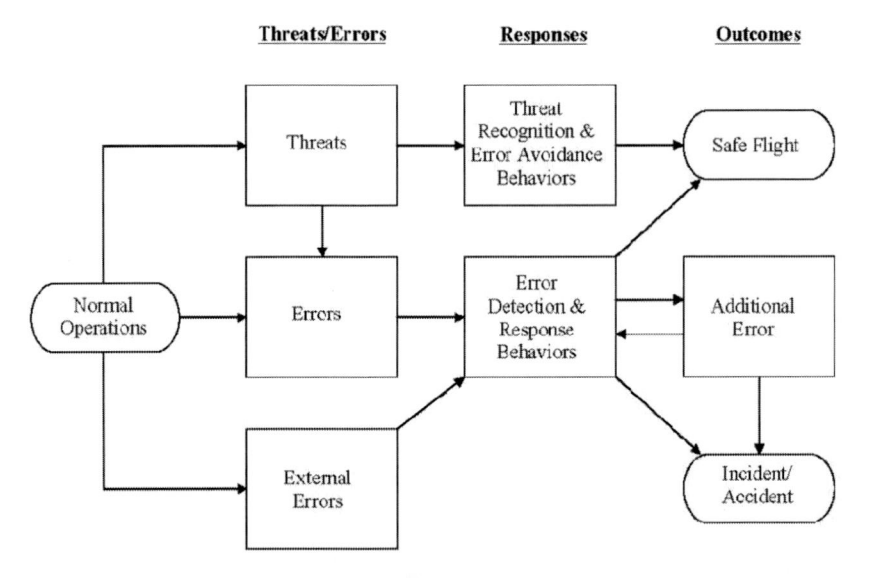

Figure 2. Delta Airline threat and error model

The figure shows how errors can be a consequence of threats and errors by others (external errors), but can also initiate from the ramp employees themselves.

Adaption of standard ramp LOSA

The Ramp LOSA tools that are available from the FAA website[*] include guidance material on threat and error management, threat and error codes, forms, software, training material, and instructions. These have been modified to suit the circumstances of KLM Ground Services at Schiphol Amsterdam Airport. A full report of the changes is available (Habekotté & Koncak 2011).

Error codes

To start observations as soon as possible the FAA provides a standard list of threat and error codes in their Ramp LOSA toolkit (FAA, 2010). While the threat codes are

---

[*] Some of these tools are currently undergoing enhancements and are reportedly available again from November 15th, 2011

generic, error codes require customization to specific company and airport procedures. Because an error is defined as a deviation from a standard operating procedure, all error codes must comply with KLM operating procedures and Schiphol requirements, and missing codes have been added. Although laborious, this task had the added benefit of identifying inconsistencies and omission in the company and airport operating procedures. In total 25 relevant LOSA error codes were initially not supported by company or airport procedures. In cases where original error codes were irrelevant to KLM operations at Schiphol they were deleted from the list, for instance concerning marshalling. To keep the KLM version of ramp LOSA compatible with the original system, error code numbering remains as close to the original as possible so that future benchmarking can be done with other handlers. However, to keep error codes in a logical order some changes were needed. Additionally, we have identified inconsistencies with language at KLM and ambiguity in the original wording and have strived to eliminate these. For example, the FAA uses the verb 'uploading' instead of 'loading' and the verb 'downloading' instead of 'offloading'. In total of the original 201 error codes, we have preserved 167 and added 12 new codes.

Checklist

The original set of observation forms consist of 41 pages of which 28 pages have to be filled in. The purpose of this form is to fill the form right after the observation. Evidently, observers are not expected to be able to fill in all 28 pages of the form for each turnaround (even though the 'Did Not Observe' check box can act as an escape). Unfortunately, the forms are published in an Adobe PDF file on the FAA Human Factors portal. The site does not provide the file to adjust the forms, like a MS Word document.

We prepared a modified observation form for the use of Ramp LOSA at Schiphol airport for KLM flights. The layout of the form has been kept as the original as far as possible, but the form has been split into a checklist to be used on the platform, and an observation form that can be filled in offline. The length of the form has been reduced from 28 pages to 8 pages for the checklist and 12 pages for the observation form. Various sections have been deleted and the demographics form has been shortened to one page. Many irrelevant checkpoints have been removed.

Observers and training

The quality of observers is a key element in the success of Ramp LOSA. This success is achieved by choosing the right observers and training them properly. To be able to collect sufficient data KLM Ground Services wishes to check a minimum of 5% of all flights on a continuous basis. This means that when the system has been fully implemented, approximately 17 observations should be conducted every day, divided over five intercontinental turn-arounds and twelve European arrival and departures. With an observer able to execute four observations a day and allowing for absences, this translates into an observer resource requirement of 4 full-time equivalents (FTE) for intercontinental operations and 2 for Europe.

It has not been easy to identify the right observers for LOSA at KLM Ground Services; several alternatives were considered. The choice has been to use operational staff (as advised by the FAA) complemented (due to cost awareness) with KLM ramp employees that temporarily require less physically demanding jobs due to medical circumstances. Rejected options included operational management, temporarily redundant workers from other departments, ramp employees from other ground handling companies, and students with experience in ramp operations.

Training for the observers has been devised. The duration is one day and the training includes modules on the basics of observing, threat and error management, threat and error codes, and use of the forms and the software. The training does not include refreshing of procedural knowledge, although it envisaged that this will form a prerequisite to being accepted into the observer training. An important consideration is "calibration" or alignment of the observers such that they judge errors equally. This is achieved by allowing the recruits to perform three audits together with the ramp LOSA coordinator/lead trainer. As an alternative, videos of actual turn-arounds are being considered as calibration material.

Data entry and analysis

For the analysis of the observations, software is required. The FAA provides software (based on MS Access) to fit standard FAA forms, but this cannot be modified to suit the KLM error codes and observation forms. For the initial observations MS Excel has been used, but a MS Access application is being commissioned that meets KLM requirements but will allow data exchange with other parties. It is envisaged that the observers will input their observation records into the package immediately upon return from the platform. A tablet application (to be used directly on the platform) has been considered but not adopted at this point because some form of data screening before entry is thought to be beneficial. It is desirable that previous entries, forms and codes can be modified by an administrator. This edit function is not available in the FAA Ramp LOSA software. Data validation sessions are planned. The analysis tool should calculate the prevalence and mismanagement index of selected errors and threats.

Communication and organisation

In addition to the modifications to the original ramp LOSA material explained above, the implementation at KLM Ground Services included the roll-out of a communication plan, the appointment of a ramp LOSA coordinator within the department of Operational Integrity.

Results

To date more than 100 safety audits have been conducted. The observations have been limited to the intercontinental / wide-body fleet of KLM at Schiphol Airport. On average, an arrival observation lasts 33 minutes and a departure observation has a duration of 56 minutes.

General

The results show that on average, 0.9 external threats and 15.8 errors are observed each arrival or departure (compared to 3 to 4 errors per flight for cockpit operations). A common discrepancy with the procedures was exceeding the speed limit at the airport (currently 6 kilometres per hour): this was observed in 70% of the assessments (vehicles for toilet servicing: 100%). Other common errors include neglecting to check for foreign objects that can damage the engines (FOD, 64%) and incorrect use of the Ground Power Units (different errors varying from 20% to 68%).

Arrivals

Coon errors in the arrival process included Ground Service Equipment not waiting in regulatory areas (68%), chocks not correctly placed (44%) and hearing protection not worn (37%). In 4% of the cases personnel was not available in time for the arriving aircraft, leading to delays, increased fuel consumption and higher workloads.

Unloading and loading

In all observed cases (i.e. 100%) the safety barriers alongside the conveyer belts leading up to the hold were not used. This led to baggage falling on employees and time-lost accidents in four specific cases. Other common errors included not checking hold after unloading (69%), setting conveyor height during driving (62%), and employees walking on a running conveyer belt (48%).

Departures

In 40% of the observations the chocks were removed before connecting the push-back vehicle. A pre-departure check was not performed correctly in 29% of the cases.

Conclusion

This paper presents the results of the introduction of Ramp LOSA at KLM Ground Services. The methodology as it has been made available by the FAA demonstrates some weaknesses, such as a questionable threat and error management framework, elaborate forms, and inflexible software. The error codes are necessarily generic. The original tools have been modified to make them consistent with KLM's procedures at Amsterdam Airport Schiphol.

The initial results show that there high number of errors are observed in platform observations compared to cockpit operations. This probably reflects the less procedural context of platform operations, the less skilled labour content, time pressure and the high turnover of staff within this environment. Many of the errors are well-known problems, which have apparently not yet been tackled successfully. The feedback of the assessment results to the responsible managers has inspired them to intensify their actions. However, some errors may reflect excessive stringency of the regulations that are not compatible with the time pressures of the turn-around process for highly utilized aircraft, for instance setting the height of conveyor belts while driving. Other errors may reflect the insufficiency of technical aids to balance safety with effectiveness, like the safety barriers alongside the conveyor belts. Many errors do not impact flight safety as much as that they lead to personal injuries (e.g. hearing aids, walking on the conveyor) and may have been underemphasized in the daily operation. None of the LOSA findings have been recorded as occurrences in the airlines safety system. This shows the potential value of this tool in revealing safety issues. Results to date show that platform LOSA is a useful tool to quantify safety performance at the bottom of the iceberg and to generate trend data. It inspired management actions and the effect of safety interventions can now be monitored.

References

Dekker, S. (2006). *The field guide to understanding human error.* Farnham, UK: Ashgate Publishing.

Delta Airlines (2007). *Ramp Operations Safety Audit,* electronic report retrieved November 2010 from https://hfskyway.faa.gov/HFSkyway/. Washington D.C.: Federal Aviation Administration / U.S. Department of Transportation.

FAA (2006). *Introduction to Safety Management Systems for Air Operator.* Advisory Circular 120-92 on Line Operations Safety Audit. Washington D.C.: Federal Aviation Administration / U.S. Department of Transportation.

FAA (2010). Line Operations Safety Assessment Website accessed November 2010. Washington D.C.: Federal Aviation Administration / U.S. Department of Transportation.

Gunther, D. (2006). Line Operations Safety Audit (LOSA) and the Safety Change Process. Paper presented at *the fourth ICAO-IATA LOSA and TEM Conference*. Toulouse: International Civil Aviation Organization.

Habekotté, R. & Koncak, B. (2011). *Ramp LOSA Implementation for KLM Ground Services*. BEng thesis. Amsterdam: Amsterdam University of Applied Sciences.

Helmreich, R.L., Klinect, J.R., & Wilhelm, J.A. (1999). Models of threat, error, and CRM in flight operations. In *Proceedings of the Tenth International Symposium on Aviation Psychology* (pp. 677-682). Columbus, OH: The Ohio State University.

Hobbs, A. and Kanki, B.G. (2008). Patterns of error in confidential maintenance incident reports. *International Journal of Aviation Psychology, 18(1),* 5-16.

ICAO (2002). *Line Operations Safety Audit (LOSA)*, Document 9803 AN/761. Montreal: International Civil Aviation Organization.

Klinect, J.R., Wilhelm, J.A., & Helmreich, R.L. (1999). Threat and error management: Data from line operations safety audits. In *Proceedings of the Tenth International Symposium on Aviation Psychology* (pp. 683-688). Columbus, OH: The Ohio State University.

Merritt, A. and J. Klinect (2006). *Defensive flying for pilots: An introduction to threat and error management.* Ausin, TX: University of Texas.

Woods, D. D., Dekker, S., Cook, R., Johannesen, & L., Sarter, N. (2010). *Behind human error*. Farnham, UK: Ashgate Publishing.

A phonetic approach for detecting sleepiness from speech in simulated Air Traffic Controller-communication

Jarek Krajewski[1], Thomas Schnupp[2], Christian Heinze[2], Sebastian Schnieder[1], Tom Laufenberg[1], David Sommer[2], & Martin Golz[2]
[1]*University of Wuppertal*
[2]*University of Applied Sciences Schmalkalden*
Germany

Abstract

The aim of the study was to develop a phonetic based instrument to estimate sleepiness within Air Traffic Controller -Communication. Thus, we conducted a within-subject partial sleep deprivation design (20.00 - 04.00 h, N = 57 participants) and recorded 356 speech samples of simulated Air Traffic Controller-Communication. During the night of sleep deprivation a well established, standardized self-report sleepiness measure, the Karolinska Sleepiness Scale (KSS), and a KSS Observer Scale (used by two experimental assistants rated each time just before the speech recordings) was applied to determine the sleepiness reference value (ground truth), which was further used for machine modeling purposes. The 170 phonetic features which have been computed partially represent auditive-perceptual concepts of prosody (pitch, intensity), articulation (slurred speech), and speech quality (breathy, tense, sharp, or modal voice). Several acoustic features show significant correlations to fusioned KSS ratings, representing e.g., a more slurred articulation and a less tense speech quality. Applying a simple linear regression method using a leave-one-sample-out cross-validation protocol reaches a mean linear error (MLE) of 1.54 KSS units (resp. r = .69) for male and 1.68 KSS units (resp. r = .65) for male speaker.

Advantages of speech based sleepiness measurement

The working conditions of Air traffic controllers (ATC) are characterized by long working hours, movement restriction, dim light levels, background noise and high level of workload. All of these factors are known to cause sleepiness, and even microsleep events (e.g. Golz et al, 2007; Horberry, Hutchins, & Tong, 2008; Sommer et al., 2005). In this manner sleepiness is a factor in a variety of incidents and accidents in road traffic (e.g. Flatley, Reyner, & Horne, 2004; Read, 2006) and work contexts (e.g. safety sensitive fields as e.g. chemical factories, nuclear power stations, and air traffic control; Melamed, & Oksenberg, 2002; Wright, & McGown, 2001). Accordingly 21% of the reported incidents mentioned in the Aviation Safety Reporting System (including pilots and air traffic controllers) were related to sleepiness. Thus, the prediction and warning of traffic employees against impending critical sleepiness play an important role in preventing accidents and the resulting human and financial costs.

Moreover, the aim to enhance joy of use and comfort within Human-Computer-Interaction (HCI) could also benefit from the detection of and automatic countermeasures to sleepiness. Knowing the speaker's sleepiness state can contribute to the naturalness and acceptance of HCI. If the user shows unusual sleepiness states, giving feedback about this fact would make the communication more empathic and human-like. This enhanced naturalism might improve the acceptance of these systems. Furthermore, it may result in better comprehensiveness, if the system output is adapted to the user's actual sleepiness-impaired attentional and cognitive resources.

Hence, many efforts have been reported in the literature for measuring sleepiness related states. But these electrode- (EOG/EEG reaching 15% error rate; Sommer & Golz, 2005) or video-based instruments (PERCLOS reaching 32% error rate; Sommer, Golz, Trutschel, & Edwards, 2008) still do not fulfill the demands of an everyday life measurement system (Golz, Sommer, Trutschel, Sirois, & Edwards (2010). The major drawbacks are (a) a lack of robustness against environmental and individual-specific variations (e.g. bright light, wearing correction glasses, angle of face or being of Asian race) and (b) a lack of comfort and longevity due to electrode sensor application.

In contrast to these electrode- or video-based instruments, the utilization of voice communication as an indicator for sleepiness could match the demands of everyday life measurement. Contact free measurements such as voice analysis are non-obtrusive (not interfering with the primary driving task) and favorable for sleepiness detection since an application of sensors would cause annoyance, additional stress and often impairs working capabilities and mobility demands. In addition, speech is easy to record even under extreme environmental conditions (bright light, high humidity and temperature), requires merely cheap, durable, and maintenance free sensors and most importantly, it utilizes already existing communication system hardware. Furthermore, speech data is omnipresent in many professional settings and thus does not interfere with primary work task ("hands-free" and "eyes-free" measurement). Given these obvious advantages, the renewed interest in computational demanding analyses of vocal expressions has been enabled just recently by the advances in computer processing speed.

Little empirical research has been done to examine the effect of sleepiness states on acoustic voice characteristics. Most studies have analyzed only single features (Harrison, & Horne, 1997; Whitmore, & Fisher, 1996) or small feature sets containing only perceptual acoustic features, whereas signal processing based speech and speaker recognition features (e.g. mel-frequency cepstrum coefficients, MFCCs; cf. Table 1) have received little attention (Greeley et al., 2007; Schuller, Batliner, Steidl, Schiel, & Krajewski, 2011; Nwe, Li, & Dong, 2006). Building an automatic sleepiness detection engine reaching sufficient precisions still remains undone. Thus, the aim of this study is to apply a state-of-the-art speech emotion recognition engine (Schuller, Batliner, Steidl, Schiel, & Krajewski, 2011; Batliner et al., 2006) on the detection of critical sleepiness states.

Phonetic feature computation

The acoustic sleepiness analysis is mainly based on speech emotion recognition research, general audio signal and computational intelligence research (e.g. Schuller, Batliner, Steidl, & Seppi, 2011; Batliner et al., 2006). Acoustic features can be divided according to auditive-perceptual concepts in prosody (pitch, intensity, rhythm, pause pattern, and speech rate), articulation (slurred speech, reduction and elision phenomena), and speech quality (breathy, whispery, tense, sharp, hoarse, or modal voice). Our approach prefers the fusion of purely signal processing based features without any known auditive-perceptual correlates and perceptual-acoustic features as e.g. the speech intensity (loudness), fundamental frequency (pitch), voiced/unvoiced duration (rhythm). Typical acoustic features used in emotion speech recognition and audio processing are (a) fundamental frequency, (b) intensity, (c) formant positions, (d) formant bandwidths, (e) mel frequency cepstrum coefficients (MFCCs), and (f) ratio of frequency bands derived from the long term average spectrum (LTAS).

Table 1. Typical phonetic features applied within speech emotion recognition

Phonetic features	Description
Fundamental frequency (F0)	acoustic equivalent to pitch; rate of vocal fold vibration; maximum of the autocorrelation function; models prosodic structure; speech melody indicator
Intensity	models intensity, based on the amplitude in different intervals; average squared amplitude within a predefined time segment; stressing structure
formant position (F1-F6)	resonance frequencies of the vocal tract (VT) depending strongly on its actual shape; represent spectral maxima, and are known to model spoken content and speaker characteristics; influenced by lower jaw angle, tongue body angle, tongue body horizontal location, tongue tip angle, tongue tip horizontal location, relative lip height, lip protrusion, velum height
formant bandwidth (Fbw1-Fbw6)	model VT shape and energy loss of speech signal due to VT elasticity (yielding wall effect), viscoelasticity of VT tissue or heat conduction induced changes of air flow (jet streams, turbulences); width of the spectral band containing significant formant energy (- 3 dB threshold)
Mel frequency cepstrum coefficients (MFCCs)	"spectrum of the spectrum"; have been proven beneficial in speech emotion recognition, and speech recognition tasks; homomorphic transform with equidistant band-pass-filters on the Mel-scale; holistic and decorrelated representation of spectrum
Ratio of frequency bands	averages out formant information; giving general spectral trends; relative amount of energy within predefined frequency bands; speech quality

Method

Procedure, subject, and speech material

In this study a within-subject partial sleep deprivation design (20.00 - 04.00 h, N = 57, 29 female, 28 male participants) containing simulated ATC Communication was conducted. Each session comprised of 40 min human-computer-interaction, followed by responding to sleepiness questionnaires (2 min), recording speech material (2 min), conducting vigilance tests (15 min), and a break (1 min). A well established, standardized subjective sleepiness questionnaire measure, the Karolinska Sleepiness Scale (KSS), was applied to determine the sleepiness reference value (ground truth), which was further used for machine modeling purposes. It was used by the subjects (self-assessment) and additionally by the two experimental assistants (observer assessment, given by assessors who had been formally trained to apply a standardized set of judging criteria). In the version used in the present study; scores range from 1–10: extremely alert (1), very alert (2), alert (3), rather alert (4), neither alert nor sleepy (5), some signs of sleepiness (6), sleepy, but no effort to stay awake (7), sleepy, some effort to stay awake (8), very sleepy, great effort to stay awake, struggling against sleep (9), extremely sleepy, cannot stay awake (10). During the night, the subjects were confined to the laboratory, conducting several human-computer-interaction simulator tasks and were supervised throughout the whole period.

The recording took place in a laboratory room with dampened acoustics using a high-quality, clip-on microphone (sampling rate: 44.1 kHz, quantization: 16 bit). Furthermore, the subjects were given sufficient prior practice so that they were not uncomfortable with this procedure. The verbal material consisted of a simulated pilot-air traffic controller communication ("County tower, Cessna nine three four five Lima, one zero miles southeast at two thousand five hundred, landing County"). These recordings represent speech turns and will be subsequently referred to as turns. The participants recorded other verbal material at the same session, but in this article we focus on the material described above. Total number of speech samples: 356 (247 female, 109 male) samples; KSS:= mean of the three KSS-Ratings; female: M = 6.20; SD = 2.36; male: M = 6.50; SD = 2.62. Note that we do not pre-select 'friendly' instances, such as instances with a high rater agreement, for evaluation. Rather, in line with recent studies in paralinguistic research (e.g., Schuller et al., 2011a), our goal is to design a system that robustly classifies all available data, as needed for a system operating in the daily work task setting ('in the wild').

We conduct a validation experiment to examine whether automatically trained models can be used to recognize the sleepiness of subjects. Our approach can be summarized in four steps (cf. Figure 1): 1. Collect individual speech data, and the associated sleepiness ratings for each participant; split the data into train and test data; 2. Extract relevant acoustic features from the speech data and build statistical models of the sleepiness ratings based on the acoustic features; 4. Test the learned models on unseen speech data. The following sections describe each of these steps in more detail.

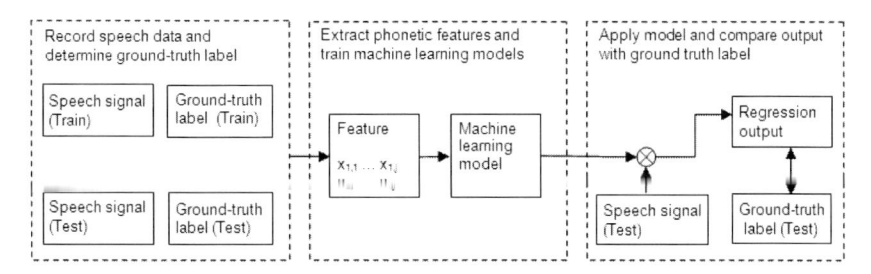

Figure 1. Procedure of 3-step construction of phonetic sleepiness measurement system according to standard procedure in speech emotion recognition

Phonetic feature extraction

To extract relevant acoustic features from the speech data, we employed Praat speech analysis software (Boersma, 2001) on the turn-wise segmented speech recordings and computed the following groups of prosody, articulation, and speech quality related features that reflect a broad coverage for paralinguistic information assessment:

(a) F0 features: The following F0 related features are computed: mean, range, minimum, maximum, 1st, 2nd, and 3rd quartile; (b) formant position: The analysis of formants uses the position of the formants 1 to 5 (F1-F5). For each formant mean, range, minimum, maximum, 1st, 2nd, and 3rd quartile are computed; (c) formant bandwidth: formant bandwidth indicates the peak width of a formant. It is defined as the frequency region in which the amplification differs less than 3 dB from the amplification at the centre frequency. For each formant bandwidth mean, range, minimum, maximum, 1st, 2nd, and 3rd quartile are computed; (d) Intensity: The intensity of a recorded sound can be measured by determining the root mean square amplitude over a short time period. Related to intensity is the popular shimmer feature, which gives an evaluation in percent of the variability of the peak-to-peak amplitude within the analyzed voice sample. It represents the relative period-to-period (very short-term) variability of the peak-to-peak amplitude. The following intensity features are computed: mean, range, minimum, maximum, 1st, 2nd, 3rd quartile and shimmer; (e) MFCC: MFCCs are a widely used variant of the cepstral coefficients that follow the Mel-frequency scale as proposed by psychoacoustics and are the standard features used in Automatic Speech Recognition (ASR). In the experiment the first 12 MFCCs, their delta and delta-delta coefficients were calculated; (f) Ratio of frequency bands: Hammarberg indices were used as another set of coarse measures of spectral properties. In sum we computed: 26 prosodic features, 108 spectral features I, and 36 spectral features II (MFCCs), resulting in a total of 170 acoustic features per speech sample. Further details have been published elsewhere (Krajewski, Batliner, & Golz, 2009; Krajewski & Kröger, 2007).

Stepwise linear regression analyses were conducted with all acoustic features to determine the most influential features and the overall measurement precision. For evaluation of regression tasks, we rely on the measures for the INTERSPEECH 2011 Speaker State Challenge mean linear error (MLE) which is the expected absolute difference, between the outputs of the regression function on the test set and the corresponding target values. In contrast to commonly applied correlation coefficients, the MLE provides a straight forward answer to the question: What is my real margin of sleepiness, when the instrument shows e.g. a value of 7.3? Due to data sparcity, a speaker-dependent approach has been chosen: a leave-one-sample-out cross-validation, i.e., in turn, one case was used as test set and all other as train. The final classification errors were calculated averaging over all classifications.

Results

For male speakers the following phonetic features show the highest correlations to KSS ratings (see Table 2): Mean formant 1 Position- formant 3 Position, $r = -.59$ Mean formant 1 Position- formant 3 Position, $r = -.59$ and Mean formant 1 Position- formant 5 Position, $r = -.58$. These features are related to sleepiness induced changed shapes of the vocal tract. Moreover, especially formant bandwidth are increased with growing sleepiness, which represents a softening of the vocal tract. For female speakers the difference between Mean formant 1 Position and mean formant 2 Position is reduced, which is indicative for a more centralized, slurred and more schwa vowel like articulation. Then, the difference between the Power Spectral Density (PSD) in the frequency bands 0 to 500 Hz and 500 to 4000 Hz is reduced, which shows a more tensed speech quality in sleepy females. Again, the formant bandwidth are increased, which demonstrates several more relaxed parts within the vocal tract.

Table 2. Correlation of phonetic features and KSS reference values separated in male and female participants

Phonetic feature (male)	r	Phonetic feature (female)	r
Mean Formant 1 Position- Formant 3 Position	-0.59**	Mean Formant 1 Position- Formant 2 Position	-0.36**
Mean Formant 1 Position- Formant 4 Position	-0.59**	Power Spectral Density 0 to 500 Hz - 500 to 4000 Hz	-0.34**
Mean Formant 1 Position- Formant 5 Position	-0.58**	Power Spectral Density 0 to 500 Hz - 500 to 1000 Hz	-0.34**
Mel-Frequency Cepstrum Coefficient 1	-0.55**	Mean Formant 1 Position - Formant 3 Position	-0.29**
Mel-Frequency Cepstrum Coefficient 2	0.56**	Slope Long term Average Spectrum	-0.27**
Median Formant 2 Bandwidth	0.56**	Mean Formant 1 Position - Formant 4 Position	-0.26**
Mean Formant 3 Position	0.56**	Power Spectral Density 125 to 200 Hz	0.25**
Percentile 25 Formant 2 Bandwidth	0.56**	Median Formant 2 Bandwidth	0.24**
Minimum Formant 4 Position	0.55**	Mean Formant 1 Position - Formant 5 Position	-0.22**
Mean Formant 5 Position	0.55**	Minimum Formant 1 Position	-0.20**

In order to determine the multivariate detection performance, we applied a simple linear regression method using a leave-one-sample-out cross-validation protocol which reaches MLE = 1.54 KSS-units (resp. r = .69) for male and MLE = 1.68 KSS-units (resp. r = .65) for female speaker (cf. Figure 2). In detail, the stepwise procedure calculates the following standardized beta coefficient model for male speaker: distance of formant 1 and formant 3 Position = -.56; mean of formant 5 position = .30; location of peak value of long-term average spectrum = -.35; mean of fundamental frequency = -.32; 25th percentile of formant 2 bandwidth = .25; small fundamental frequency perturbation (jitter) = -.19; maximum of formant 3 position = -.20; time of minimum of formant 2 position = -.15; maximum of formant 1 position = .13. For female speaker we calculated the following beta coefficients: distance of formant 1 and formant 2 Position = -.21; PSD differences of [0 500] Hz minus [500 4000] Hz = .03; formant 4 position = -.10; 25th percentile of formant 3 bandwidth = .14; PSD of [125 200] Hz band = .34; regression slope of fundamental frequency = -.47; regression slope of long term average spectrum = -.59; ratio of PSD of [0 1000] Hz and [1000 4000] Hz = -.39; standard deviation of fundamental frequency = .25. All beta coefficients (except for "PSD differences of [0 500] Hz minus [500 4000] Hz) reach the two-sided significance level of 0.05.

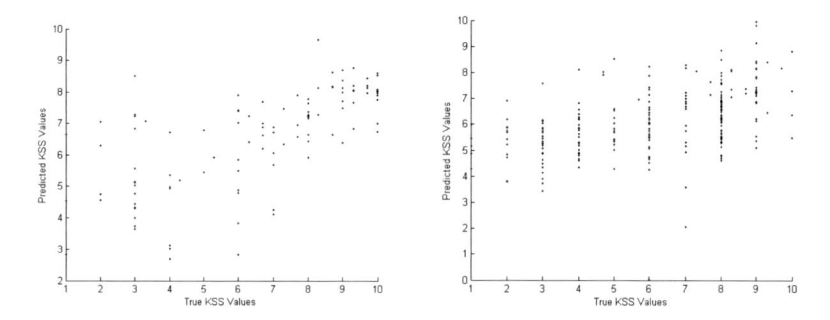

Figure 2. Distribution of the output of linear regression based combination of phonetic features and their KSS reference values for speech samples of male (left) and female (right) speakers. Each dot represents values for a single speech sample computed from ca. 5 sec. lasting ATC phrase. X-axis: True KSS Values, as resulted from fusion of self and observer reports; Y-axis: Predicted KSS Values, as computed from phonetic models

Discussion

The aim of the study was to construct and validate a non-obtrusive sleepiness detection instrument based on simulated ATC communication. The advantages of this ambulatory monitoring measurement approach is that in ATC settings obtaining speech data is objective and non-obtrusive, and it allows multiple measurements over long periods of time. The main findings may be summarized as following. First, acoustic features, that were extracted from speech and subsequently modeled with machine learning algorithm, contain a substantial amount of information about the speaker's sleepiness state. Our acoustic measurements showed differences between alert and sleepiness speech in male speaker especially in changes of vocal tract shape, in a reduced pitch rate, in a softening of the vocal tract, in less small perturbation of pitch. In female speaker we found the following sleepiness induced changes of speech: a more slurred articulation (smaller differences of formant positions), a steeper decrease of the intonation contour, a less tense voice quality (steeper decrease in LTAS), and a higher deviation of the pitch. These results are mainly consistent with the predictions of the cognitive-physiological mediator model of sleepiness (Krajewski, Batliner, & Golz. 2009). Nevertheless, the results clearly indicate different effects of sleepiness on male and female speakers.

Secondly, we found that applying a large phonetic feature set of 170 features on male and female speaker separately yielded a precision of +/- 1.54 KSS units (resp. 1.68 KSS units) on unseen data but known speakers. This result indicates for a male speaker that a phonetically measured sleepiness value of 8, corresponds in average to a real KSS value of [6.46 9.54]. In sum, our regression performance for both male and female speaker is in the same range that has been obtained for comparable tasks, e.g. for emotional user-state classification (cf. Schuler et al., 2010), which are usually based on much larger databases (over 9000 turnes). Thus, it seems likely that sleepiness detection could be improved by collecting similar-sized speech databases, containing emotional speech samples from different types of speakers and speaking styles.

Limitations and future work

The validity of our results is limited by several facts. First, the major criticism refers to the choice of the applied ground truth. The used fusion of one self-report and two observer-report measures could be criticized because of its (semi-)subjective nature lacking an involvement of "objective" physiological ground truth measures. Until now, many studies found associations between physiological data (e.g. EEG or EOG) and sleepiness. Nevertheless, they still remain in a premature developmental stage without offering a standardized, commonly accepted scaling as it is realized in the KSS. Furthermore, the KSS has proven its validity in several studies, particularly when the application context is unlikely to provoke self or observer deception. Despite of this currently premature status of "behavioral and physiological sleepiness instruments" they offer a promising potential as future "gold standard" measures. Secondly, sleepiness might be confounded by annoyance states due to the multiple repetition of speak task. Thus, the results obtained in the current study with a within subject design should be replicated with a between subject design.

References

Batliner, A., Steidl, S., Schuller, B., Seppi, D., Laskowski, K., Vogt, T., Devillers, L., Vidrascu, L., Amir, N., Kessous, L., & Aharonson, V. (2006). Combining efforts for improving automatic classification of emotional user states. In T. Erjavec & J. Z. Gros (Eds.), *Language Technologies, IS-LTC 2006*, (pp. 240-245). Ljubljana, Slovenia: Infornacijska Druzba.

Boersma, P. (2001). PRAAT, a system for doing phonetics by computer. *Glot International, 5,* 341–345.

Flatley, D., Reyner, L.A., & Horne, J.A. (2004). Sleep-Related Crashes on Sections of different Road Types in the UK (1995–2001). *Road Safety Report No. 52* (pp. 4-132). London: Department of Transport.

Golz, M., Sommer, D., Holzbrecher, M., & Schnupp, T. (2007). Detection and Prediction of Driver's Microsleep Events. In RS4C (Ed.), *Proceedings 14th International Conference Road Safety on Four Continents*. Bangkok, Thailand.

Golz, M., Sommer, D., Trutschel, U., Sirois, B., & Edwards, D. (2010) Evaluation of Fatigue Monitoring Technologies. *Journal of Somnology, 14*, 187-189.

Greeley, H.P., Berg, J., Friets, E., Wilson, J., Greenough, G., Picone, J., Whitmore, J., & Nesthus, T. (2007). Sleepiness Estimation using voice analysis. *Behaviour Research Methods*, *39*, 610-619.

Harrison, Y., & Horne, J.A. (1997). Sleep deprivation affects speech. *Sleep, 20*, 871-877.

Horberry, T., Hutchins, R., & Tong, R. (2008). Motorcycle rider fatigue: A review. *Department of TransportRoad Safety Research Report, 78*, 4-63.

Krajewski, J., & Kröger, B. (2007). Using prosodic and spectral characteristics for sleepiness detection. *Interspeech Proceedings, 8,* 1841-1844.

Krajewski, J., Batliner, A., & Golz, M. (2009). Acoustic sleepiness detection – Framework and validation of a speech adapted pattern recognition approach. *Behavior Research Methods, 41*, 795-804.

Melamed, S., & Oksenberg, A. (2002). Excessive daytime sleepiness and risk of occupational injuries in non-shift daytime workers. *Sleep, 25*, 315-322.

Nwe, T.L., Li, H., & Dong, M. (2006). Analysis and detection of speech under sleep deprivation. *Proceeding of Interspeech, 9*, 17-21.

Read, L. (2006). Road safety Part 1: Alcohol, drugs and fatigue. In Department for Transport (Ed.), *Road safety Part 1* (pp. 1-12). London: Department for Transport.

Sommer, D., & Golz, M. (2005). *Clustering of EEG-Segments Using Hierarchical Agglomerative Methods and Self-Organizing Maps*, Proc International Conference on Artificial Neural Networks (ICANN2001), (pp. 642-649) Vienna, Austria: Springer.

Sommer Golz, M., Trutschel, U., & Edwards, D. (2008). *Assessing driver's hypovigilance from biosignals.* Proceedings of the 4th European Conference of the International Federation for Medical and Biological Engineering (ECIFMBE-2008): pp. 152-155. Berlin: Springer.

Schuller, B., Batliner, A., Steidl, S., & Seppi, D. (2011). Recognising Realistic Emotions and Affect in Speech: State of the Art and Lessons Learnt from the First Challenge. *Speech Communication 53,* 1062-1087.

Schuller, B., Batliner, A., Steidl, S., Schiel, F., & Krajewski, J. (2011). The Interspeech 2011 Speaker State Challenge. *Proceedings Interspeech, 12*, 3201–3204.

Sommer, D., Chen, M., Golz, M., Trunschel U., & Mandic, D. (2005). Fusion of state space and frequency domain features for improved microsleep detection. In W. Dutch et al. (Eds.), *Proceedings International Conference Artifical Neural Networks (ICANN 2005)* (pp. 753-759). Berlin: Springer.

Whitmore, J., & Fisher, S. (1996). Speech during sustained operations. *Speech Communication, 20,* 55–70.

Wright, N., & McGown, A. (2001). Vigilance on the civil flight deck: Incidence of sleepiness and sleep during long-haul flights and associated changes in physiological parameters. *Ergonomics, 44*, 82-106.

Surface Transportation

The Lane Change Test: United Kingdom results from a multi-laboratory calibration study

Terry C. Lansdown
Heriot-Watt University, Edinburgh
UK

Abstract

Dual task scenarios have long been used to evaluate the demands imposed by secondary in-vehicle information systems. International effort has been expended to develop standardised methodologies for valid, reliable, and efficient system assessment. The 'Lane Change Test' is an example of one of these protocols. A multi-laboratory coordinated data calibration exercise was undertaken to explore the utility of the Lane Change Test. This paper reports on UK data which contributed to this exercise. The Lane Change Test encompasses a primary task representative of some control aspects of the driving task. It was evaluated in the context of several secondary tasks, with easy and difficult variations. These tasks were an auditory one, a visual one and an integrated audio, visual and manual route guidance task. Results indicate broadly the same directionality of findings for all laboratories. For example, mean lane deviations were found to be smallest for the auditory tasks, followed by the visual only tasks, with the integrated tasks being most disruptive of primary task lane deviation. However, some laboratory-specific findings were identified.

Introduction

The analysis and evaluation of task difficulty may be undertaken by pushing users to the point at which the performance fails. For example, Wickens & Hollands (2000) discuss the use of performance resource functions to characterise the relationship between effort and performance. However, when secondary or additional tertiary tasks are undertaken to achieve a goal, performance on the additional, typically secondary task, may provide insight into both the individual's capabilities, and potential performance in the primary and secondary tasks. There has been an ongoing debate (Salvucci & Taatgen, 2008) as to whether task execution is distributed (Wickens, 2002) or unified in the imposed cognitive demands (Tombu, et al., 2011). In many applied contexts, regardless of the processing mechanism, it is pertinent to understand the functional performance relationship between tasks. For example, it has been suggested (Wickens & Hollands, 2000) that for a primary task with a relatively constant level of difficulty, an increasingly difficult secondary task may 'mop up' spare cognitive capacity, until performance on it or the primary task begins to degrade (setting aside any limits in the availability of data, rather than resources, for the tasks).

Driving provides a familiar and safety-relevant domain in which to consider dual task performance. Of course, it is rather arbitrary to use the term 'dual-task', as both the primary (presumably vehicle control) and secondary tasks (e.g., use of a route guidance system) may be rather difficult to define in absolute terms. Further, a typical driver will be performing many other activities which may also be defined as tasks that could potentially have bearing on performance in other areas (for example, eating, conversing with a passenger, thinking about non-journey-related matters). However, the dual-task paradigm has been frequently adopted to investigate the impact of additional in-vehicle systems on performance in the driving task.

Numerous empirical methodologies have been developed to facilitate investigation of the dual-task paradigm. Application areas have included use of mobile telephones, route guidance systems, and other OEM in-vehicle human-machine interfaces, e.g., in-vehicle entertainment or climate controls. Several methods that have been produced to assist in the development and evaluation of secondary in-vehicle systems are noteworthy with respect to findings reported in this paper, i.e., the EU statement of principles (SOP, 1999), the Occlusion task (ISO, 2007), the HMI Checklist (Stevens, Board, Allen, & Quimby, 1999), the 15 second rule (Green, 1999), the Peripheral Detection Task (Olsson & Burns, 2000), and more recently the Lane Change Task (ISO, 2010). Each approach has its strengths and limitations. For example, the Peripheral Detection Task is inherently visual in nature, thereby requiring caution in its application for the evaluation of auditory distractors.

The Lane change task is a primary tracking task which is representative of some of the aspects of the driving task. It was developed to assist in the evaluation of in-vehicle human machine interfaces (HMI) with specific consideration for their usability while the vehicle is in motion. Such a task must be based upon a sound evaluation methodology and therefore needs to be valid, diagnostic and reliable. To obtain objective data on these parameters an international research effort was co-ordinated to determine reliability of findings and calibrate results (Bengler, Mattes, Hamm, & Hensel, 2009). The task consists of a personal computer running a simulation of a three lane roadway. The driver is required to manoeuvre the virtual vehicle between lanes as a result of the roadside signage while performance is being measured with or without a secondary task. This paper reports on findings from trials in the United Kingdom contributing to the inter-laboratory collaboration exercise. It was pertinent to contribute statistical data with regard to the suitability and effectiveness of the 'standard' secondary tasks and their impact on the task performance. Thus, the study has three hypotheses:

i. that the introduction of the secondary tasks would compromise lateral control,
ii. the secondary tasks requiring visual attention will produce the largest performance decrement, and
iii. the Hard variants of the secondary tasks will produce relatively greater performance decrement than their Easy variants.

Method

Design

The experimental protocol for the LCT is defined in ISO (2010). This was followed as closely as possible with one exception. One of the four secondary tasks, the Critical Tracking Task, was not presented to participants. This was omitted as the researcher did not consider an embedded secondary tracking task to be reflective of any reasonable task that may be safely expected to be performed during the driving task. In summary for the data reported here, the study had a repeated measures design, in which all participants experienced three secondary tasks. Each of these secondary tasks had two levels of difficulty, relatively 'easy' or 'hard'. The order of presentation of the secondary tasks was counterbalanced according the schedule in the published documentation (ISO, 2010). Additionally, there were pre-test and post-hoc baseline drives undertaken without a secondary task. The primary dependent variable was mean lane deviation (MDEV). The study was approved by the Heriot-Watt University, School of Life Sciences Ethics committee.

Participants

There were sixteen participants, eight of whom were female. Average age was 37 years ($SD = 11$). The participant's average driving experience was 19 years ($SD = 14$). Two participants reported driving less than 10,000km per year, eleven drove between 10,000-20,000km per year, with three participants reporting driving more than 20,000km per year . Their Vision was normal or corrected to normal for driving and fifteen of the participants were right-handed.

Procedure

Participants were welcomed to the laboratory, and after completing informed consent forms, had their demographic information collected. They were then provided with training to familiarise themselves with the characteristics of the primary driving task, and secondary tasks. Participants then undertook the experimental conditions, as indicated above. The experimenter was present in the laboratory during all trials.

Primary driving task

The primary task utilised standardised equipment to represent a three-lane roadway, see Figure 1. The driver was instructed to accelerate to 60km/h and maintain the speed. Initially, two start signs are presented for each condition, then periodically, lane change indicators would appear on each side of the road, corresponding to the lane in which the downward arrow was presented, see Figure 1. The participants were then required to switch lane quickly and efficiently. The protocol specifies that this process is repeated during baseline training until a MDEV value of 1.2m or less is established. The lane changes are repeated over a series of tracks, lasting approximately 3 minutes at 60km/h.

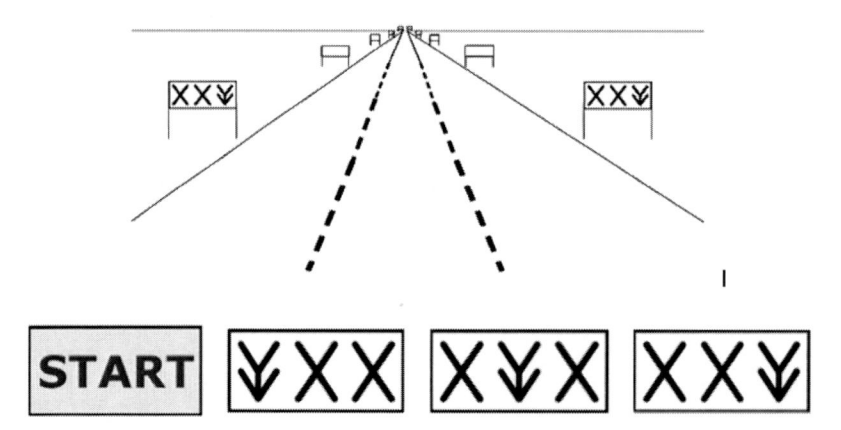

Figure 1. The LCT simulated roadway and lane change signage

The auditory secondary task

This task is based on the Sternberg (1966) auditory discrimination test. When attempting this activity, participants were required to listen to a sequence of three (the easy variant) or six (the hard variant) spoken numbers. After 15 seconds, participants heard a further single digit. They were then required, as quickly as possible, to declare 'yes' if the digit was part of the previous sequence; or 'no' if not. There were six presentations of auditory stimuli for both the easy and hard auditory conditions.

The SuRT visual secondary task

In this task, the secondary visual display (see Figure 2) is filled with circles, one of which is larger than the others. The visual search task requires selection of either the left or right side of the screen in the easy condition, or one of six regions, three to the left and three to the right in the hard condition. Participants must select the region with the larger circle, using buttons corresponding to left or right (once for the easy version, from one to three presses for the hard), at which point, a grey selection area appears on the screen, then press another button to confirm their choice. In the easy version of the task the non-target stimuli subtended 26 arcmins (approximately 6mm for the participants), and the targets 61 arcmins (approximately 14mm). For the hard variant, the non-targets subtended 53 arcmins (12mm) and the targets 61 arcmins.

The Integrated Route Guidance Task

A route guidance task was selected as it required both visual, verbal and manual interaction. A standalone system was used, the 'TomTom Go'. This device, is operated using a touch screen interface. In the easy task, participants were required repeatedly to adjust the volume to pre-determined percentages, throughout the experimental 'drive'. The difficult task was destination entry. In this activity, the participants were required to enter the city, street and house number using the touch screen. When completed, they were required to repeat the exercise with a different address, again and again, throughout the drive. The addresses were balanced in terms of the number of required characters in each element.

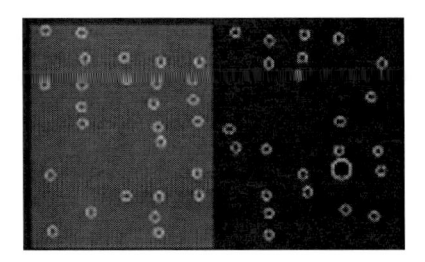

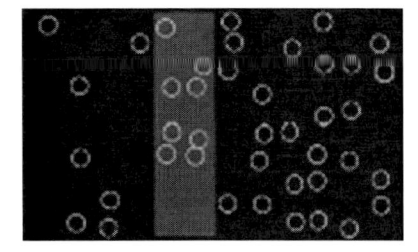

Figure 2. The SuRT secondary task, with Easy (left) and Hard (right) variations

Equipment

Experimental equipment for this study was specified in ISO (2010). Figure 3a depicts the arrangement of apparatus. The primary lane change task was displayed using a 21" LCD monitor with the resolution set to 1600x1200 pixels. The SuRT task display was a 19" monitor with the resolution set to 1280x1024 pixels, this was switched off when not in use. Control interaction with the lane change task was facilitated using a Logitech Momo force-feedback steering wheel. A TomTom Go navigation system was used for the integrated route guidance task. The position of the route guidance system with respect to the primary task is shown in Figure 3b.

Figure 3a (left). Example experimental setup. Figure 3b (right). Route guidance system position

Results

Primary task performance

Mean lane deviation (MDEV) was analysed according to ISO (2010). Two metrics are calculated following this protocol, Normative MDEV and Adaptive MDEV. Calculation of Normative MDEV compares the participant's performance to an idealised path for lane change behaviour. The Adaptive MDEV additionally considers the participant's baseline performance in the calculation of their MDEV.

Analysis of variance of the Normative data revealed a significant difference in primary task performance for the secondary tasks (F(3.14,47.1) = 9.08, p <.0001), see Figure 4. Post-hoc comparisons revealed the Easy Visual, and both Integrated tasks produced significantly greater MDEVs than the Pre-baseline. For the Post-Baseline condition, all secondary tasks resulted in significantly greater lane deviation, except the Easy Auditory one. The Easy Auditory task was found to produce significantly less lane deviation to all the other secondary tasks, except the Hard Auditory one, which was not significantly different. The Easy Visual task was significantly different to the Post-Baseline, the Hard Visual and both Integrated tasks. The Hard Visual task was significantly different to The Post-Baseline, and both Integrated tasks. Both Integrated tasks imposed significantly greater lane deviation than any of the other conditions, although they were not significantly different from each other. Figure 4 shows the order of increasing lane deviations with the Baseline drives having the lowest deviation, followed by the Auditory, Visual, and Integrated secondary task, with the highest MDEVs. Paired t-tests were performed on each Normative Easy-Hard secondary task pair, e.g., Easy Visual vs. Hard Visual. No significant differences were found for lateral control MDEV values for any pair.

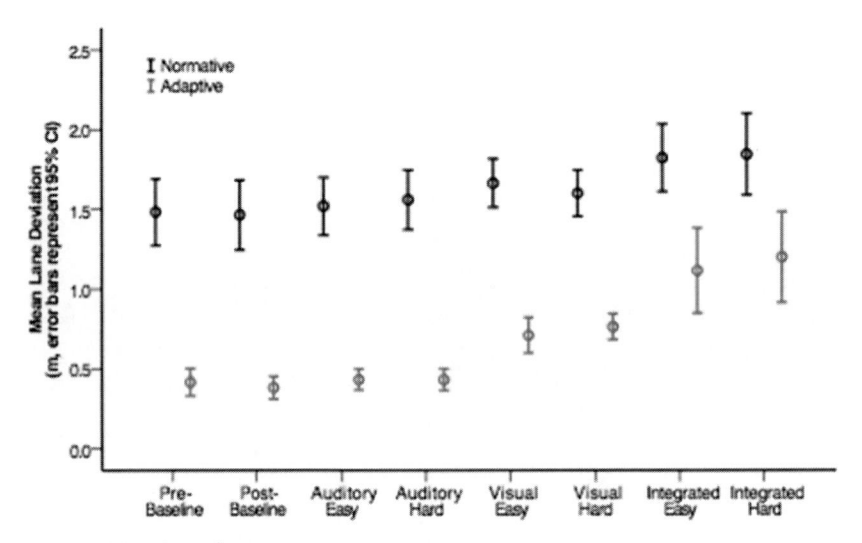

Figure 4. Mean Lane deviation

Analysis of variance of the Adaptive data revealed significant differences for MDEV between secondary tasks (F(1.88,29.13) = 31.5, p <.0001), see Figure 4. The same general pattern of differences emerged as identified in the Normative data. Pre-Baseline significant differences were found between all the Visual and Integrated secondary tasks, which all resulted in greater MDEV values. The Post-Baseline condition was found to have significantly lower MDEV than all other conditions, except the Pre-Baseline, with which it was not significantly different. Both the Auditory tasks were significantly different to all conditions except the Pre-Baseline and the each other. Similarly, both the Visual tasks and both the Integrated tasks were significantly different to all other conditions except the other variant of that task type. Figure 4 presents data indicating that the addition of a secondary task significantly reduces lateral control. Further, that auditory, visual, and integrated visual-auditory tasks will significantly degrade lateral control, in that order. Paired t-tests were performed on each Adaptive Easy-Hard secondary task pair. No significant differences were found for lateral control MDEV values for any pair.

Secondary tasks

Auditory secondary task
All participants completed 100% of the auditory task presentations. Four of the sixteen participants made one error each in the auditory digit discrimination task.

SuRT visual secondary task
Paired sample t-tests were conducted on the visual secondary task performance data, see Figure 5. They revealed significantly fewer items processed ($t(15) = 8.98$, $p<0001$ (two-tailed)) in the hard condition ($M = 80.8$, $SD = 33.1$) than the easy one ($M = 19.8$, $SD = 8.4$). Further, mean task duration was significantly longer ($t(15) = -6.3$, $p<0001$ (two-tailed)) in the hard condition ($M = 10.4$ secs, $SD = 5.9$) than the easy one ($M = 2.6$ secs, $SD = 1.4$). No significant difference was found for the percentage of correct items.

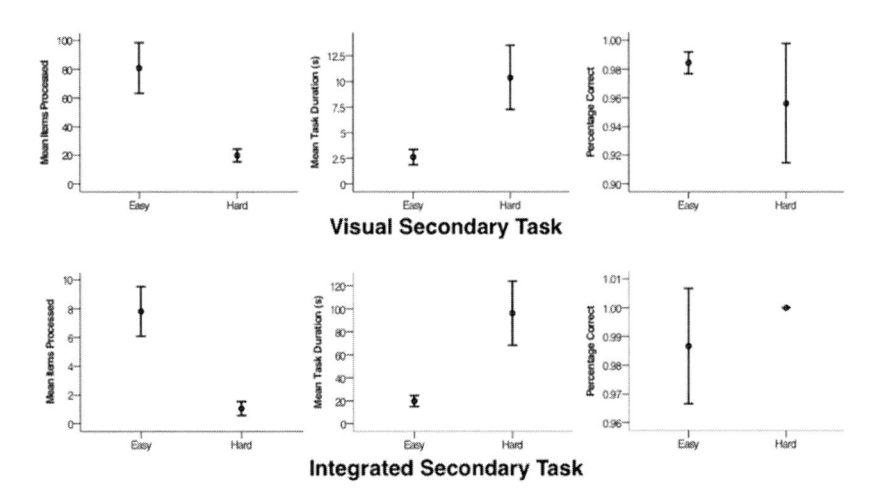

Figure 5. Visual and Integrated secondary task performance

Integrated route guidance task
Paired sample t-tests were conducted on the integrated route guidance secondary task performance data, see Figure 5. Much like the visual task data, they revealed significantly fewer items processed ($t(15) = 10.6$, $p<0001$ (two-tailed)) in the hard condition ($M = 7.8$, $SD = 3.2$) than the easy one ($M = 1.1$, $SD = 0.9$). Further, mean task duration was also significantly longer ($t(10) = -6.4$, $p<0001$ (two-tailed)) in the hard condition ($M = 96.0$ s, $SD = 41.5$) than the easy one ($M = 19.8$ s, $SD = 7.2$). Five participants were excluded from this analysis as they did not complete any of

the route guidance items, and entering zero scores would skew the data. No significant difference was found for the percentage of correct items.

Discussion

This paper reports on finding from a UK study contributing data to a multi-laboratory calibration activity. Sixteen drivers participated in an experiment which followed the LCT protocol presented in ISO (2010). Results present a picture in which primary task performance (lateral deviation) is predictably impaired by the introduction of secondary tasks, as has been shown many times previously. The nature of the primary task performance decrements is consistent with the existing literature on secondary task complexity (Kantowitz, 1996; Lansdown, 2002; Nieminen & Summala, 1994; Santos, Merat, Mouta, Brookhuis, & de Waard, 2005; Summala, Nieminen, & Punto, 1995). Secondary task performance clearly demonstrates that the visually-based tasks were further degraded by the hard variants, but this had no significant effect on primary task performance. No support was found for the effectiveness of the easy-hard secondary task variants as candidate metrics for the LCT. Results suggest merit in the visual, auditory and integrated tasks, but not in the additional effort of two levels for each modality.

Results support rejecting the null for hypothesis i). Both for Normative and Adaptive MDEV values were found to significantly increase with the introduction the candidate secondary tasks. The influence of the auditory secondary tasks was rather limited, and only revealed a limited impact on either Normative or Adaptive MDEV values, particularly for the hard variant of the auditory tasks. The visual and integrated tasks were found have a significantly more pronounced negative effect of lateral vehicle control, resulting in greater MDEVs. Thus, the null is also rejected for hypothesis ii), in that both the visual and integrated secondary tasks frequently produced significantly greater MDEVs than the baseline conditions. This was especially so, in the Adaptive MDEV analysis which took greater account of the individual differences when calculating lateral control, than the Normative analysis.

Empirically, it would appear that the Adaptive method for calculation of MDEV values offers more promise for the reduction of experimental variance, than the Normative analysis. Indeed, it does appear to draw the data apart, as evidenced by the higher F values revealed by the ANOVAs, 9.08 for the Normative vs. 31.5 for the Adaptive respectively. However, the different analysis method did not result in meaningful differences in results when considering post-hoc testing of the MDEV values. Further, only three participants in this study achieved the 1.2m MDEV training criteria after multiple attempts in the laboratory, as specified in the ISO standard. It is unclear why this may have been the case, perhaps participants were uniquely unfamiliar with the lane change task, or this study was the 'one in twenty' which probabilistically would produce inaccurate results with a $p \geqslant 0.05$ alpha criterion. If this were the case, one might expect the relative order of significance for the other findings to be substantially different from that in the published data (Bengler, et al., 2009). The same trend was identified across all available data, with the baseline conditions producing the lowest MDEVs, the auditory higher, visual higher, and integrated tasks with the highest MDEV values. Thus, this study supports the relative validity of the LCT tasks, but not their absolute validity. Such findings are commonplace the in the driving simulation literature (Blana, 1996; Blana, 1997; Wang, et al., 2010). Data from other contributors (Bengler, et al., 2009), did reveal some significant differences for secondary task difficulty that were not found in this study. However, there was only agreement for the critical tracking task which was not adopted in this study. The data presented for the auditory, visual and integrated tasks does not present agreement between contributing laboratories.

Hypothesis iii) is rejected in that no significant MDEV differences were found when comparing the easy and hard versions of the secondary tasks in this study. Practically, data presented here provides no encouragement for the adoption of both easy and hard variants of the secondary tasks utilised here for the LCT. Performance on the secondary tasks, shows clearly that regardless of the metric considered, the participants found the secondary tasks to be significantly different in their demands on driver's resources. It is interesting, then, that these secondary task performance decrements did not result in significantly different primary task failures. The drivers may have been protecting primary task performance, in the manner that one might reject an unnecessary secondary distraction when vehicle control demands more resources. For example, engaging in conversation when approaching a complex junction; drivers will often rationally and naturally disengage with conversation to concentrate on the more important primary task. Alternatively, the demands imposed by the secondary tasks may not have been high enough to statistically influence primary task performance, although this seems unlikely as there is widespread agreement that, for example, route guidance destination entry is a highly taxing secondary task.

Observation of participants during the experiment suggests that there may be some scope to refine the primary and secondary task engagement in the LCT protocol. It could perhaps require minimum levels of secondary task performance in addition to the primary task training requirement. Such measures, might address the potential situation where a participant disregards secondary task performance in order to favour primary task performance. While this reflects a rational and, one might suggest, sensible approach to driving; it rather undermines the purpose of the protocol. Therefore, it is suggested that measures be put in place to reduce the scope for individual strategies by LCT participants.

Conclusions

This study utilised auditory, visual and integrated tasks to establish performance when engaging in a simulated driving lane change task. The lane change task offers promise for efficient evaluation of in-development in-vehicle human machine interfaces. Results from this study offer some support for the protocol in that the same general pattern of significant findings was established as has been reported previously. The selection of different types for the candidate secondary tasks was shown to be effective for differentiating both primary and secondary task performance. However, the secondary tasks could not be differentiated in terms of their effect on the primary task. Utilisation of the Adaptive MDEV calculation had the practical impact of reducing the absolute MDEV values by considering the baseline performance of the participants. However, it did not have a meaningful effect on the pattern of significant differences found for the candidate secondary tasks.

Acknowledgements

The author would like to thanks Stefan Mattes for assistance in analysis of the MDEV values.

References

Bengler, K., Mattes, S., Hamm, O., & Hensel, M. (2009). Lane Change Test: Preliminary Results of a Multi-Laboratory Calibration Study. In G.L. Rupp (Ed.), *Performance Metrics for Assessing Driver Distraction: The Quest for Improved Road Safety* (pp. 243-253). Warrendale, Pennsylvania: Society of Automative Engineers International.

Blana, E. (1996). A survey of driving research simulators around the world. ITS Working Paper No. 481. Leeds, UK: Institute for Transport Studies, University of Leeds.

Blana, E. (1997). The pros and cons of validating simulators regarding driving behaviour. In A. Kemeny (Ed.), *DSC 97 Driving Simulation Conference* (pp. 125). Lyon - France: ETNA.

Green, P. (1999). Estimating compliance with the 15-second rule for driver-interface usability and safety. In *Proceedings of the Human Factors and Ergonomics Society 43rd Annual Meeting* (pp. 987-991). Santa Monica, CA: Human Factors and Ergonomics Society.

ISO 16673 (2007). Road vehicles - Ergonomic aspects of transport information and control systems - Occlusion method to assess visual distraction due to the use of in-vehicle systems. International standard, International Organization for Standardization.

ISO 26022 (2010). Road vehicles - Ergonomic aspects of transport information and control systems - Simulated lane change test to assess in-vehicle secondary task demand. International standard, International Organization for Standardization.

Kantowitz, B.H., Hanowski, R.J., & Tijerina, L. (1996). Simulator Evaluation of Heavy-Vehicle Driver Workload II: Complex Secondary Tasks. In *Proceedings of the Human Factors and Ergonomics Society 40th Annual Meeting* (pp. 877-881). Santa Monica, CA: Human Factors and Ergonomics Society.

Lansdown, T.C. (2002). Individual differences during driver secondary task performance: verbal protocol and visual allocation findings. *Accident Analysis & Prevention, 34*, 655-662.

Nieminen, T., & Summala, H. (1994). Novice and experienced drivers' looking behavior and primary task control while doing a secondary task. In *Proceedings of the Human Factors and Ergonomics Society 38th annual meeting* (pp. 852 - 856). Santa Monica, CA: Human Factors and Ergonomics Society.

Olsson, S., & Burns, P.C. (2000). Measuring distraction with a peripheral detection task. In D. Salvucci and N. Taatgen. (2008). Threaded cognition: an integrated theory of concurrent multitasking. *Psychological Review, 115*, 101-130.

Santos, J., Merat, N., Mouta, S., Brookhuis, K., & De Waard, D. (2005). The interaction between driving and in-vehicle information systems: Comparison of results from laboratory, simulator and real-world studies. *Transportation Research Part F: Traffic Psychology and Behaviour, 8*, 135-146.

SOP. (1999). Commission recommendation of 21st December 1999 on safe and efficient in-vehicle information and communication systems: A European statement of principles on human machine interface. *Official Journal of the European Communities, L 19*.

Sternberg, S. (1966). High-Speed Scanning in Human Memory. *Science, 153*, 652-654.

Stevens, A., Board, A., Allen, P., & Quimby, A. (1999). A safety checklist for the assessment of in-vehicle information systems. A user's manual. Wokingham, Berkshire, UK: Transport Research Laboratory.

Summala, H., Nieminen, T., & Punto, M. (1995). Maintaining lane position with peripheral vision during in-vehicle tasks. *Human Factors, 38*, 442-451.

Tombu, M.N., Asplund, C.L., Dux, P.E., Godwin, D., Martin, J.W., & Marois, R. (2011). A Unified attentional bottleneck in the human brain. *Proceedings of the National Academy of Sciences, 108*, 13426-13431.

Wang, Y., Mehler, B., Reimer, B., Lammers, V., D'Ambrosio, L.A., & Coughlin, J.F. (2010). The validity of driving simulation for assessing differences between in-vehicle informational interfaces: A comparison with field testing. *Ergonomics, 53*, 404-420.

Wickens, C.D. (2002). Multiple resources and performance prediction. *Theoretical Issues in Ergonomics Science, 3*, 159-177.

Wickens, C.D., & Hollands, J.G. (2000). Attention, time-sharing, and workload. *Engineering psychology and human performance* (3rd ed.). New York: Harper Collins.

Electric vehicles: an eco-friendly mode of transport which induces changes in driving behaviour

Elodie Labeye[1], Myriam Hugot[1], Michael Regan[2], & Corinne Brusque[1]
[1]French Institute of Science and Technology for Transport, IFSTTAR, Bron, France
[2]University of New South Wales, Sydney, Australia

Abstract

The electric vehicle (EV) represents a new eco-friendly mode of transport which involves different kinds of constraints to use that are likely to affect driving behaviour. In order to study the impact of electric technology on driving behaviour in France, the BMW Group in Munich contracted IFSTTAR to lead and undertake scientific research for the MINI E France Project. In this study, 25 "private users" from Paris drove for 6 months a MINI E (an advanced electric vehicle prototype). The study focused on how drivers used the EV, via analysis of their trips and the behaviours they adopted in their daily routine, using questionnaires, travel and charge diaries. This paper describes the outcomes of the study, in particular how EV characteristics can induce changes in driving behaviours that affect different levels of control of the driving task, at the strategic, tactical and operational levels. The important connection between eco-friendly driving and safe driving behaviour in electric vehicles is discussed.

Introduction

Faced with current energy conservation issues, the reduction of CO2 emissions is a global priority. In the field of transport, this is reflected in the push for sustainable urban mobility. In this sense, unprecedented financial support has been provided by governments for the development of new energy, new mobility projects and low-carbon emission vehicles.

The electric car is, in this context, a potentially effective and practical alternative to traditionally fuelled vehicles that can play an important role in reducing the environmental impact of transport. Advanced lithium battery technologies also enable the car to help improve the quality of ambient air while supporting the reduction of transport noise.

Use of an electric vehicle, however, has potential to induce in drivers changes in daily vehicle usage patterns. The aim of this study was to identify these changes and their impact on the activity of "traditional" driving, as described in the literature (e.g., Michon, 1979, 1985).

In D. de Waard, N. Merat, A.H. Jamson, Y. Barnard, and O.M.J. Carsten (Eds.) (2012). *Human Factors of Systems and Technology* (pp. 171 - 180). Maastricht, the Netherlands: Shaker Publishing.

In a model proposed by Michon (Michon, 1979, 1985; see also Van der Molen & Bötticher, 1988), driving is modelled as a hierarchical categorization of all the tasks the driver must make throughout a trip. Michon asserts that there are three main levels in this hierarchy, which are differentiated according to the time and cognitive requirements of tasks.

Trip planning, for example, is held to occur at the strategic level and the tasks performed here require high cognitive investment. According to Michon (1979, 1985), trip characteristics are selected according to the objectives of the trip while taking into account all the fluctuations of the environment. Then, the driver evaluates the quickest ways (based on roadwork, traffic jams, etc.), and the potentially dangerous situations (driving at night…). Planning for the trip will be also determined by the time the driver has available for the trip, the costs involved, the choice of vehicle and the road. There is potential, however, for all of these criteria and their respective weight in the planning of a trip to be changed by driving an electric car. Indeed, new issues arise in relation to the unique properties of this type of vehicle. Because the electricity stored in the EV is more limited than a tank of conventional fuel in terms of energy available, trips of the driver may change according to its charging and energy retention capabilities. The travel planning undertaken by the driver must therefore take into account the constraint of range which is more salient than that for a conventional car, and this additional task can make the driving activity more complex.

Michon (1979, 1985) also identifies in his model of driving a tactical level of control. Here, it is the planning of actions and "Maneuver control that allowing the driver to negotiate the directly prevailing circumstances" (Michon, 1985, p. 489). At the tactical level, the interaction with other road users and the traffic environment is controlled in real time and decisions need to be made quickly. The latter would include overtaking decisions, decisions to give way and to stop at a pedestrian crossing, the choice of speed; and all depending on traffic density, speed of vehicles, and other situational factors. In the scenario of driving an electric car, the silent nature of the EV may have a significant impact by changing the perception of the car and the means that other users have to identify the presence of it (Garay-Vega et al., 2010; Stefan, 2006). Tactical treatment of the environment will therefore have to take account of this new component to optimize silent driving during the trip, obliging the driver to be more attentive to other road users in driving the EV.

Finally, some inherent properties of the EV may have an impact on the third level of driving control proposed by Michon (1979, 1985) - the operational level. At this latter level, control of the stability of the vehicle (speed and lateral control, etc.) is held to occur automatically. However, the EV has a feature absent in conventional cars and related directly to the operational level: regenerative braking. When the user releases the accelerator pedal the car slows quickly (as if the driver has very lightly braked) and a certain amount of kinetic energy is recovered and transferred to the battery. This new control mechanism offers new possibilities in addition to the different manoeuvres possible in conventional cars, which may have a direct impact on the operational activities of the driver.

Objectives

As previously described, the objective of the present study was to identify the changes induced by the use of electric cars and their impact on the activity of driving. We believe that certain characteristics the EV have the potential to induce changes in driving behaviours that affect the different levels of the driving task previously defined: the strategic level, in terms of route planning due to the limited range of the vehicle; the tactical level, in terms of interactions with other road users to deal with the silent nature of the EV; and the operational level, in terms of braking behaviour to master regenerative braking.

To address these objectives, a study already conducted in Germany (Cocron et al., 2011a, 2011b), USA and England for the manufacturer BMW was replicated in France with 25 drivers in Paris. These users utilized for six months - from December 2010 to July 2011 - a MINI E electric car. Data for the study were derived from a set of questionnaires, and travel and charge diaries. The aim was to study the acceptance of electric vehicles and the new behaviours of the users. For this paper, we highlight the changes in behaviour and attitudes that occurred after 6 months of use of EVs; the results at 3 months were essentially identical.

Methodology

Participants

More than 700 people applied online (via the MINI.fr site) to participate in the study. Driving such an energy saving vehicle would not be desirable for all types of potential applicants and, hence, the 700 who applied were not a representative sample of the French population, but rather more representative of those that might eventually buy electric vehicles. From the vehicle manufacturer's perspective, this sample was of more relevance in this study. A first selection was made to represent the overall composition of the 700 applicants and it was also based on the following logistic criteria: residence in the Paris area; having a garage or a dedicated parking place for the car; acceptance of all the conditions of the experiment (payment for leasing the vehicle: 475 euros/month insurance included + 25 euros/month for a wall box setting by the energy provider EDF; a commitment to respond to questionnaires; etc.); and having access to suitable electrical power and other technical apparatus. Finally, other criteria (see below) were used to select the final sample of users from among the participants which met the above requirements. Twenty-five subjects were chosen principally based on the number of miles they were driving each day (we wanted drivers who would drive enough to really experience the vehicle), their age (it was roughly representative of the original sample of applicants), and their gender (we tried to oversample women because men were totally overrepresented among applicants in previous studies in other countries involving the MINI E; we wanted to collect the opinions and attitudes of both sexes.). Moreover, in order to observe any driving changes with use of the EV, we tried to oversample for drivers who had had no experience with pre-electric or hybrid vehicles.

Nevertheless, of the final 25 participants selected, only 7 were women and 18 were men, with an average age of 44.5 years, and 24% had experienced an electric vehicle and 12% a hybrid vehicle. The number of people per household for each driver selected was on average 3.32; 80% of selected participants had a university level qualification and 36% were driving more than 70 km per day. At the beginning of the study, 60% of participants described their style of driving as calm and balanced and 52% as sporty and dynamic; 32% had already had a MINI. Finally, 95% of them had one vehicle at home or more (32% of users had only one vehicle and 68% had more than one vehicle). The demographic characteristics of the final sample were not entirely as planned, given the difficulties in recruiting enough people who met the practical requirements for participation in the study (e.g., garaging capabilities; access to suitable power sources, etc). Moreover, it should be noted that the sample was not representative of the French population but rather of potential early adopters of EV, which explains their high experience of electric or hybrid vehicles. Their involvement in the study was based on their interest in the brand and the technology of EV, as well as for the ecological impact of it.

Electric vehicles

The electric vehicles used in this study are prototypes similar in external appearance to the MINI Cooper, but with two-seats and equipped with a lithium-ion battery. The average range of the vehicle is 160 km and the car has a regenerative braking system that allows the driver to lightly brake (without applying brake pressure) while regenerating energy from the moment s/he releases the accelerator pedal. Of course, braking in the conventional way is still possible. To charge the vehicle, each participant had a wall box of 12 amps installed by the French electricity provider EDF in his or her home. Drivers could also charge their vehicles from Parisian public charging stations. A full charge lasts for about 9 hours.

Material

Data collection for the study derived from three different sources: questionnaires, travel diaries and charge diaries. To assess changes in behaviour and acceptability occurring over time with use of the EV, data were collected from participants at several points during the study: at T0, T3 months and at T6 months.

Questionnaires

Four questionnaires addressing different dimensions of use were administered during the study. Questionnaires were conducted on the Internet or administered face-to-face and contained between 225 and 356 items measured on a Likert scale of six points, ranging from 1 "very strongly disagree" to 6 " very strongly agree", or a 1 "never" to 6 "always". The first questionnaires administered at the beginning of the study concerned the prospective views and expectations of future users about the electric vehicle, their considerations about the ecological aspects and techniques of EVs and also their driving habits with a traditional car. After 3 months of using the EV, participants were asked to complete two questionnaires with items which were either existing items, from T0, or new items. The new ones related to the experience and appreciation of participants about the use of the MINI E on a daily basis. Several topics were covered: the drive, charge, displays inside the car, the absence of noise of the vehicle, regenerative braking, and critical situations. After 6 months, participants again completed a questionnaire where the majority of items were identical to items from previous questionnaires. Redundancy of items at the middle and the end of the study allowed us to be sure that the behaviours observed were not the result of the attraction for the novelty of the vehicle but, rather, appropriate behaviours in the EV, which persisted over time.

Travel patterns

During the study, participants also completed travel and charge diaries, each for a week (Golob & Gould, 1998). Before they utilized the EV, they had to register in a travel diary all their trips, detailing the trip distance, means of transport taken, the purpose of the trip, and departure and arrival times for their own conventional car. After 3 months and at the end of the study they completed travel diaries which concerning the travels with the MINI E. Users were also required to complete charge diaries detailing all charges that they made during a one-week period. Here, users reported on the following characteristics: place of charge, charge status at the beginning and the end of the charging process, and the reasons for the charge. Obviously, a daily record keeping would have also been a solution, but the rhythm of life of participants could not allow them to be truly exhaustive in their notes for the entire duration.

Results

The results were derived from the qualitative treatment of open-ended questions and the quantitative treatment of Likert scales. For the quantitative treatment, accumulated responses to "very strongly disagree", "strongly disagree" and "disagree" were used and the same for "very strongly agree", "strongly agree" and 'agree" was made. As noted already in the introduction, driving is a complex cognitive activity which includes a set of hierarchically structured tasks (Michon, 1979; Van der Molen & Bötticher, 1988). When driving an electric vehicle, the nature of these tasks has potential to be modified because new constraints appear in connection with the particular characteristics of the EV. We present below the changes in behaviour that occurred after 6 months of use of EVs (the results at 3 months were, as noted, essentially identical).

The strategic level

Range was perceived by users of the MINI E as a major constraint in managing the use of the car. The results indicate that this problem leads them to adopt new behaviours such as developing a strategy for planning their charges, reconsidering their choice of using different modes of current transport and trying to re-define their routes to optimize their range. In addition, MINI E users expressed the need for tools to help them in these various planning tasks that are new to them.

Analysis of charge diaries allowed us to describe the strategies and behaviours that drivers adopted to charge their vehicles. The results show that participants charge their vehicle on average 5.2 times per week, mainly at night (in 73% of cases) and at home (86% of cases). Although drivers claimed they charged their vehicle when it did not have enough range to make the next trip, or when the battery was discharged, the diaries show that vehicles were charged when they had on average at least 42% of range remaining. Users seemed to be afraid to run out of battery power and, thus, to risk breaking down.

The results reveal that only 3% of all charges made by the participants were made in public charging stations. This low percentage is mainly due to the fact that places are regularly occupied by traditional cars and the fact that public charging points are not in places where users visit daily. Finally, charging the MINI E became a daily routine for 81% of participants after 6 months of driving and 96% of users appreciated being able to charge at home and therefore no longer needing to go to the gas station.

In addition to bringing about in drivers planning strategies to manage the charging process, the use of the MINI E also urged users to review their choice of using different modes of transport. Indeed, use of the MINI E did not amount to a simple replacement of the conventional car. Use of the vehicle caused the driver to review and reassess the role of walking, of cycling, of the conventional car, and of public transport in assessing mobility requirements. Overall, the analysis of questionnaires revealed an increase in the number of short trips by MINI E users: for 76% of drivers, the electric car is more useful for their daily trips than the traditional car and 43% think they have used their MINI E for trips for which they had not previously used their normal car (the participants state that is usually trips for "small shopping"). More specifically, 76% of users agree with the following item: "For short trips, I used more the MINI E than my traditional car ".

Analyses of travel diaries confirm that the number of trips involving only the car increases between T0 and T6 months (from 60% to 87%) – and, in parallel, the average distance of trips reduced (95% of trips only done by car were less than 68 km before the MINI E study while they were less than 42 km at T6 months). Interestingly, 43% of users thought they had used the MINI E for trips they usually did by foot (or by bike) before the study (against only 19% for trips they usually did by public transport before the study). However, the decrease of the walking is consistent with the increased use of EV for short trips. From the environmental point of view, the low energy cost of EV compared to the car engine and the fact that the EV emits less CO_2 appear to be the main reasons which remove any guilt of participants in using more their electric vehicle. Of course, it is also possible that the increase in use of the EV can be due to the novelty of the vehicle and the impact of the monthly fixed payment schedule or the cheap charging process. But we think that at the end of six months of use, participants entered into a routine of trips, and by then the effects of the novelty would have worn off.

Finally, the limited range of the EV also induced for drivers a redefinition of their routes to optimize the power consumption of the car. The analysis of questionnaires revealed that almost half of the drivers (48%) say that, when they can, they do trips which allow them to use less energy. But the questionnaires also show that there are still trips that participants would have liked to do in the EV but which could not be realized due to the limited range of the vehicle (and for which they had to use another mode of transport.) Indeed, users believe that they have not been able to do with the MINI E 21% of the daily trips they have usually done with their conventional car; and for 62% of them, the limited range of the MINI E does not permit them to do all of their normal driving. More than half of participants (52%) believe that the MINI E has changed their mobility behaviour significantly. Finally, although users are aware of the parameters that influence range, they still seek navigation tools enabling them to handle the charge: whether it is information on the location of the closest public charging station (95%) or navigation systems which direct them to more eco-friendly routes (88 %).

In summary, these results show, collectively, that the activity of charging vehicles and the adaptation of their mobility patterns to the range of the EV are new tasks that participants include in the planning of their trips. The strategic level of the activity of driving an electric car still relies on the evaluation of trips in terms of timing, of danger, travel cost, etc; however, also, and even more so, on the optimization of vehicle energy consumption.

The tactical level

Another main feature of the electric car is that, unlike a conventional vehicle, it emits no engine noise. At low speed (<30 km/h) EVs are silent and their approach may not be perceived by pedestrians. By contrast, beyond 30 km/h the rolling noise of the vehicle on the road is dominant and other users have no problem in hearing it approaching. The lack of noise at low speed implies the need for a change in the interaction between road users, since usually their perception of the traffic environment is based on information from both the visual and auditory modalities. Action planning and manoeuvre control that allow the driver to negotiate through traffic in real time will be modified when using an electric car - and the tactical level of driving, described by Michon (1979, 1985), which allows for proactive behaviours (Van Zomeren et al., 1988), will have to bear in mind that the pedestrian does not hear the car coming.

Overall, 38% of users reported that they had to change their driving behaviour due to the lack of outside noise of the MINI E, especially in order to pay more attention to pedestrians. Only 18% of users report that the lack of engine noise caused problems, but 50% think the lack of noise is potentially dangerous. In practice, 28% of users identified critical situations concerning the acoustic properties of the MINI E. Such situations involved bicyclists, pedestrians, pedestrians with a Walkman, and occurred primarily in parking areas where pedestrians did not perceive that the car had started (drivers had to honk to warn them). Other problem situations relating to the acoustic properties of the MINI E included driving in underground garages, on driveways, and during snow days (when pedestrians only looked at their feet and were therefore less aware of cars) and when other people got out of their car and did not perceive the MINI E. However, despite the potential risks and the need for increasing attention in their anticipatory behaviour, the study shows that the absence of noise is perceived as an advantage for drivers: 73% of participants do not want outside vehicle noise, even though the provision of artificial noise could strengthen the perception of low-speed vehicles for other road users, especially pedestrians.

The operational level

In the hierarchy of driving activity described by Michon, the last level (the "operational" level) is concerned with the implementation of decisions taken at higher levels and corresponds to the achievement of sub-tasks of the driving activity. It is also responsible for reactions to unexpected and dangerous events. As will be seen, the use of regenerative braking when driving an electric vehicle changed driving behaviour at the operational level, and this was also modulated by energy issues established at the strategic level of driving activity.

Many users (86%) say that their driving style changed as a result of using the regenerative braking function in the MINI E. They reported that they anticipated more their braking distances and tried to drive more slowly. For nearly 90% of participants, the main challenge is to not use the brake pedal, in order to maximize the regenerative braking energy. These results show how the operational and strategic levels of driving are dependent on each other: on the one hand the new feature (regenerative braking) alters the activity of driving in an operational way since the users only use a single pedal (i.e., the accelerator pedal) to accelerate and decelerate; and on the other hand, the regenerative braking is also involved in the implementation of strategic energy issues since its use enhances energy. Overall, 57% of users think that, with their MINI E, they drive in a more flexible way than with a conventional car. The development of new driving behaviours, such as no longer using the brake pedal, or using it less often, can be problematic. While 76% of users report that the MINI E made them a safer driver when they drive it and that they much appreciate the regenerative braking, all of them try to use it whatever the braking situation. They estimate that they use it instead of the brake pedal in 92% of braking situations, and 91% report that the regenerative braking system is not strong enough. Users in this study report that they would like somehow to have a regenerative braking more powerful for use in all braking situations. In case of emergency braking, for example, regenerative braking could be used out of habit and be less effective than the traditional braking process with the brake pedal. Nevertheless, 96% of users declared in this experiment that they were always able to brake in time and 91% of them agreed that the regenerative braking always worked reliably.

Discussion and conclusions

The aim of this study was to examine the impact of the MINI E electric vehicle on self-reported driving behaviours. We postulated that use of such vehicles would induce changes in drivers in their daily vehicle usage patterns. Overall, the findings from the study confirmed that certain features and properties of the MINI E changed driving activity by generating new tasks, and modifying others.

Indeed, the results show that the electric vehicle modified certain aspects of driving activity, leading drivers to critically review their choice of transport mode and to rethink their trips according to the range of their vehicle. Thus, at the strategic level of control, a new planning task evolved to enable drivers to manage the optimization of vehicle energy consumption. Although we studied only 25 drivers, they used an EV for 6 months for most of their trips. This long period of exposure presumably gave them enough time to understand the characteristics of this new type of vehicle and to adapt to its constraints to best deal with their mobility needs - although further research is needed to determine quantitatively how long it takes for drivers to fully adapt to EVs. The charging solution chosen by most drivers was to charge almost every night, at home. However, the limited distribution of charge infrastructure and the duration of the current charge appear to have had an impact on charging behaviour: drivers did not charge when they had the need to do so, but rather took advantage of opportunities they had to charge. It is possible, however, that this strategy will change when charging infrastructure becomes more widespread and when charging times are faster.

Moreover, two more key findings emerged from the study: MINI E drivers reportedly developed strategies at the tactical level of control to anticipate risky situations in response to the silence of the MINI E; and they reportedly acquired new behaviours at the operational level of control to manage the regenerative braking function. Given that pedestrians reportedly perceived the EV with more difficulty because of its silent nature (and, hence, could be vulnerable to risk), the driver of the EV must be more attentive, decrease his speed and achieve a more intensive scan of the environment. But, despite reportedly paying more attention to pedestrians, more than 70% of drivers appreciated the silence of the car and didn't want any artificial noise generated outside the vehicle. For them, the more electric vehicles there are on the street, the more pedestrians will be alerted to them. Concerning the regenerative braking function, users need to relearn to drive with a new pedal, but it is easy and fast to acquire. Finally, the large majority of users reported that they anticipated more their braking distances and drove more slowly with the MINI E. Thus, this electric vehicle function encouraged drivers to drive in a safer, more eco-friendly manner.

Finally, the study reported here had nevertheless some limitations, the primary one being that it relied on the collection of self-reported data (questionnaires, charging diaries, and travel diaries). Thus, it is not known to what extent the self-reported behaviours recorded here are reflective of actual behaviours. The next phase in this line of research, which commenced late in 2011, has been to study drivers' actual behavioural adaptation in response to MINI E characteristics using a more experimental approach, involving instrumented vehicles equipped with video, accelerometers and other sensors. This will enable us to determine, in real driving situations, how the reduced sound of the vehicle actually changes drivers' behaviour, how the need to conserve battery energy impacts on driver behaviour in terms of speed, acceleration, deceleration and other measures of behaviour and vehicle performance; and how driving behaviour actually changes as a result of using the regenerative braking function.

The findings from the present study suggest, nevertheless, that driving a MINI E encourages drivers to drive in a manner that is more safe, and eco-friendly. The next phase of research, using instrumented vehicles, will seek to prove that this is so.

Acknowledgements

We are grateful to BMW Germany and BMW France for providing us with the opportunity to conduct this study. In particular, we thank Michaela Luehr, Roman Vilimek, Michael Hajesch and Jean-Michel Cavret and his colleagues for their support, and for their input to the project. We thank also colleagues from CEESAR, for their important role in recruiting participants and collecting the data for this study, in particular Julien Adrian, Annie Langlois and Reakka Krishnakumar. Finally, we thank Julien Delaitre, Magalie Pierre (EDF) and Julien Augerat (Veolia) for their support.

References

Cocron, P., Bühler, F., Franke, T., Neumann, I., & Krems, J. (2011a). The silence of vehicles- blessing or curse? *Proceedings of the 90th Annual Meeting of the Transportation Research Board.*

Cocron, P., Bühler, F., Neumann, I., Franke, T., Krems, J.F., Schwalm, M., & Keinath A. (2011b). Methods of evaluating electric vehicles from a user's perspective – the MINI E field trial in Berlin. *IET Intelligent Transport Systems, 5*, 127-133

Garay-Vega L., Hastings A., Pollard J.K., Zuschlag M., & Stearns M.D. (2010). *Quieter Cars and the Safety of Blind Pedestrians: Phase I* (Report No. DOT HS 811 304).Washington, D.C.: U.S. Department of Transportation

Golob, J. & Gould, T.F. (1998). Projecting use of electric vehicles from household vehicle trials'. *Transportation Research. Part B: Methodology, 32*, 441-454.

Michon, J.A. (1979). Dealing with danger: *Report of the European Commission MRC Workshop on physiological and psychological factors in performance under hazardous conditions* (Report VK 79-01). Haren (The Netherlands): Traffic Research Centre, University of Groningen,

Michon, J.A. (1985). A critical view of driver behavior models: What do we know, what should we do? In L. Evans and R.C. Schwing (Eds.), *Human behavior and traffic safety* (pp. 485-520). Plenum Press, New York.

Stefan, K. (2006). The importance of Vehicle Exterior Noise Levels in Urban Traffic for Pedestrian – Vehicle Interaction, *ATZ – Worldwide, 108* (07-08), 19-21

Van der Molen, H.H., & Bötticher, A.M.T. (1988). A hierarchical risk model for traffic participants. *Ergonomics, 31*, 537-555.

Van Zomeren, A.H., Brouwer, W.H., Rothengatter, J.A., & Snoek, J.W. (1988). Fitness to drive a car after recovery from severe head injury. *Archives of Physical Medicine and Rehabilitation, 69*, 90-96.

Interaction between driver and infotainment system using a touchpad with haptic feedback

Andreas Blattner, Roland Spies, Klaus Bengler, & Werner Hamberger
Technische Universität München
Germany

Abstract

Modern car infotainment systems contain increasingly more functions, which the driver must be able to handle easily even while driving in order to reach the destination stress-free and particularly safely. Therefore, several types of control elements like, for example, rotary push buttons, touchscreens, and/or joysticks, are offered by different car manufacturers. In addition to these control elements, the search for new ways of facilitating the interaction between system and driver still continues. In light of this, the use of a touchpad is proposed as a facile and intuitive, alternative control element. Based on the findings of Hamberger (2010) and Spies et al. (2009, 2010), the development of a new touchpad with haptic feedback is accomplished. The current paper presents the results of a field experiment that compared a touchpad with realistic haptic feedback via sensible and operable elements to a touchpad with haptic feedback via vibration of the touchpad surface in a real driving situation.

Introduction

With an increasing number of features available in modern car infotainment systems, many problems arise in terms of what is required between the human and the device in order to interact efficiently. There are more and more functions integrated into the infotainment system, which the driver must be able to handle with a minimum of distraction using a limited number of control elements even while driving. The various car manufacturers in the world try to solve this problem with different types of control elements like touchscreen, joystick, voice control, rotary push button, etc. A new approach to actualize a facile and intuitive interaction between driver and infotainment system is the usage of a touchpad, like the Audi AG applies it in the Audi A8.

According to Hamberger (2010), an in-car-touchpad offers several potentials. It is familiar to the users because of the accustomed usage with computer touchpads and enables handwriting recognition. Furthermore, a touchpad is a multipurpose control element, which perpetuates the possibility of splitting up display and controls. Thus, the display remains in the ideal field of vision and the touchpad can be positioned in the ideal reaching distance; robustness, optics, and the ease of use, are additional positive arguments. The results of an experiment in a driving simulator prove that a

---

In D. de Waard, N. Merat, A.H. Jamson, Y. Barnard, and O.M.J. Carsten (Eds.) (2012). *Human Factors of Systems and Technology* (pp. 181 - 187). Maastricht, the Netherlands: Shaker Publishing.

touchpad reduces track deviation relative to a rotary push button or a touchscreen. Additionally, text entry via touchpad has been shown to decrease gaze diversion times in comparison to a rotary push button (Bechstedt et al., 2005). This trend is mirrored in how customers feel about the touchpad, as they report to prefer to use the touchpad rather than the touchscreen (Hamberger, 2010).

In dual task situations, i.e. a driver interacting with an infotainment system while driving, interferences occur if tasks draw upon the same mental resources (Wickens, 1984 & Wickens et. al, 2004). In accordance with Bubb (1992), it is beneficial that during interaction with technical systems, a redundant feedback is given via multi sensory channels. Therefore, because the visual channel is most often the primary sensory channel in use, it is reasonable to relieve the load placed on this channel by using a touchpad with an additional haptic feedback. Spies et al. (2009) developed an automotive touchpad with haptic feedback based upon Braille technology in order to map display elements (e.g. buttons and sliders), as sensible and operable elements onto the touchpad (see figure 1 left). This touchpad with haptic feedback was compared to a conventional touchpad in a driving simulator experiment. The results were that a touchpad with haptic feedback decreases time for interaction and improves driving performance in a dual task situation significantly (Spies et al., 2009 & Spies et al., 2010).

The touchpads with haptic feedback

Based upon the findings of Hamberger (2010) and Spies et al. (2009), the development of a new touchpad with haptic feedback is accomplished, which enables an easy handling of the infotainment system. Investigations on technical realisation possibilities for a haptic feedback on touchpads resulted in two basic variants that are fit for automotive applications.

Figure 1. Touchpad with real haptic feedback (left), touchpad with simulated haptic feedback (right)

The first alternative is a touchpad with real haptic feedback (see figure 1 left) like the Braille touchpad of Spies et al. (2009). Real haptic feedback means that the displayed contents (e.g. buttons) in the menu screen can be imaged on the touchpad as sensible and operable elements by deploying the appropriate pins. Thus, the user is able to feel and press every elevated element on the touchpad representing concrete menu content, for instance the selectable button "Navigation". This principle is shown by the cross section of the touchpad on the left hand side in figure 2.

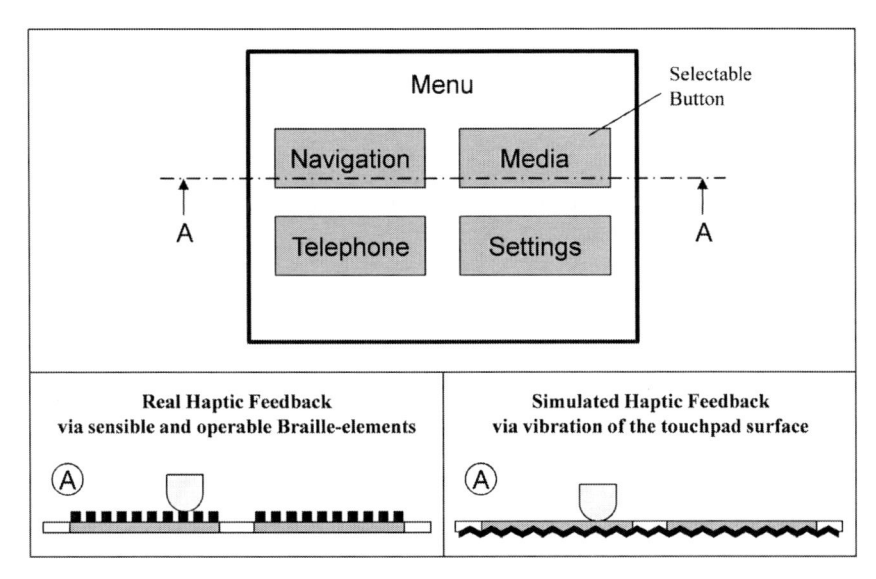

Figure 2. Technical realisation of the two different haptic feedback types

The second variant, a touchpad with simulated haptic feedback via vibration, is displayed in figure 1 (right). In order to realise this simulated haptic feedback, the configuration of the menu screen with its operable items (e.g. list elements) is virtually reproduced on the touchpad surface.

If the user touches a marked area, for example the button "Navigation", he will get a haptic feedback via constant vibration of the touchpad surface as long as his finger stays within the virtual region of the button (see cross section on the right hand side in figure 2). This vibration feedback is generated by four electromechanical oscillators. Choosing a menu function by pressing the planar touchpad surface in the area of a selectable menu element is indicated by another brief vibration pattern. Thus, a haptic feedback via vibration is given for both finding and selecting available menu functions. An additional visual feedback accompanies the haptic feedback of the two touchpad variants by highlighting the currently touched graphical widget on the screen and thus demonstrates the effective finger position on the touchpad.

For technical reasons, the realisation of a touchpad with simulated haptic feedback is less complicated than a touchpad with realistic haptic feedback. Hence, research would need to prove the assumption that a touchpad with haptic feedback via vibration leads to comparably good interaction, driving and gaze deviation results, as a touchpad with realistic haptic feedback in a dual task situation. The current experiment was conducted in order to compare the two aforementioned touchpad variants with different haptic feedback and to clarify the following hypotheses:

H1. A touchpad with haptic feedback via vibration leads to menu operation times that are equal or shorter than using a touchpad with real haptic feedback.

H2. The gaze behaviour while using a touchpad with haptic feedback via vibration is at least equal to a touchpad with real haptic feedback.
H3. Giving a haptic feedback via vibration yields an equal or better driving performance than using a touchpad with real haptic feedback.

Method

A within subjects usability test with 30 test persons (23M, 7F, ∅=38.8 years, SD=8.5 years) was conducted. The experiment was arranged in a real driving situation where the test persons had to drive on the autobahn A9 between Lenting and Denkendorf, without any junctions, at a constant speed of 100 km/h. This drive was performed without the use of any other technical devices, like speed control for instance. While performing the described driving task and paying attention to the other cars on the road, the subjects had to perform the following three menu interaction tasks:

T1. Entering a given destination using the last destinations
T2. Calling a specified contact member of the telephone book
T3. Reducing the loudness of the navigation system

These menu tasks were integrated into a specially implemented infotainment system and the test persons handled the three different tasks using both aforementioned touchpads with real and simulated haptic feedback in alternating order. The three tasks consisted of handling list and matrix compositions of functions, and using sliders for scrolling in lists or adjusting concrete values. During the experimental session, driving performance and gaze deviation data were measured while performing the various tasks and driving. Therefore, the Dikablis system (Lange et al., 2006) from Ergoneers GmbH was used as eye tracking device. The dependent variables were operation time, maximum glance time, percent total eyes off road time, standard deviation of speed and standard deviation of steering angle. Following the experiment, subjects were additionally asked to indicate which touchpad variant they prefer and to evaluate the usage of the two different touchpads. The objective and subjective results of the described usability test are presented in the following section.

Results

All data comparing the two touchpad variants with different haptic feedback was statistically evaluated by several t-tests. In the following, the findings for interaction, gaze deviation and driving performance are exemplified by the results for interaction task T1 "entering a given destination using the last destinations" because the tasks T2 and T3 yielded similar results.

The comparison of the control speed for completing the defined menu task by using both touchpad variants showed significantly shorter menu operation times for the touchpad with real haptic feedback (see figure 3). Accordingly, hypothesis H1 could not be verified in the conducted experiment.

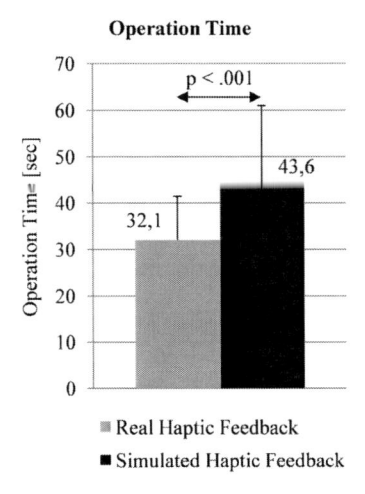

Figure 3. Operation time of the test persons for entering a given destination by using the touchpad with real haptic feedback and the touchpad with simulated haptic feedback

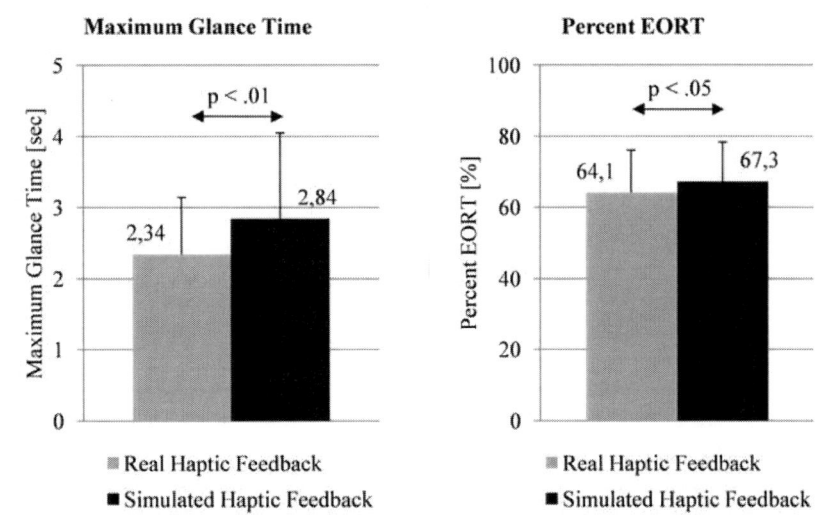

Figure 4. Maximum glance time (left) and percent total eyes off road time (right) of the test persons for entering a given destination by using the two different haptic touchpad variants

Considering the menu task of entering a given destination, the analysis of the gaze behaviour led to the following outcome: the maximum glance time during the menu task, was significantly longer when the touchpad with simulated haptic feedback via vibration was used (as can be seen in figure 4 on the left side). Furthermore, the percent total eyes off road time (Percent EORT) was significantly higher when the touchpad with simulated feedback was used (see figure 4 right). This parameter is defined as Percent EORT = (Total Eyes Off Road Time) / (Total Task Time) = EORT / TTT. Eyes were considered "off road" when the test person was not looking forward onto the road. Consequently, using the touchpad with real haptic feedback resulted in a significant better gaze deviation performance and hypothesis H2 could not be supported.

Hypothesis H3 was supported because the equation of the two touchpad variants with different haptic feedback led to identical driving performance results. Figure 5 shows that there was no significant difference in the standard deviations of speed (see figure 5 left) and steering angle (see figure 5 right) between the two touchpads with different haptic feedback.

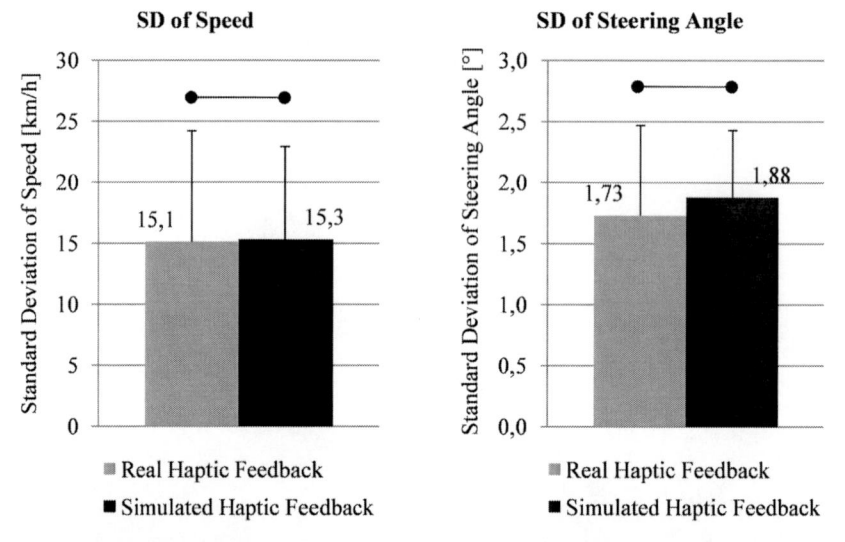

Figure 5. Standard deviations of speed (left) and steering angle (right) of the test persons for entering a given destination by using the touchpad with real haptic feedback and the touchpad with simulated haptic feedback

On the basis of the objective findings, the assumed equivalence between the touchpad with haptic feedback via vibration and the touchpad with real haptic feedback could not be shown for interaction and gaze deviation performance. Additionally, the subjective questionnaire regarding interaction reliability, traceability or selection of menu elements, and operation comfort, further supported the objective findings in that the touchpad with real haptic feedback performed significantly better. Moreover, 90 percent of the test persons preferred the real haptic feedback in the final ranking of the touchpad variants.

Discussion

In conclusion, it can be said that the touchpad with real haptic feedback had significant advantages over the touchpad with simulated haptic feedback. This was even more the case regarding menu interaction time and gaze behavior in a dual task situation. Based solely on driving performance, the simulated haptic feedback led to equal results and, thus, the assumed equivalence between the two haptic feedback variants could not be shown in the conducted usability test. There are many different approaches and technologies with great potential available on the market, which add a simulated haptic feedback to touchscreens or touchpads. Presently, these technologies are not technically advanced enough for usage in an ambitious dual task situation, like the interaction between driver and infotainment system while driving. In the future it might be possible to create a simulated haptic feedback on a touchpad that is as good as a realistic haptic feedback. However, at the moment, a touchpad with real haptic feedback has to be preferred for a driving situation as it was shown in the conducted usability test. Thus, the results described above provide the basis for the further development and technical concept of a new touchpad with real haptic feedback. This will additionally ensure a facile and intuitive interaction between driver and infotainment system even while driving.

References

Bechstedt, U., Bengler, K., & Thüring, M. (2005). Randbedingungen für die Entwicklung eines idealen Nutzermodells mit Hilfe von GOMS für die Eingabe von alphanumerischen Zeichen im Fahrzeug. In L. Urbas, and C. Steffens (Eds.), *Fortschritt-Berichte VDI Reihe 22(22), 6. Berliner Werkstatt MMS – „Zustandserkennung und Systemgestaltung"* (pp. 125-130). Düsseldorf: VDI-Verlag.
Bubb, H. (1992). Menschliche Zuverlässigkeit, Definitionen - Zusammenhänge - Bewertung. Landsberg/Lech, Germany: EcoMed-Verlag.
Hamberger, W. (2010). MMI Touch – new technologies for new control concepts. *IQPC – Automotive Cockpit HMI 2010*. Stuttgart: Steigenberger Graf Zeppelin.
Lange, Ch., Wohlfarter, M., & Bubb, H. (2006). Dikablis, engineering and application area. In R.N. Pikaar, E.A.P. Koningsveld, and P.J.M. Settels (Eds.) Proceedings IEA2006 congress (CD). Elsevier.
Spies, R., Peters, A., Toussaint, C., & Bubb, H. (2009). Touchpad mit adaptiv haptisch veränderlicher Oberfläche zur Fahrzeuginfotainmentbedienung. In H. Brau, S. Diefenbach, M. Hassenzahl, K. Kohler, F. Koller, M. Peissner, K. Petrovic, M. Thielsch, D. Ullrich, and D. Zimmermann (Eds.), *Usability Professionals 2009*. Stuttgart: Fraunhofer Verlag.
Spies, R., Hamberger, W., Blattner, A., Bubb, H. & Bengler, K. (2010). Adaptive Haptic Touchpad for Infotainment Interaction in Cars – How Many Information is the Driver Able to Feel? In *AHFE International – Applied Human Factors and Ergonomics Conference 2010*. Oxford: Wiley-Blackwell.
Wickens, C.D. (1984). Attention, Time-Sharing and Workload. In C.D. Wickens (Ed.), *Engineering psychology and human performance* (pp. 291-334). Columbus: Merril.
Wickens, C.D., Lee, J., Liu, Y., & Becker, S.G. (2004). Cognition. In C.D. Wickens, J. Lee, Y. Liu, and S.G. Becker (Eds.), *An introduction to human factors engineering* (pp. 145-159). Upper Saddle River, NJ: Pearson Prentice Hall.

Behaviour of deck officers with new assistance systems in the maritime domain

Albert Kircher[1], Fulko van Westrenen[2], Håkan Söderberg[1], & Margareta Lützhöft[1]
[1]Chalmers University of Technology, Göteborg, Sweden
[2]Umantec, The Netherlands

Abstract

The maritime domain is seeing new developments in systems aimed at increasing safety and efficiency of transport. These systems are tested for norm compliance and technical capability, but human factor aspects are not always part of the evaluation. The article argues that this is required, and exemplifies factors of interest with data from a simulator study. Two new systems related to collision warning and speed management were tested on 32 participants in full mission bridge simulators. The factors of main interest were the officers' experience, attitudes and workload. Data on the officers' behaviour were collected for complete runs and for different events related to possible collisions for the one system, and areas where speed should be limited for the second system. Differences in the officers' behaviour were observed in relation to several variables. The study claims that in addition to ensuring the technical capability of systems, a number of human factor issues have to be taken into account. Important aspects of the evaluation are highlighted.

Introduction

Ship's officers face an increasingly high number of novel systems to make navigation safer and more efficient. Theoretically, a nearly complete picture of their own ship, the surrounding ships, the navigation area, water depth and weather conditions is available, which could make navigation fundamentally safe. Radar systems are able to scan the surroundings and give the position and movement of other ships even at night. The Automatic Identification System (AIS) offers even more complete information about other ships by transmitting and receiving a variety of data related to ship type, voyage, position, and many more to all ships equipped with AIS. Electronic maps contain precise nautical charts, and can be integrated with Radar and AIS systems. Nonetheless, serious accidents happen, and a common claim is the involvement of the "human element". What the human factor is all about takes into account how humans are, and how they work with technical systems; this simple consideration entails several aspects, and the need for evaluating technical systems accordingly. This addition to function and compliance testing of technical systems may sound self-evident, but reality provides another picture. Human factor aspects can, and in fact should, be incorporated from the very early design stages of products or systems. We elucidate some of the factors which play a role in the way humans interact with systems (from here on systems will be synonymous with technical systems), and show how these were analyzed in a scientific way for two different new systems. The study was part of a larger EU funded project ("ITERATE") aiming at validating a model of driver behaviour for different transport means, i.e. shipping, train, and car. Technical details and implementation of the assistance systems (described below) were not evaluated directly, but instead behaviour of different groups of officers while navigating with the systems.

There are many aspects which compose a human factor evaluation, and the variety of methods is growing steadily. For the presented study, factors considered of importance of how professionals working in a certain environment deal with technology are: experience in the field, a person's attitude and personality and mental effort. These factors were part of the study, and clearly represent a sub-selection of possibly important factors; their pertinence is described below. Gender, fatigue and culture are factors which were not included in the study, because the study design did not allow for more factors to be incorporated.

Mental effort

In shipping, mental effort can vary greatly during a voyage. It was expected that mental effort would have an effect on the behaviour of the navigator. Mental effort is related to mental workload. The measurement of mental workload can be accomplished by a number of subjective, physiological and performance measures (O'Donnell & Eggemeier, 1986). The underlying theories of mental processing will not be described here, as this is outside the scope of the paper. In the present study the method had to have very low application and training time demands, allow straightforward data collection and analysis, and be validated. This led to choosing the Rating Scale of Mental Effort (RSME) (Zijlstra, 1993) to assess mental effort. It is a uni-dimensional scale, where the participant indicates the perceived mental effort on a continuous line ranging from 0 to 150 with verbal anchors (ranging from "almost no effort" to "extreme effort"). Note that the invested mental effort is not directly comparable to mental workload, but can be described as a two-folded proportion of cost of cognitive processing related to the processing of information and compensatory effort (Mulder, 1986). The RSME has been found to be sensitive (de Waard, 1996), and sometimes even superior to other scales (e.g. the NASA-TLX), as reported by Veltman & Gaillard (1996). In the present study, mental effort was prepiloted to create a scenario with three distinct levels (on the RSME scale).

Attitude and sensation seeking

Attitudes are complex factors involving a person's values, feelings, inclinations and principles, and affect how a situation is reacted to. Sensation seeking (as a subcomponent of personality) received attention already in the 1970s. In the present study a short questionnaire was sought, and the choice fell on Hoyle's Brief Sensation Seeking Scale (BSSS), which according to the developers (Hoyle, et al., 2002) is based on the same contents as the Zuckerman's sensation seeking scale (SSS-V). The response scale is a five-point Likert-type ranging from *strongly disagree* to *strongly agree*. Hoyle et al. found the scale to be significantly correlated with expected behaviour, a recent study by Litvin (Litvin, 2008) found the BSSS to be an effective alternative to the SSS-V.

The factor sensation seeking has been found relevant in several areas, but in the maritime field only little research was found, and usually related to the military research (see Biersner & LaRocco, 1983 for example). For car drivers sensation seeking predicts violations (Schwebel, et al., 2006; Machin & Sankey, 2008), aggressive and risky driving (Dahlen & White, 2006), keeping high speeds and disregarding rules (Iversen & Rundmo, 2002). A hypothesis is that these effects found for car drivers are present for the maritime field as well.

Experience

Experience is considered an important factor in the context of the study, and in shipping it is usually coupled with age, as certain qualifications (and levels of responsibility) take time to acquire. Experience, defined as accumulated skills and knowledge from exposure to a task, is measured e.g. by counting the number of days a seafarer has worked at sea in the last year, or how many months or years were spent working at sea in total. No agreed upon definition of experience in the shipping sector was found in the literature.

Experience, or the lack of it, has been reported to relate to occupational hazards and injuries in chemical factories (low experience resulting in higher injury occurrence, see Saha, Kumar, & Vasudevan, 2008), for seafarers (Nielsen, 2001; Jensen, et al., 2004). For car drivers the risk in being involved in accidents is highest just after taking the drivers license, as statistics point out. There is a limited number of studies on the effect of experience in the handling of technology for the maritime sector. Dahlman et al. (2008) reported that inexperienced navigators relied more on navigational aids, while experienced navigators made more use of environmental cues and paper charts. Age of participants, and thus familiarity with technical aids which often is associated with young (and consequently inexperienced) navigators, could not explain the difference between the groups in the study by Dahlman.

Common systems in merchant vessels

Collision warning is already present in professional and many leisure Radar systems on ships, and is a part called Automatic Radar Plotting Aid (ARPA). However, these systems have their limitations, especially in narrow waterways, estuaries and archipelagos. New technology is expected to allow for better warnings in these situations in the near future. Speed warning systems are not present in merchant vessels.

Method

The tested assistance systems related to collision warning and to speed management. In this experiment two rudimentary instruments are used to see the possible effect on the navigators: a maximum-speed warning and a collision warning tool. The speed warning is given for small, precisely defined areas. The collision warning is given based on the expected passing distance and time to other ships. The warning is provided using a visual and sound display. The warning tool is implemented on a laptop not requiring any interaction by the user.

Participants and design

A total of 32 navigators participated, their mean age was 35.75 years (SD=11.7). Two types of participants volunteered: novices and professionals. The 15 novices were students in their final year of professional education to become deck officer and one year practical training on board, recruited from the Chalmers Master Mariner education program. The 17 professionals hold a master license and have at least two years experience at that level. The novices were on average younger than the professionals: novices were on average 27.7 years old (SD=7,4) and professionals 42.9 (SD=10.1). Students were compensated with cinema tickets, professionals with 650 SEK (70 Euro).

A two (type of navigator) times two (level of sensation seeking) times three (level of workload) design was used. The type of navigator was novice vs. professional. The level of sensation seeking was low vs. high. The level of workload was low, medium, and high. The type of navigator and sensation seeking was a between participants design, the workload was a within participant design. Both assistance systems were evaluated with the same study design.

Equipment

Two full mission bridge simulators (Transas NAV 5000) were used for the simulator runs (see figure 1 below). One bridge has a field of view of 260°. Two monitors provide a rear view, giving a total field of view of almost 360°. The other bridge has a field of view of 220°. Both bridges contain a complete set of ship controls and instruments, identical to those found in real vessels. Conning, Radar/ARPA, ECDIS, GMDSS/communication, etc. are present (see figure 1). The simulator is a fixed-base, the environment is projected on a round screen. A multichannel sound system provides realistic, directional sound effects.

The area selected for this experiment is the archipelago west of Sweden, 18 nautical miles NW of Gothenburg (around N57° 55', E011° 35'), see figure 2. The participants navigated a small, water-jet driven vessel used by the rescue services. It is 12m long, capable of 37kn, and is highly manoeuvrable. Weather conditions were the same for all participants: good weather, little wind, good visibility.

Procedure

The participants were scheduled in pairs, as two simulators were available. They were briefed, provided with written instructions, and introduced to the bridge. The training was done using three scenarios of about 12 minutes each. There were two experimental scenarios: a speed management scenario ("speed") and a collision avoidance scenario ("traffic"). All subjects did both runs, but the order was balanced. Both scenarios used the same route. The route consisted of three parts: low workload, medium workload, and high workload (in that order). The workload was the result of environment (low at open sea), traffic (medium due to traffic in the middle part), and a secondary task (high due to traffic and counting in the last part).

Figure 1. Bridge simulator used for the study. Participants have Radar, assistance system screen and electronic chart available

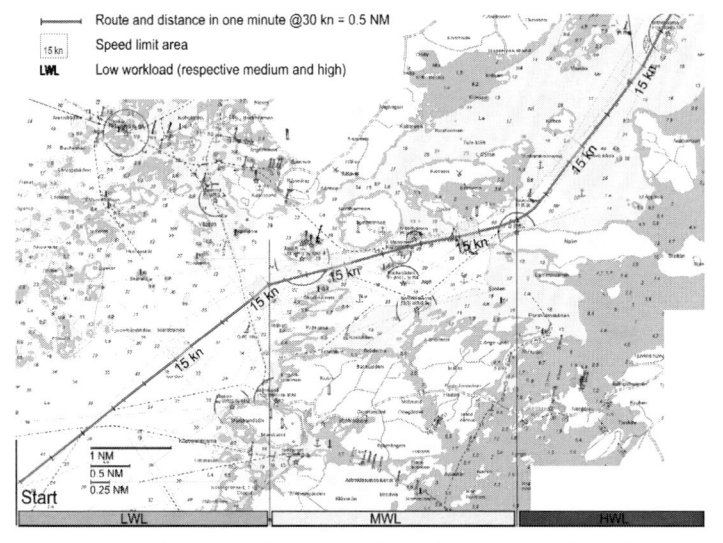

Figure 2. Area navigated. Workload sections and speed limits are marked. The collision avoidance system used the same area

For the secondary mental task participants were asked to count backwards from 948 in steps of seven, and this raised mental effort. This was practised beforehand (with a different starting number) by all participants; the number of errors was not recorded, although participants were not aware of this. The workload was estimated by navigational experts during design using the RSME workload scale (Zijlstra, 1993), and the targeted RSME score was 25 for low effort, 35 for medium, and > 70 for high effort. The three workload parts had roughly equal length, and a scenario took 20-25 minutes to complete. The scenarios were evaluated by navigation-training experts. In between the two experimental scenarios there was a coffee break and the first part of a computer-based questionnaire about sensation seeking. Before and after each scenario the participants were asked about their preferences for the warning system timing (on paper).

In the speed scenario the participant had to adjust speed according to the speed warning instrument. In each of the three workload parts two "speed boxes" were defined on the map, not visible to the participant. This resulted in six speed boxes in the speed scenario. Before entering the box the participant received a warning for the reduced maximum speed, about 15 seconds before the speed limit came into effect. In the box, an alarm was given when the participant was speeding.

In the traffic scenario the participant had to resolve conflicts fully, helped by the conflict warning instrument. A warning was given based on closest point of approach (CPA) and time, the CPA<0.1 nautical miles and the time<60s. In the low workload part easy to resolve conflicts were created, in the medium workload part the participant was confronted with a rather complex traffic situation, and in the high workload part the participant had conflicts and the secondary task. The number of conflicts was designed to be the same in each of the three parts. When done, the participants received their compensation. The entire experiment took ca. 2.5 hours. Simulator data logging was automated and most questionnaires PC-based.

The following hypotheses were focussed on:

- Sensation-seeking operators select assistance system settings which are more forgiving.
- Experienced operators will receive fewer warnings than inexperienced operators.
- Operators will receive most warnings in the low workload area.

More "forgiving" system settings imply that an operator has more time (or distance) for taking measures before the situation related to the warning occurs. For the speed warning system this means longer distance from warning to actual start of the speed limit area; for the collision warning system it means more distance or time from warning to the point where a collision could occur. The reason that operators could receive more warnings in the low workload area is being under-loaded and not paying enough attention.

Results

Data were first pre-processed and extracted for sections of the scenario; analyses were carried out in SPSS (TM) 19.0, confidence level: 95%, p-values are only reported as above or below .05.

Experts were significantly older than novices (mean difference = 15.2 years, $t(30)$=-4.8, $p<.05$). All were Swedes, and gender distribution was 84.4 % (27) males versus 15.6% (5) females. 15 participants were novices, and 17 experts. All participants had at least some experience of navigating a boat, with the highest prevalence for leisure boats and passenger vessels. Mainly relevant for the level of experience was the time spent working at sea: here experts had on average 88.47 (SD=99.8) months in total as deck officer, and 27.47 (SD=62.9) months as master, while novices had 2.2 months (SD=6.5) and 1.6 (SD=6.2) months as deck officer and master respectively. From the BSSS scores 14 participants were classified as high sensation seekers (9 novices and 5 experts), and 18 as low sensation seekers (6 novices and 12 experts). This classification was based on a participant's median value for the respective questions (see Hoyle, et al., 2002) and group median value: participants having a median value above the group median were classified as high sensation seekers, and those having a value equal or below the group median as low sensation seekers. There was no significant correlation between the calculated median sensation seeking score and age of participants or experience level of participants. Pre-piloting the mental effort fixed levels on five participants led to RSME mean scores of 17.3, 32.1 and 53.4 for low, medium and high workload respectively.

The first tested hypothesis was that sensation seekers would select more forgiving assistance system settings. For the speed management system there was no significant difference in the preferred warning distance (before entering an area with speed limitation), anyway, the reported preferred warning distance was well above the standard system setting of 300 meters i.e. 475.4 m (SD=209.7) for novices, and 425 m (SD=266.3) for experts. There was a significant correlation between the "before" and "after" system settings (Spearman's rho = 0.71, p<.05).

For the collision avoidance system the settings were asked in relation to warning distance ("CPA" related to the closest point of approach) and warning time ("TCPA" related to time to reaching the closest point of approach). Here differences were evident (see figure 3 and 4 below). The difference was statistically significant for CPA reported after the run (Mann-Whitney U(32)=73.5, p<.05) and TCPA before the run (U(32)=81.5, p<.05). CPA "after" values were 206.8 m for low sensation seekers and 350,6 m for high sensation seekers, while TCPA "before" values were 48.3. for low sensation seekers and 80.8 for high sensation seekers.

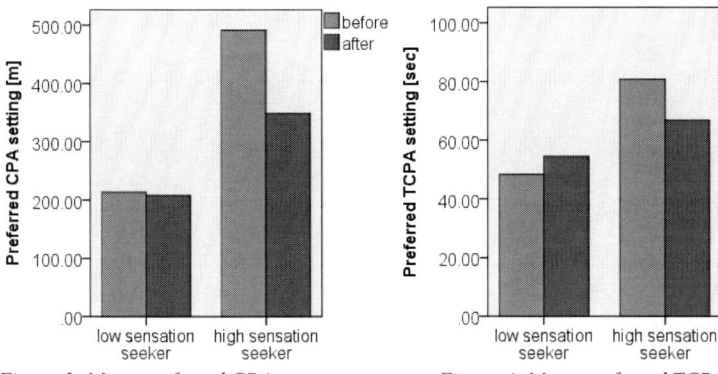

Figure 3. Mean preferred CPA system settings before and after the run

Figure 4. Mean preferred TCPA system settings before and after the run

Looking at the correlations between the combined preferred settings for CPA and TCPA gives significant correlations between CPA before and after (r=.925, p<.05) and for TCPA before and after (r=0.5, p<.05), but no significant correlations between CPA and TCPA.

Related to the second hypothesis, the mean number of received speeding warnings for novice navigators was 9.33 (SD=10.99) and for experts 6.82 (SD=5.67), however, not statistically significant. Figure 6 shows the mean number of speed warnings. For the collision warnings novices received on average 36.2 warnings (SD=38) and experts 39.6 (SD=37) warnings, without being statistically significant.

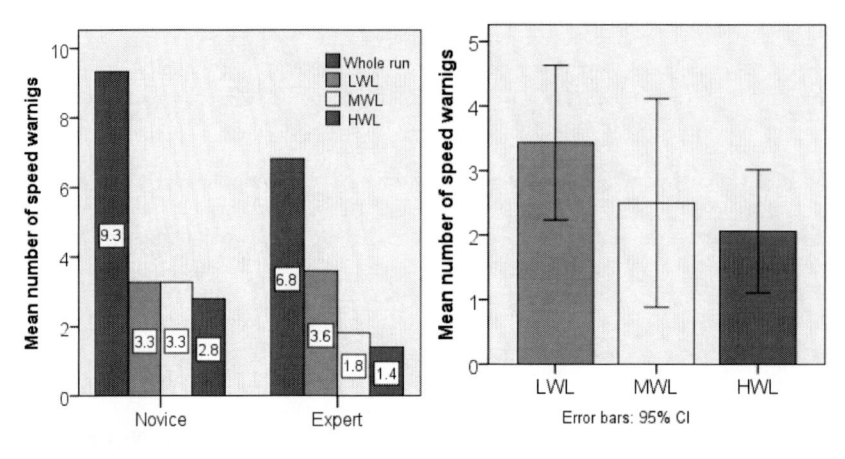

Figure 5. Mean number of speed warnings for novices and experts, in total and for each workload section

Figure 6. Mean number of speed warnings for the three workload sections

A third hypothesis was that the highest number of warnings would be visible in the low workload section; this was again tested for both the speed and the collision system. The mean number of speed warnings for the three workload levels was as follows: low workload (LWL) area: 3.44 (SD=3.3), medium workload (MWL) area: 2.5 (SD=4.5), high workload (HWL) area: 2.1 (SD=2.7). Friedman's test showed a significant difference in the groups (χ^2=6.24, p<.05). The mean number of collision warnings for the low, medium and high workload sections are 13.4 (SD=6.6), 11.5 (SD=5) and 13.1 (SD=5.8), however, the difference for the mean number of collision warnings was not statistically significant.

Conclusions and discussion

For the speed management system the sensation seeking scores failed to discriminate the preferred warning distance to the speed limit area, however, the mean preferred warning distance was larger than the default system settings of 300 meters. During the run with the collision warning system participants classified as sensation seekers reported larger CPA and TCPA preferred system settings. This seems to indicate that such navigators want to prepare well ahead for oncoming situations, probably to allow a higher degree of freedom of action.

The number of received system warnings (in both systems) was not influenced by the sensation seeking scores, but it appears that higher variances are present in novices. The reason for this could be that some novices are very cautious and try to avoid warnings, while others are too inexperienced and receive many warnings, leading to higher variance in the data. On the one hand it can be argued that people working in certain trades such as sea rescue services, where hazardous situations have to be dealt with on a daily basis, are more likely to score higher in sensation seeking. On the other hand, professional trades with high level of responsibility and strict rules may be less interesting for high sensation seekers. A third alternative is that the sensation seeking notion itself is less relevant for the professional shipping trade. The data sample with this group of participants does not confirm either assumption, although there was a tendency of more low sensation seekers in the expert navigator group.

The experience of operators did not have any influence on the number of warnings received (for both systems), although trends in figure 5 seem to point out experts as being more influenced by the workload level in relation to the number of received speeding warnings. It should be noted that no assertions for long term effects and adaptation to the systems can be made, a fact which is general for simulator studies. Quantifying the experience of navigators deserves a thorough discussion during the study planning phase, as the terms are multifaceted.

In the low workload area the highest number of warnings was observed. This could be explained by a person being under-loaded not paying enough attention the speed restrictions. For the collision warnings this was not the case: no significant variation was observed in the three workload sections.

Limitations of the study

The number of participants was limited to 32. This number is seen as sufficient for the tested hypotheses, but a larger number of participants would have allowed the introduction of additional factors. Furthermore, most novices came from the same education institution. Anyway, there is a limited likeliness that companies testing new systems in a similar fashion would invest in controlled studies with a large number of participants, thus the number of factors analyzed (levels) has to be limited.

Workload, or more precisely mental effort, was varied at three levels, but it was not actually assessed during the simulation run, but piloted beforehand. This means that the mental effort exerted by navigators may have varied in a different way than incorporated in the scenario design.

Conclusion

The study pointed out possibilities of integrating human factor aspects in evaluating technical systems before their introduction on ships. It is not suggested that a simulator study as the one presented here is sufficient for a human factor evaluation of new ship systems, but it can be a significant part.

Acknowledgements

The study was part of a project financed by the European Union 7th Framework Programme. All participants of the study are gratefully acknowledged.

References

Dahlen, E.R., & White, R.P. (2006). The Big Five factors, sensation seeking, and driving anger in the prediction of unsafe driving. *Personality and Individual Differences, 41*, 903-915.

Dahlman, J., Forsman, F., Sjörs, A., Lützhöft, M., & Falkmer, T. (2008). *Eye tracking during high speed navigation at sea: Field trial in search of navigational gaze behaviour.* Paper presented at the 6th annual meeting of the Society for Human Performance in Extreme Environments, September 21-22, 2008, USA.

De Waard, D. (1996). *The Measurement of Drivers' Mental Workload* (Ph.D. Thesis). Haren, The Netherlands: University of Groningen, Traffic Research Centre.

Hoyle, R.H., Stephenson, M.T., Palmgreen, P., Lorch, E.P., & Donohew, R.L. (2002). Reliability and validity of a brief measure of sensation seeking. *Personality and Individual Differences, 32*, 401-414.

Iversen, H., & Rundmo, T. (2002). Personality, risky driving and accident involvement among Norwegian drivers. *Personality and Individual Differences, 33*, 1251-1263.

Jensen, O.C., Sørensen, J.F.L., Canals, M.L., Hu, Y.P., Nikolic, N., & Thomas, M. (2004). Incidence of self-reported occupational injuries in seafaring - An international study. *Occupational Medicine, 54*, 548-555.

Litvin, S.W. (2008). Sensation Seeking and Its Measurement for Tourism Research. *Journal of Travel Research, 46*, 440-445.

Machin, M.A., & Sankey, K.S. (2008). Relationships between young drivers' personality characteristics, risk perceptions, and driving behaviour. *Accident Analysis & Prevention, 40*, 541-547.

Mulder, G. (1986). The concept and measurement of mental effort. In G.R.J. Hockey, A. Gaillard, and M. Coles (Eds.), *Energetics and human information processing* (175–198). Dordrecht, The Nerherlands: Martinus Nijhoff.

Nielsen, D. (2001). Seafarers' accidents: does age, rank or experience matter? *International maritime health, 52*, 27-38.

O'Donnell, R.D., & Eggemeier, F.T. (1986). Workload assessment methodology. In K. Boff, L. Kaufmann, and J. Thomas (Eds.), *Handbook of perception and human performance.* New York: John Wiley.

Saha, A., Kumar, S., & Vasudevan, D.M. (2008). Factors of occupational injury: A survey in a chemical company. *Industrial Health, 46*, 152-157.

Schwebel, D.C., Severson, J., Ball, K.K., & Rizzo, M. (2006). Individual difference factors in risky driving: The roles of anger/hostility, conscientiousness, and sensation-seeking. *Accident Analysis & Prevention, 38*, 801-810.

Veltman, J.A.H., & Gaillard, A.W.K. (1996). Pilot workload evaluated with subjective and physiological measures. In K. Brookhuis, C. Weikert, J. Moraal and D. de Waard (Eds.), *Aging and Human Factors*: Proceedings of the Europe Chapter of the HFES Annual Meeting 1993 (pp. 107-128). Haren, The Netherlands: University of Groningen.

Zijlstra, F.R.H. (1993). *Efficiency in Work Behavior. A Design Approach for Modern Tools*. Ph.D. thesis. Delft, The Netherlands: Delft University of Technology.

Investigating visually distracted driver reactions in rear-end crashes and near crashes based on 100-car study data

Henrik Lind[1,2], Selpi[2], & Marco Dozza[2]
[1]*Volvo Car Corporation, Göteborg*
[2]*Chalmers University of Technology, Göteborg*
Sweden

Abstract

Rear-end crashes are common accident scenarios and account for approximately 30% of all police-reported accidents in the United States. Previous studies have shown that driver inattention just before the rear-end crashes or near crashes is a major contributing factor. To improve the development of active safety systems, which take into account driver inattention, it is important to understand when and how the driver reacts; from the moment the driver shifts state from being distracted to attentive and becomes aware of the potential threat. This paper investigates the reaction (braking, steering) selection that visually distracted drivers make in a critical rear-end crash situation using the 100-car naturalistic driving study data. A simple model describing the driver's perception reaction time (PRT) based on headway distance to the forward vehicle is presented. The results presented in this paper can assist further studies where more data will be available and help elucidate how people react to critical near crashes and crashes on road.

Introduction

Rear-end crashes have been shown as the most frequent among all crash types, accounting for 29% of all police reported crashes in the United States, summing up to approximately 1.8 million annually (Najm et al., 2007). Society and industry have found crashes to be a serious problem and several solutions have emerged over time. Between the 60s and 80s safety belts emerged followed by airbags. In the 90s new active safety systems, exemplified by anti lock brakes and electronic stability control, supported the driver-vehicle system from loosing stability. From the year 2000 and onward the focus has shifted towards active safety systems assisting the driver in a non-attentive situation. For example, a forward collision warning (FCW) system behaves like an ever vigilant guardian alerting the driver if the driver response to a forward braking vehicle is delayed or missing. If the driver is estimated to be incapable of reacting in time the vehicle may apply autonomous emergency brake.

One important input in designing forward collision warning systems, roads, and assessment of the benefits of new driver assistance systems is perception reaction

In D. de Waard, N. Merat, A.H. Jamson, Y. Barnard, and O.M.J. Carsten (Eds.) (2012). *Human Factors of Systems and Technology* (pp. 201 - 211). Maastricht, the Netherlands: Shaker Publishing.

time (PRT), i.e. the time from an external stimulus to brake activation (Georgi et al., 2009). Investigation of PRT in critical situations has traditionally been studied in simulator studies, bench tests, or on test tracks. In such studies, the length of the experiments is often short. Often the driver is instructed to look for a certain event and asked to respond by braking. Thus the driver's attention level and level of surprise compared to a critical situation in normal driving can be seriously questioned. However, the benefit of such studies lies in their ability to control the experimental environment.

PRT estimation is relevant for active safety functions in order to estimate the driver reaction that is an outcome of a relative state change between drivers, vehicles and the environment. An active safety function should not interfere unless the driver is inattentive or immediate action is required. The function includes the predicted driver reaction as part of the algorithm where an over estimation of the expected driver reaction time may lead to false or nuisance alerts and an under estimation leads to late intervention and a less efficient system performance. Currently active safety functions use an estimated driver reaction time that may be changing based on a set of fixed values since the influencing factors for on road critical events are still not well understood. In comparison, an accurate model of the driver reaction time, including the error distribution for the model, may potentially allow better adapted thresholds for activation of active safety systems while reducing nuisance and false activations to a minimum level. This approach may increase the effectiveness of active safety systems. It is therefore of interest to investigate, and if possible to model, the PRT and its distribution for critical situations on road.

Many studies have investigated human response time from different aspects. For instance in a laboratory situation, MacAdam (1995) investigated how long it takes for a human to respond from the time a stimulus is given and found that visual response times are about 180 ms, whereas auditory and tactile response times are about 140 ms. Specifically related to driving tasks, response time have also been studied in a number of studies. For example Taoka (1989) and Sohn and Stepleman (1998) investigated total braking time (TBT). TBT is often defined as the time duration from the activation of an external event, like forward vehicle braking, until the brake pedal is pressed. Taoka (1989) combined the TBT results from three studies and performed a meta analysis to review if the total braking time estimate of 1 second is an accurate representation of the 85 percentile of TBT for all drivers. From his study, Taoka recommended to use 1.8 seconds for the 85 percentile value for TBT. It is worth noting here that none of the situations investigated by Taoka involved full emergency braking. Sohn and Stepleman (1998) performed a meta-analysis on 26 studies investigating the TBT and the influencing factors. They created a mental model to estimate the TBT and its distribution. It was found that both the log-normal and normal models have the same significant characteristics for both the mean and variance of TBT at 5% level of significance. The factors found significantly influencing the variance of TBT were: the headway distance to the brake stimulus (object or event triggering the brake reaction), the awareness level of the driver, the type of brake stimulus, and the country in which the experiment took place (USA or elsewhere).

The field of psychology has also investigated the driver reaction time in controlled experiments. However, there are issues around the measurement and the interpretation of the reaction time. The model of the time from perception to action by steering or braking is influenced by a number of stages and mental processes (see Figure 1). Although the current research cannot describe the mental processes in detail, it is possible to measure the time from onset of a stimulus to reaction and statistically analyse the result. However, the measurements of the stimulus activation and the threshold for the start of reaction depend on the design of the experiments and thus vary in between studies. As a result, different studies cannot be fully compared.

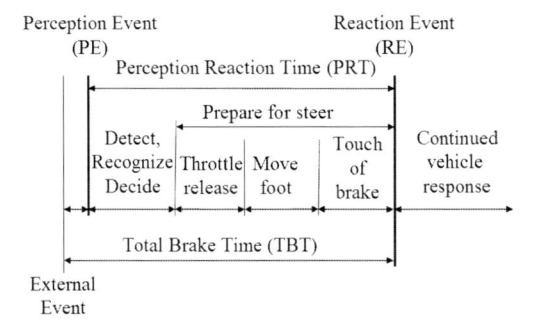

Figure 1. Perception reaction time in relation to total brake time

This study aims to investigate the PRT and the influencing factors for distracted drivers in rear-end crash and near crash events using the 100-car naturalistic driving study data (Dingus et al., 2006). The choice of steering and/or braking response is also considered.

Method

Definitions used in this study

Perception Event (PE): the point in time when the driver regains gaze on road from an off road visual distracted state.

Reaction Event (RE): the point in time when the host vehicle initiates a braking and/or steering manoeuvre to avoid an impending crash.

Perception Reaction Time (PRT): the time difference in between the Perception Event and the Reaction Event.

Lognormal Perception Reaction Time (logPRT): the natural logarithm of PRT, The conversion to logPRT=ln(PRT) is used to convert a skewed response to normal distributed response (Taoka, 1986).The conversion table from PRT to logPRT is listed in Table 1.

Distribution of logPRT: the probability density function of logPRT (log seconds).

Table 1. Conversion table from PRT in seconds (s) to logPRT in log seconds

PRT (s)	0,27	0,33	0,41	0,50	0,61	0,74	0,90	1,11	1,35	1,65	2,01	2,46
logPRT	-1,3	-1,1	-0,9	-0,7	-0,5	-0,3	-0,1	0,1	0,3	0,5	0,7	0,9

Host Vehicle (HV): the vehicle with the subject driver and is the host for the measurement.

Forward Vehicle (FV): the lead vehicle directly preceding the host vehicle.

Headway Distance: the distance (m) in between the front of the host vehicle and the rear of the forward vehicle.

Headway Time: the Headway distance divided by host vehicle velocity (s).

Precipitating event: the event that initiates the crash, near-crash, or incident.

Distraction Time: the distraction time (s) preceding the Perception Event.

Time to Collision (TTC): headway distance divided by relative velocity (s). As proposed by Lee (1976), Cavallo and Lauren (1988).

Data

The base of the analysis presented in this paper is the 100-car naturalistic driving study data as published by Virginia Tech Transportation Institute (VTTI) (Neale et al., 2005). In the 100-car study, 100 cars collected naturalistic driving data for one year. The study included 109 primary drivers and 132 additional drivers. A goal of the study was to maximize the potential to record crash and near-crash events through the selection of participants with higher than average crash or near-crash risk exposure. Events were defined according to a set of triggers followed by manual classification. All events were annotated by accident researchers at Virginia Tech. Annotations include a textual description as well as a classification of the event in the form of descriptive factors. Furthermore every event published by VTTI includes a time sequence of vehicle signals. In the absence of an eye tracking camera the investigation team at Virginia Tech performed gaze analysis by manual annotation of recorded video of the drivers face. The data also includes information of the forward traffic from a radar sensor mounted in the front of the cars. However the videos are not publicly available due to privacy reasons.

Data pre-processing

The 100-car public dataset in Matlab format (Dozza, 2010) was used as the basis for this analysis. The dataset is available on the VTTI website (Virginia Tech data warehouse).

A closer look into the 100-car data-set reveals that the velocity signal of the host vehicle is sampled with approximately 3.5 Hertz compared to 10 Hz for the rest of the time sequence signals. Thus, every velocity value is repeated during 3 to 4 cycles. Additionally, the velocity information is partly missing in a number of the rear-end crash and near crash events. The missing data seems to be correlated to harsh braking manoeuvres of the host vehicle. Fortunately, the acceleration is independently measured and can be used for reconstruction of the velocity signal. A tuned Kalman filter (Kalman, 1960) was successfully applied using host vehicle velocity and host vehicle acceleration (10 Hz) as input and provided an estimation of the host vehicle velocity as output.

The radar information is often lacking during high deceleration of the host vehicle. The reason might have been missing vehicle velocity leading to confused radar target tracker. The missing radar data cannot be restored, leading to a large number of events lacking the information of the forward vehicle target after the reaction event (host onset of brakes).

The time sequence radar information did not include target selection of the forward vehicle. Therefore, a complex and a simple algorithm were designed for the selection of the forward vehicle (see Figure 2). The complex algorithm performed target selection within a predicted path based on estimation from the filtered yaw rate (angle per second) and velocity of the host vehicle. The simple algorithm is a zero azimuth based algorithm. The width of the target selection was set to 2 meters. Investigation shows that at close and medium headway distance the azimuth based target selection provides a more stable target selection due to less influence of dynamic signals like yaw-rate and velocity. Since most rear-end crash events are initiated at short to medium headway distance and at zero azimuth the simple target selection algorithm was selected.

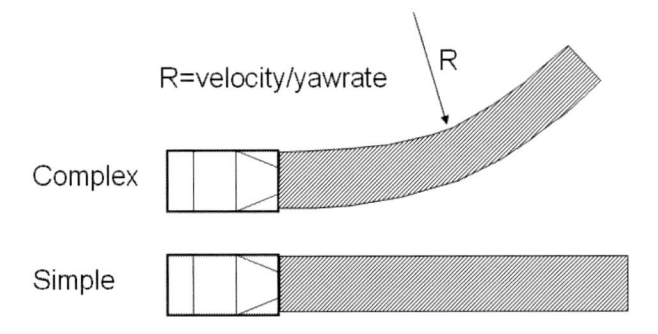

Figure 2. Host vehicle with simple and complex target selection area in grey

Evaluation process

The 100-car time series data allows an investigation of the FV state change synchronised with the host driver and HV state change. An analysis related to PRT was performed on several events consisting of recorded rear-end crash and near crashes.

Pre-selection of events. The 100-car data consists of incidents and events. The events constitutes of 760 near crash events (92% of the total number of events) and

68 crash events (8% of the total number of events). Rear-end crash situations are defined in the dataset as incident type='Rear-end striking'.

The precipitating event is one of the three following classes: 'Other vehicle ahead - but decelerating', 'Other vehicle ahead - slowed and stopped 2 seconds or less', or 'Other vehicle ahead - stopped on roadway more than 2 seconds' leaving 186 events (22% of the total number of events).

Final selection of events. It is of importance to select a robust PE (see Figure 1). PE is the point in time where an alert and vigilant driver would normally detect the approaching vehicle. This study defines the perception event to the point in time the driver reacquires gaze on the road from a visual distracted state. The selection of events is further reduced by discarding PRT values lower than 0.1 seconds (Luce, 1986), and discarding events where a forward vehicle does not exist at time PE. At time PE, the angular expansion rate of the forward vehicle must be detectable by the driver (>0.0035 *rad/s*) (Mortimer, 1990) and the driver must have reacquired look on road from distracted state within 3 seconds in advance of the driver reaction. The final selection of events rendered 19 events for the final analysis. One out of 19 events is a crash and the remaining 18 are near-crashes. The 19 selected events are from 14 drivers. One driver produced 4 events, two drivers produced 2 events each and the remaining 11 drivers produced one event each.

Selection of candidate factors. The candidate factors for modelling the PRT are a mix of derived and measured signals at time PE. Table 2 lists the factors used in the correlation analysis to define significant candidate factors that are suspected to influence the PRT.

Table 2. Candidate factors used in correlation analysis

Factor name	Description	Unit
D_PRT	Driver PRT	s
logD_PRT	natural logarithm of PRT	log s
DriverID	Driver ID	number
D_BrakingInEvent	Driver Braking to avoid accident	boolean
D_SteeringInEvent	Driver Steering to avoid accident	boolean
HV_Velocity_PE	Host Vehicle Velocity at PE	m/s
HV_Braking_PE	Host Vehicle Braking at PE	boolean
HV_ThrottleAct_PE	Host Vehicle Throttle Active at PE	boolean
FV_HeadwayTime_PE	Headway Time at PE	s
FV_HeadwayDistance_PE	Headway Distance (Range) at PE	m
FV_Range_PE	Forward vehicle range at PE	m
FV_TTC_PE	Forward Vehicle TTC at PE	s
FV_InverseTTC_PE	Forward Vehicle Inverse TTC at PE	1/s
FV_Velocity_PE	Forward Vehicle Velocity at PE	m/s
FV_RelVel_PE	Forward Vehicle Relative Velocity at PE	m/s
FV_InvRelVel_PE	Forward Vehicle Inverse relative velocity at PE	s/m
FV_AngularRate_PE	Forward Vehicle Angular Expansion Rate at PE	rad/s
FV_RelLongAcc_PE	Forward Vehicle Relative Longitudinal acceleration at PE	m/s^2
FV_LongAcc_PE	Forward Vehicle absolute Long acceleration at PE	m/s^2

The dependent factor and independent factors in Table 2 were analysed using Pearson correlation analysis. Significant factors ($r>0.5$ and $p<0.05$) are identified.

Cross factor correlation can be expected due to the non-random event selection, however it was not specifically investigated in this study. Several of the factors investigated are partly based on similar information and thus dependent. For example, FV_TTC_PE, FV_AngularRate_PE, FV_HeadwayTime_PE are linked to FV_HeadwayDistance_PE. The factors FV_TTC_PE, FV_AngularRate_PE and FV_inverseTTC_PE include velocity information and are thus correlated.

The factors which have high correlation coefficient and low p-value are the prime candidates for modelling the predicted PRT. Due to the low number of events in the dataset, only one factor can be used to model logPRT at a time.

Results

Distribution of PRT. The histogram indicates a skew. A fit to logPRT with 99% confidence provides M=-0.0522 (corresponds to mean PRT 0.6 seconds), SD=0.577. The histogram for PRT with scaled lognormal distribution is presented in Figure 3.

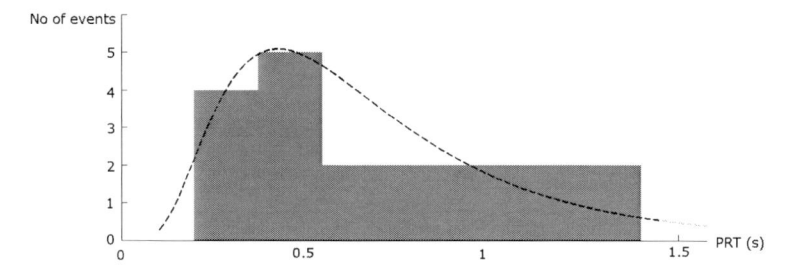

Figure 3. PRT histogram and estimated scaled probability distribution function

Correlation analysis. Pearson correlation analysis in between logPRT and each of the factors listed in Table 2 suggests that FV_HeadwayDistance_PE (r=0.633, p=0.004) and FV_TTC_PE (r= 0.522, p=0.022) are the two most contributing factors to logPRT. Inverse TTC (FV_inverseTTC_PE) also shows a high correlation with logPRT (r=0.447, p=0.055). These three factors are not independent and only one should be selected at a time for a regression model.

Regression analysis. The TTC formula includes PV_HeadwayDistance_PE, thus the correlation of TTC may be caused by this common factor. A test was performed where logPRT was modelled using PV_HeadwayDistance_PE yielding F>11 and R^2adj=36.6% (R^2 adjusted is used when studying low number of samples). The single factor FV_HeadwayDistance_PE is therefore the dominant contributing factor in the 19 events. It is worth noting here that a larger number of samples may give a different result.

The proposed regression model is:
logPRT = -0.979+0.0405 x FV_HeadwayDistance_PE

The standard deviation on error residuals, Figure 5, is reduced from 0.58 (SD no model) to 0.46 with proposed model. The unexplained variance is still high. In case of low headway distance the proposed model, Figure 4, gives a higher estimation of the perception reaction time. A more accurate model may be based on an exponential curve.

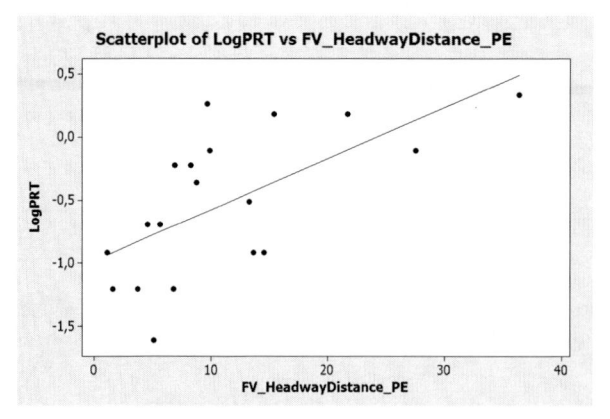

Figure 4. Scatter plot including regression equation logPRT(FV_HeadwayD_PE) for dataset

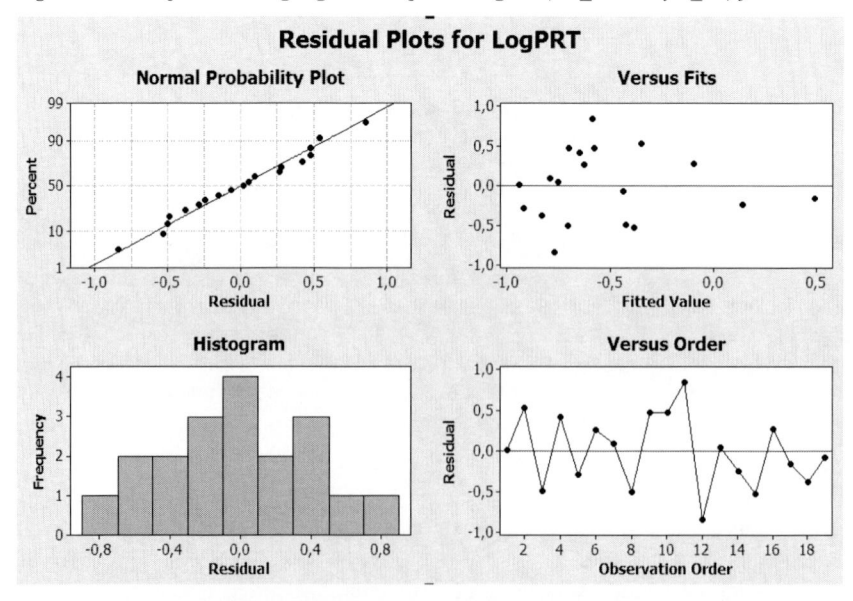

Figure 5. Residual plot for logPRT (FV_HeadwayDistance_PE)

Steering and braking reactions. In all of the 19 events, the drivers react by activating the brakes. In 4 out of the 19 events, the drivers complemented the braking manoeuvre with avoidance steering. The brake activation preceded steering avoidance action in all steering events. A significant correlation (r=0.482, p=0.036) was found between relative velocity to the forward vehicle at PE and avoidance manoeuvre with steering. This indicates that at higher relative velocities there is an increased probability for a steering manoeuvre. This is partly consistent with the vehicle dynamics since at higher relative velocities a steering manoeuvre is more efficient than braking (Brännström et al., 2011). However, since the number of events in this set of rear-end crashes from the 100-car data is very small, it is difficult to draw any conclusions regarding which factors influenced the selection of braking and/or steering actions.

Discussion and conclusions

This study models PRT using events from the 100-car naturalistic driving study. 19 events where the drivers were distracted just before the critical event are examined. Given the low amount of samples, the result points towards headway distance as primary factor for modelling logarithmic perception reaction time. The time to collision shows a high correlation due to contribution of the headway distance factor. There is a small correlation to relative velocity related factors. No significant correlation is found between the change of headway distance before and after distraction, and logPRT. The result of the investigation shows that drivers are able to react fast in a close to crash situation. The primary driver reaction is braking in 100 percent of the events, supported by steering in 22 percent of the investigated events. The steering selection was positively correlated to a higher relative velocity.

Sohn and Stepleman (1998) found four significant characteristics that influence the variance of the mean total braking time. They are the distance away from the brake stimulus, the awareness level of the driver, the type of brake stimulus, and the country in which the experiment took place. Since the latter three factors are fixed in the 100 car study, this leaves the headway distance factor as the influencing factor; this is consistent with the current study. Related to the time from looking on the road to driver reaction Shutko (1999) performed a simulator study involving drivers of commercial vehicles. The drivers were instructed to drive a truck passing a trailer when a number of barrels hidden behind the trailer suddenly rolls out on the roadway. In the baseline condition (no warning) the driver was distracted at the onset of the event. The study showed a mean time from the first glance to throttle reaction of 0.39 seconds. In comparison, this study suggests a mean PRT of 0.6 seconds; this time incorporates brake activation. The explanation of the lower mean time in Shutko's case is that the measure was performed from the external event to throttle release and that the drivers were exposed to a set-up test situation.

Of the 19 studied crash and near crash events only one ended in a crash. This indicates that drivers in this study were able to respond as fast as required. A model of predicted PRT in close to crash situations may require a different model than normal driving where closure rate is a significant factor. It is of interest to further study when and how a change of models takes place. This could be the point of the reactive barrier as proposed by Engström (2011).

In the dataset the possibility for a steering avoidance manoeuvre is unknown. This may prove to be significant for the driver's selection of steering. The low TTC value may as well limit the driver's choice of action leading to a bias in the results. The driver's choice of manoeuvre is an interesting area for future study.

Analysis of naturalistic driving data poses a number of problems. First, it is hard to judge, from the data, if the driver was warned by external factors like for instance brake sound or a stop light ahead. Second, the manual (video) annotation might be noisy due to its subjective nature. Third, the sensor data is not perfect. Fourth, similar to controlled studies, the mental occupation of the driver is unknown. Despite of this imperfection, study of naturalistic data is important and naturalistic driving is the only way to get access to unbiased surprise reaction of drivers on road in a normal environment.

Overall, the results from this study can only be seen as an indication. First because the number of events studied is small (19). Second because the 100-car study includes a set of higher risk drivers where the skill level can be questioned. Third because of a possible sampling bias i.e., only the events from 100-car data were studied; limited knowledge about the events from the 100-car data that has similar initial conditions as the events but did not lead to either near crash or crash events. Thus no comparison between the statistical properties of the events and the non-events with similar starting conditions were made. In the future a more sophisticated model may be estimated when larger sets of on road crash and near crash events have been collected.

The results from this study can be used as input to the modelling of driver reaction in critical situations for use in the development of active safety systems and simulation assessment of active safety systems. Future studies will investigate driver response from warning to braking using field operational test data.

Acknowledgements

This work is supported by Vinnova through the IVSS project "ASIS- Algorithms and Software for Improved Safety" and in cooperation with the FFI project "QUADRA-Quantitative Driver Behaviour Modelling for Active Safety Assessment". Part of the work was carried out at SAFER – Vehicle and Traffic Safety Centre at Chalmers, Sweden. We thank Professor Olle Nerman (Chalmers-Mathematical Sciences) for valuable input on statistical issues discussed here.

References

Brännström, M., Coelingh, E., & Sjöberg, J. (2011). Decision Making on when to Brake and when to Steer to Avoid a Collision. *Future Active Safety Technology toward zero-traffic-accident FAST-zero '11*, Japan: JSAE Japan Society of Automotive Engineers.

Cavallo, V., & Lauren, M. (1988). Visual information and skill level in time-to-collision estimation. *Perception, 17*, 623-632.

Dingus, T.A., Klauer, S.G., Neale, V.L., Petersen, A., Lee, S.E., Sudweeks, J., Perez, M.A., Hankey, J., Ramsey, D., Gupta, S., Bucher, C., Doerzaph, Z. R., Jermeland, J., & Knipling, R.R. (2006). *The 100-car naturalistic driving study, Phase II – results of the 100-car field experiment* (Report DOT HS 810 593), Washington D.C., USA: NHTSA National Highway Traffic Safety Administration.

Dozza, M. (2010). *SAFER100Car: a toolkit to analyze data from the 100 Car Naturalistic Driving Study*, proceedings of the Second International Symposium on Naturalistic Driving Research, Blacksburg, USA: VTTI Virginia Tech Transportation Institute.

Engström, J. (2011). *Understanding attentions selection in driving- from limited capacity to adaptive behaviour* (Report : ISBN 978-91-7385-545-7), Göteborg, Sweden: Chalmers University of Technology.

Georgi, A., Zimmermann, M., Lich T., Blank, L., Kickler, N., & Marchthaler, R. (2009). *New approach of accident benefit analysis for rear end collision avoidance and mitigation systems*, Enhanced Safety Vehicles – ESV'21, 09-0281, Washington D.C., USA: NHTSA National Highway Traffic Safety Administration.

Kalman, R.E. (1960). A new approach to linear filtering and prediction problems, *Journal of Basic Engineering 82*, 35–45

Lee, D.N. (1976). A theory of visual control of braking based on information about time to collision. *Perception, 5,* 437-459.

Luce, R.D. (1986). *Response times: Their role in inferring elementary mental organization*. New York, USA: Oxford University Press.

MacAdam, C.C. (2003). Understanding and Modeling the Human Driver. *Vehicle System Dynamics, 40, (Nos. 1–3),* 101–134.

Mortimer, R.G. (1990). Perceptual factors in rear-end crashes. *Proceedings of the Human Factors Society 34th Annual Meeting* (pp. 591-594). Santa Monica, USA: Human Factors Society.

Najm, W.G., Basav, S., Smith, J.D., Campbell, B.N. (2007). *Analysis of Light Vehicle Crashes and Precrash Scenarios based on the 2000 General Estimates System* (Report DOT HS 809573), Washington D.C., USA:NHTSA National Highway Traffic Safety Administration.

Neale, V.L., Dingus, T.A., Klauer, S.G., Sudweeks, J., & Goodman, M. (2005). *An overview of the 100-car naturalistic study and findings.* Enhanced Safety Vehicles – ESV19', 05-0400, Washington D.C., USA: NHTSA National Highway Traffic Safety Administration.

Shutko, J. (1999). *An Investigation of Collision Avoidance Warnings on Brake Response Times of Commercial Motor Vehicle Drivers*, Master Thesis, Blacksburg, USA: VTTI Virginia Tech Transportation Institute.

Sohn, S.Y. & Stepleman, R. (1998). Meta-analysis on total braking time, *Ergonomics, 41*, 1129-1140.

Taoka, G.T. (1989). Brake reaction times for unalerted drivers. *ITE Journal (March)*, 19-21.

VTTI data warehouse: http://forums.vtti.vt.edu/index.php?/forum/13-100-car-data/

Contributing factors to driving errors in trucking industry: drivers' individual, task and organisational attributes

Ya Li & Kenji Itoh
Tokyo Institute of Technology
Japan

Abstract

The present study explored contributing factors to traffic accidents involved by occupational truck drivers by analysing 10-year incident records collected from 18 trucking companies in Japan. A total of 1292 drivers' records were collected, each of which specified a driver's individual attributes – including his/her task- and organisation-related factors – and history of incident cases for the recent ten years. As the driver's individual attributes, each record included age, gender, work experience, penalty points on driving license, and scores of an "aptitude test" which comprised five personality/attitude traits and four sensorimotor functions. Examples of task-related variables were vehicle type, vehicle's carrying capacity, driving area and annual working days. From results of the Mann-Whitney test with the recent three-year incident data, it was suggested that truck drivers' emotional stability and safety attitudes contribute to safety performance on the road. As other influential factors, results derived from the Chi-square test revealed lower risk of traffic accidents for drivers having the following characteristics: no penalty point on the driving license, longer occupational experiences, shorter driving distance per year, engaging in short-distance delivery and working in a smaller-sized company.

Introduction

Traffic safety has been recognised as one of the most important social issues to address in Japan although the number of fatalities by traffic accidents has decreased by half during the last two decades, i.e., 11,227 in 1990 and 4,863 persons in 2010 (TPDTA, 2011). In addition, like in western countries (Clarke et al., 2005), occupational drivers are involved in more road accidents compared to ordinary, non-occupational drivers in Japan. The fatal accident rate per 10,000 vehicles for commercial vehicles, e.g., trucks, buses and taxicabs, was three times higher than that of private automobiles for the last ten years (IATSS, 2010). Therefore, it is of great importance to explore contributing factors of occupational drivers to safety performance.

In the present paper, we uncover important driver factors contributing to safe driving performance in the trucking industry. Driver attributes are collected from three major aspects of occupational truck driving: driver himself/herself, task and organisation. For this purpose, incident records were collected from approximately 1,300 occupational drivers working for 18 trucking companies in Japan. Each record included not only a 10-year history of incidents involved by the driver but also his/her attributes data. Dichotomising all drivers into "incident-involving" and "no-incident" drivers in terms of the number of incidents involved, influential factors for identifying these two groups is explored for drivers' individual, task-related and organisational attributes included in the incident records collected in this study.

Methods: incident records

Collected sample

We collected a total of 1,292 drivers' incident records from 18 trucking companies, most of which were located in Tokyo and its metropolitan regions. Each incident record included two sorts of information: (1) descriptions of a driver's attributes and (2) a history of incident cases for the recent ten years or shorter. After completing an electronic file of the incident records for data analysis, driver names were removed from the records for anonymity.

Profile of the collected sample is shown in Table 1 in terms of drivers' age and their experience as occupational truck drivers. Driver age ranged between 19 and 68 years old (mean=39.5; SD=9.8) with a mean driver experience of 10.5 years (SD=8.9). The mean age of the entire Japanese truck drivers was 43.1 years (JTA, 2010), and therefore the age distribution of our sample was not greatly different from the national data. There were very few female drivers collected, i.e., 22 records (2%) – and 1,270 (98%) male records – although this female ratio was slightly higher than that of the entire driver population in Japan (0.8%; JTA, 2010).

Table 1. Profile of drivers in the collected incident records

Driver age (years old)	<=10 years experience	>10 years experience	Not described	Total
19-29	110 (9%)	4 (0%)	62 (5%)	176 (14%)
30-39	275 (21%)	158 (12%)	115 (9%)	548 (42%)
40-49	145 (11%)	154 (12%)	69 (5%)	368 (28%)
50-59	70 (5%)	48 (4%)	38 (3%)	156 (12%)
60-68	30 (2%)	9 (1%)	3 (0%)	42 (3%)
Not described	-	-	2 (0%)	2 (0%)
Total	630 (49%)	373 (29%)	289 (22%)	1292 (100%)

Safety outcome data

In each driver's incident record, a history of incidents was recorded for the recent ten years, i.e., 1997 to 2006, or in a period from the starting year for the current company to 2006. Incident cases were recorded as classifying into injury and non-injury incidents as well as their financial damage (money lost by incidents in Japanese yen; 1 Euro was equivalent to about 106 JPY in October 2011). No driver in the 18 trucking companies was involved in a fatal accident during those ten years, and all injury incidents were rather minor. In addition, more than 80% of drivers were not involved in even a non-injury incident each year. For instance, in the most recent year of our data, i.e., in 2006, there were only 189 drivers (15%) who were involved in a single incident – 20 drivers (2%) were involved in multiple incidents and a majority (83%) of drivers were involved in no incident. Thus, using a one-year incident rate or number as a safety outcome measure, we can classify our subjects to two classes: a majority of safe (or perfect) drivers and only a few risky (or dangerous) drivers. It is evident that this index does not properly express an individual driver's safety performance. Taking into account longer years, e.g., ten years in this study, as the other extreme, this may cause a problem of statistically low reliability due to a small valid sample since there were not many drivers who had worked more than ten years for their current company. Another problem of long year incident data may be due to changes in the driver's situation and condition in such a long interval.

For these reasons, we used incident records for the recent three years, i.e., 2004-2006 for analysis. There were 997 valid driver records, i.e., working for the current companies for three years and longer. Among them, 351 drivers (35%) were involved in a single (24%) or more incidents (11%) for these three years. The other 646 drivers (65%) did involve no incidents during this period. For analyses examined effects of driver attributes on safety, we primarily applied the Chi square test to driver attributes having two or more nominal categories such as gender, employee conditions, vehicle type and driving area. For attributes having rank based scores such as personal traits in an aptitude test, the Mann-Whitney test was employed to examine differences in scores of specific items between the "incident-involving" and the "no-incident" driver group.

Driver attributes

Driver attributes specified in the incident records were divided to the following three levels: (1) individual driver, (2) task-related, and (3) organisational attributes. A complete list of these attributes is shown in Table 2. It should be noted that as individual driver attributes, scores of a driving "aptitude test" were included in the data sample. There are several instruments for assessment of driver aptitude, which were authorised by the ministry, and the one developed by the NASVA (National Agency for Automotive Safety and Victim's Aid) is one of the most widely used instruments in Japan. In this instrument, the driving aptitude was measured in terms of five personality/attitudinal traits and four sensorimotor functions. The personal/attitudinal traits were emotional stability, cooperativeness, generosity, trustfulness, and safety attitudes, and each trait was assessed in terms of a five-rank score, which was measured by responses to 12-30 questionnaire items with which a subject rated their agreements/disagreements on a five-point Likert scale (Maruyama, 1981).

Regarding the sensorimotor functions, risk perception was assessed by responses to questionnaire items. The other three functions, i.e., attention distribution, motion accuracy, and decision/motion time, were assessed using a computer-based driving simulator or by measurement programmes on a computer display. First, attention distribution measures drivers' abilities to distribute their attentions properly under a succession of changing situations and the abilities to concentrate one's attention on driving lastingly. Respondents were asked manipulate a steering wheel in order to drive a clockwise moving "car" through the required green area while keep the "car" away from the red area. Second, motion accuracy assesses the ability to deal with a succession of occurring circumstances speedily and accurately. Respondents were required to light 16 lamps with different colours one by one.

Table 2. Factors examined their contributions to safety outcomes

	Factors	Descriptions
Individual attributes	Age	Subject's age, which was classified into two groups: (1) under 39 years old, and (2) 40 years or older.
	Gender	Driver's sex: male or female. The sample collected included 98% of male and 2% female drivers.
	Occupat. drv. experience	Subject's total experience as an occupational truck driver, classified to (1) 10 years or shorter, and (2) longer than 10 years.
	Employment conditions	Employment condition where the subject is expected to work permanently or to leave the employer within a certain limited period, i.e., (1) regular employee, and (2) temporary employee.
	Penalty points on driving license	Penalty points imposed a subject for offence against the traffic rules each year. Using accumulated points for the recent three years, the subjects were divided into two groups: (1) no penalty point incurred, and (2) penalty points incurred at least 1.
	Aptitude test scores	Scores for assessment of a subject's driving related aptitude in terms of personal/attitudinal traits (five items) and sensorimotor functions (four items) on a five-point rank.
Task related attributes	Carrying capacity	Maximum carrying capacity of the vehicle, which a subject usually drove, classified to (1) smaller than 13 tons, and (3) greater than 13 tons truck or trailer.
	Vehicle type	Four types of vehicles, regular type truck, cooler truck, trailer, and other types, were fused the former two (regular truck/cooler truck) and the latter two (trailer/others).
	Driving area	Delivery areas for daily work were divided to two types: (1) in-town or short range delivery, i.e., within 200km, and (2) medium or long distance delivery, i.e., longer than 200km.
	Annual driving distance	The total number of kilometers a subject has driven his/her truck for an entire year of 2006. Two groups were divided at 75,000km, which was near the mean distance: (1) ~75,000km, and (2) over 75,000km.
	Annual working days	Mean of annual waking days of drivers in the collected sample was 261 days (SD=49.7) in 2006. The two groups were classified by dividing at a near point of the mean: (1) ~260 days, and (2) more than 260 days.
	Percentage of waiting time	Waiting time is referred to duration that a subject waits in his/her vehicle for loading/unloading tasks. Percentage of waiting time over hours in duty was classified to: (1) less than 10%, (2) 11-20%, (3) 21-30%, and (4) higher than 30%.
Organisational attributes	Company size	Company size was classified in terms of the number of vehicles owned into three groups: (1) ~50 trucks, (2) 51~200 trucks, (3) 201~400 trucks.
	Mean age of drivers	Mean age of drivers was calculated for each company and classified it into two groups at the age of 40: (1) 40 or younger, and (2) older than 40 years.

When a certain colour was given, respondents should respond correspondingly at once. If a colour was given with a buzzing sound, respondents should keep keys and the pedal pressed. In this test, both hands and right foot were in charge of two keys and a pedal respectively. Reaction time and accuracy were recorded to evaluate motion accuracy. Third, decision/motion time examines drivers' judgment on the time of an action. In this test, respondents were required to gaze at a moving light dot. Then the light dot went into a ditch (disappeared) and respondents should still gaze at it as if it was still moving at the same speed and press the key when they estimated the light dot to go through the ditch. From the light dot disappeared to the key was pressed the time was assessed as the decision/motion time.

Results

Effects of individual driver factors

Regarding the subjects' demographic and personal attributes, a significant difference in the proportion of incident-involving drivers was observed only by the length of occupational driver experience (χ^2=6.519, $p<0.05$). The percentage of no incident involved during the three-year period was higher by drivers having longer than ten-year occupational experiences than those having shorter experiences. There were no significant differences by most of other demographic or personal attributes such as gender, age class, and employment conditions.

Table 3. Mean scores of each factor for incident-involving and no-incident drivers

(a) Personality/attitudinal traits

Factors	No-incident	Incident	p
Emotional stability	3.56	3.38	*
Cooperativeness	3.48	3.43	
Generosity	3.44	3.52	
Trustfulness	3.64	3.51	
Safety attitudes	3.07	2.85	*

(b) Sensorimotor functions

Factors	No-incident	Incident	p
Risk perception	3.91	3.77	
Attention distribution	3.97	3.98	
Motion accuracy	4.25	4.20	
Decision/motion time	3.73	3.67	

(c) Behavioural factor

Factors	No-incident	Incident	p
Penalty points	1.35	2.10	***

* $p<0.05$, ** $p<0.01$, *** $p<0.001$.

Results of differences in five-point rank-based scores of psychological personality and attitudinal factors, sensorimotor functions – all of which were assessed by the aptitude test – and behavioural factors between incident-involving and no-incident drivers are shown in Table 3 in terms of mean scores of the both groups and significance levels derived by the Mann-Whitney test. Among personality/attitudinal traits assessed by the use of a driving aptitude test, emotional stability and safety attitudes exhibited significant impacts on the proportion of no-incident versus incident-involving drivers, as can be seen from Table 3(a). Schematic demonstrations of these factors' effects on safety performance are provided in

218   Li & Itoh

Figure 1. As indicated in Figure 1(a), no-incident drivers had higher scores of emotional stability, compared to the incident-involving group. This may suggest that drivers who are likely to be composed and emotionally stable exhibited less risky against traffic accidents. Similarly, there were significant a greater number of drivers having positive safety attitudes in the no-incident group than in the incident-involving group.

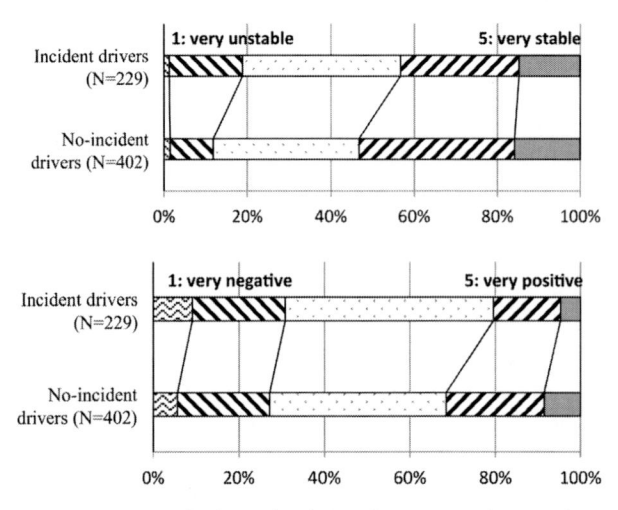

Figure 1. Differences in personality/attitudinal traits between incident-involving and no-incident drivers top: Emotional stability, bottom: Safety attitudes

We could obtain other evidence that driver behaviours are crucial indicators of accident risk besides the result of the aptitude test. The penalty points on the driver license may be associated with risk-taking behaviour. Significant differences were observed between the no-incident and the incident-involving group for both attributes. These associations between drivers' scores and incident/no incident classifications are shown for these attributes in Figure 2. As can be seen from Figure 2, percentage of no-incident drivers was much higher for those who incurred no penalty points for the last three years than those who did so. This result implies that drivers' risk-taking behaviour is likely to increase a possibility of traffic incidents or accidents.

Impacts of task-related factors

Regarding task-related factors, Chi-square test was applied to the collected sample to examine a difference in proportion of incident-involving and no-incident drivers between two groups of each attribute. We identified significant differences for two task-related attributes related to driving distance: driving area ($\chi^2=3.896$, $p<0.05$) and annual driving distance ($\chi^2=4.121$, $p<0.05$), as shown in Figure 3. Drivers who usually engaged medium or long distance delivery, i.e., daily driving distance was longer than 200km, were more likely to be involved in incidents than those driving their trucks for in-town or short distance delivery, i.e., within a 200km distance. A similar effect of the annual driving distance was seen: employees driving longer than

75,000km per year had significantly more chances to involve incidents than those whose annual driving distance was shorter. There was no effect on frequency of incidents identified of vehicle type, carrying capacity of the vehicle, annual working days and percentage of waiting time from the sample collected in the present study.

Figure 2. Differences in drivers' penalty points between incident-involving and no-incident drivers

Regarding task-related attributes, driving areas and annual driving distance were suggested to associate with frequency of traffic incidents. Drivers who are engaged in long distance delivery are likely to operate their vehicles long hours. Annual driving distance is also tied to total operation hours of the vehicle. Such situations have caused high workload to truck drivers as well as severe exposure to hazards on the road. Therefore, our results of significant effects of driving area and annual driving distance can be evidence of association of workload with traffic accidents.

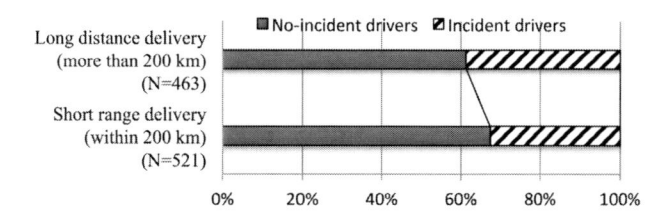

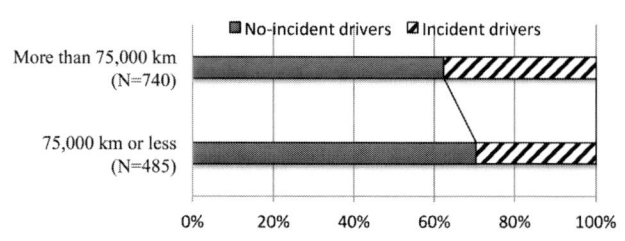

Figure 3. Differences in task-related factors between incident-involving and no-incident drivers. Top: Driving areas, bottom: Annual driving distance

Effects of organisational factors

Of two organisational factors examined in this study, we identified a significant difference in safety performance by the company scale (χ^2=17.019, p<0.001). Percentage of no-incident drivers decreased with company's scale, as depicted in

Figure 4. Approximately 70% of drivers in the smallest company group – i.e., less than 50 trucks – were involved in no incidents for the recent three years while this proportion was about 55% for the largest group drivers. The trend of lower incident rate was shared by the results of a questionnaire-based survey from 371 companies conducted by the Japan Trucking Association (JTA, 2008) – we calculated "no-incident company" rate based on the data obtained in this survey. There was no significant difference in the proportion of no-incident drivers by the other organisational factor, i.e., mean driver age of the company, when we made two groups by dividing companies at their mean age of 40 (χ^2=0.061). Since drivers in a smaller company are more likely to have short-range delivery, this result is consistent with the results mentioned in the task-related factors.

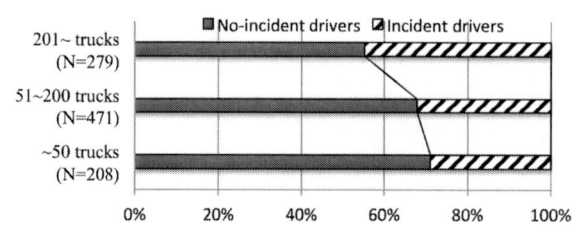

Figure 4. Effects of organisation size on safety performance

Conclusion

The present study detected several factors of occupational truck drivers contributing to safe driving performance from actual incident records. First, it can be seen that two personality/attitudinal traits assessed by "aptitude test", i.e., stable emotionality and safety attitudes are reflected to reduce a possibility of incident involvement. However, according to our follow-up analysis – though there is no space for describing this issue here – since personal trait scores may vary across years, the results of aptitude test should be carefully checked from other evidence derived by other methods when they are used, e.g., for driver assessment. Second, the present study also revealed that penalty points on the driving license represent the degrees of risk-taking behaviour, and that drivers imposed no penalty points were less likely to be involved in incidents on the road. Third, we identified positive impacts of the occupational driver experience on traffic safety, indicating that less experience could be a promising estimator of the liability to traffic incidents. Fourth, driving areas, annual driving distance and organisation size indicated significant effects on safety outcomes. Drivers who are usually engaged in medium/long distance delivery, annually long distance driving or working in a smaller-sized company have more risks to cause accidents/incidents than those engaged in in-town/short-range delivery or short distance driving. Both of the former factors are tied to drivers' workload. Therefore, we can build a hypothesis about a relationship of driver workload with safety performance: driver workload is a crucial risk factor to driving accidents.

With findings of the present study, we can establish safer and more effective work conditions in a trucking firm, e.g., an employment system to determine annual days off and work schedule based on an appropriate level of workload, an effective supervisory system to control drivers accident/incident involvement, and effective safety training not only for technical skills but also for safety attitudes and behaviours.

Acknowledgements

We are grateful to Keiichi Higuchi, President of the National Trucking Business Cooperative, for his support and cooperation throughout this study. We would like to acknowledge Yukiko Sato and Hidetoshi Furuno of the same organisation, and managers of trucking companies for their arrangement of the data collection. We are grateful to Jiro Gyoba, Professor, Tohoku University, Tomoki Ohashi, Professor, Miyagi Gakuin Women's University and Shinya Yoshida, Professor, Tohoku Gakuin University for variable information and discussion about the aptitude test for driving.

Reference

Clarke, D.D., Ward, P., Bartle, C., & Truman, W. (2005). *An in-depth study of work-related road traffic accidents. Road Safety Research Report No. 58.* London: Department for Transport.

IATSS (2010). Statistics Road Accidents Japan (1997-2007). Tokyo, Japan: International Association of Traffic and Safety Sciences Accessed on 15 October, 2011: http://iatss.or.jp/english/statistics/statistics.html..

JTA (2008). A survey report on relationships of company scale and business contents with the number of traffic accidents/incidents, Japan Trucking Association, (in Japanese). Accessed on 3 October, 2011:
www.jta.or.jp/kotsuanzen/report/jigyokibo_jiko_report200803.pdf

Maruyama, S. (1981). A guidance of psychology aptitude diagnosis. Textbook for diagnosis. National Organization for Automotive Safety & Victims' Aid, Tokyo, Japan.

TPDTA (2011). Traffic facilities during 2010. Transportation Planning Division of Transportation Authority (In Japanese). Accessed on 19 Sep., 2011:
http://www.e-stat.go.jp/SG1/estat/List.do?lid=000001069279.

Don't be upset! Can cars regulate anger by communication?

Sabine Wollstädter[1], Hans-Rüdiger Pfister[1], Mark Vollrath[2], & Rainer Höger[1]
[1] Leuphana University Lüneburg
[2] Braunschweig University of Technology
Germany

Abstract

In a study conducted online we explored several strategies targeted to regulate drivers' anger via different kinds of regulation strategies. First, participants were asked to imagine several traffic related events from a drivers' perspective, presented as written text scenarios, and intended to induce anger. Participants took the perspective of a driver who was unable to arrive as planned at his or her destination due to other drivers' obstructive driving behaviour. Each driving event was followed by different emotion regulation strategies presented via audio files. One strategy attempted to shift the participants' attention to a non-driving related topic (distraction). A second strategy attempted to modulate participants' affective state using a relaxation technique (suppression). The last strategy induced an alternate interpretation of the emotion-eliciting event (reappraisal). All strategies were aimed at down-regulating emotions that typically have a negative valence such as anxiety, fear, sadness, and anger. Participants rated their emotion via the Self Assessment Manikin. Results from these hypothetical scenarios indicate that it is conceivable to regulate emotions within a technical system such as a car. Furthermore, we find that individual preferences for the specific kind of regulation exert a significant impact on emotion regulation.

Emotion and emotion regulation in Driver-Vehicle-Interaction

Driving in rush-hour traffic is probably not everyone's cup of tea. But when progress is blocked due to a particularly slow driving car in front, anger is likely to occur. Emotions arise when physical or mental events compel attention, having particular meaning relevant to individual goals, needs or issues. Typically, emotions trigger physical, experiential and behavioural responses (Campos et al., 2004; Gross & Thompson, 2009; Lazarus, 2001). Accordingly, anger leads to changes in the driver's thinking, feeling and in driving behaviour, in some cases to maladjusted driving (Deffenbacher et al., 2003; Parker et al., 2002). It is assumed that accidents frequently occur as a result of aggressive and risky driving behaviour (Martinez, 1997; Snyder et al., 1997). In particular, anger, often provoked by others, may lead to risky driving (Roidl et al., 2010; Mesken et al., 2007; Stephens & Groeger, 2006; Lajunen & Parker, 2001; Parkinson, 2001; Scherer et al., 2001; Shinar, 1998; Arnett et al., 1997).

There is a widespread agreement that the same set of processes which are responsible for the elicitation of emotions are involved in their modification (Campos et al., 2004). Gross & Thompson (2009) categorised various strategies, used consciously and unconsciously, which intensify, maintain or down regulate emotion (Gross, 2001). Strategies which help us down-regulate emotions that typically have a negative valence such as anxiety, fear, sadness, and anger (John & Gross, 2009) are self-assertion, distraction, reappraisal and suppression.

Self-assertion relies on the idea that changing the current situation can make us happy. A change can imply problem solving supported by others or leaving a given situation (Gross & Thompson, 2009). In a driving situation this would mean e.g. leaving a traffic jam and taking another route recommended by a navigation system, or passing a slow driver. These types of modifications aren't always possible and depend on the given situation. Hence, in this study we focus on distraction, suppression and reappraisal.

Distraction refers to attempts to shift an individual's attention to a different topic or different aspects of a situation; thinking of other things than the emotion-eliciting event (John & Gross, 2009). Previous research yields evidence that distraction reduces experienced anger (Rusting & Nolen-Hoeksema, 1998), and has a major impact on decreasing aggressive behaviour (Bushman et al., 2005). Of course, while driving, a drivers' attention should be predominantly directed to his/her driving task. However, if a given situation leads to anger, shifting drivers' attention to the radio broadcasts or listening to audio books could help or constitute affective regulation. Preliminary research shows that auditory stimuli have minor effects on driving performance (Hatfield, 2008) and thus should be safe to use as a distraction.

Suppression attempts to modulate a persons' affective state by using a relaxation technique, e.g. listening to music (John & Gross, 2009; Thaut & Wheeler, 2010). Music is known to influence affect to change mood and decrease arousal (Juslin & Sloboda, 2010), and is successfully applied in music therapy for reducing negative emotion (Thaut & Wheeler, 2010; Wheeler, 2005; Sidorenko, 2000). This effect is found to be even stronger when listening to favoured music (Allen & Blascovich, 1994; Anderson et al., 1991; Stratton & Zalanowski, 1984). Current research concerning the influence of music in unpleasant driving situations suggests that music moderates anger, aggression (van der Zwaag et al., 2011; Wiesenthal et al., 2003) and stress (Wiesenthal et al., 2000). Concerning music preference, Wiesenthal et al. (2003) found that mild aggression is lower when listening to self-selected music.

The third strategy we focus on is *reappraisal*. It works by inducing an alternate interpretation of the emotion-eliciting event (John & Gross, 2009). Clinical research shows, that altering maladjusted cognitions is an effective treatment for anger-related disorders (Campbell-Sills & Barlow, 2009; Ray et al., 2008; Deffenbacher, 2006). A couple of driving simulation studies have already explored the effects of therapy techniques for cognitive behaviour therapy on reducing driving anger. Deffenbacher and colleagues (2002) taught drivers cognitive-relaxation coping skills. In group therapy sessions they learned to reappraise a situation by taking a different perspective on anger eliciting events. They found evidence that cognitive coping reduces anger and risky driving behaviour (Deffenbacher, 2009; Deffenbacher et al., 2002; Deffenbacher et al., 2000). Drivers, though, who do not recognize their anger as a problem, require different approaches (Deffenbacher, 2009).

When comparing emotion regulation strategies, individual differences need to be considered as well. Age and situational experiences have shown to lead differences in experience and regulation of anger. Gross et al. (1997) found age-related decreases in the experience of anger. Further on, aggressive driving has been reported to be the highest among young drivers (Wickens et al., 2011; Herrero-Fernandez, 2011). In summary, these findings suggest an improvement of regulation competence across the life span can be assumed. But still, we are far from understanding individual differences on the subject of anger regulation. More studies are needed examining the arousal or complexity of emotional stimuli and taking the situational context into account when interpreting age differences (Turk Charles & Carstensen, 2009). Additionally, real-life experience may alter the meaning and interpretation of the emotional stimuli (Kunzmann et al., 2005); in terms of driving, elicit more anger in a rush-hour situation.

The present study was designed to explore emotion regulation strategies, which enable technical devices to initiate emotion regulation in order to de-escalate maladjusted driving behaviour. We constructed a setting, where participants imagined themselves driving while different strategies were tested. Radio broadcast was assumed to distract attention (distraction). Music was assumed to suppress emotion as a result of relaxation (suppression). Providing specific information was assumed to lead a reappraisal of the sources of anger (reappraisal). First, we were interested whether different regulation strategies lead to specific affective responses. Second, we were interested whether the affective impact of regulation strategies is moderated by individual preferences.

Method

Participants

Two-hundred-twenty-six participants (N = 226, 143 female, 83 male, from 19 to 63 years of age with a mean of 27.55 years) completed the questionnaire. Part of them were recruited via snowball sampling and received lottery tickets to win a prize of € 50 if they agreed to participate. Twenty-five percent (N = 57) of the participants were students who could obtain credits to fulfil study requirements. Fifty-one percent (N = 115) of the participants use a car several times a week. Commute or go on errands are the main driving purpose in 61 % (N = 138) of the cases.

Design

In an online questionnaire participants were asked to imagine themselves driving in specific anger eliciting traffic situations, presented as text vignettes. Subsequently, certain strategies were tested in their ability to cause emotion regulation. A mixed within/between-subjects design was applied. Regulation strategy was used as a within-subject factor. Each participant received different types of regulation strategies targeted to regulate emotion: distraction, suppression and reappraisal. Reappraisal was divided into two sub-strategies, thus regulation strategies varied on four levels. As between-subject factor, preference-term was varied on two levels. Participants received pre-set regulations either assumed to be a) liked or helpful or, b) not liked or unhelpful (see Figure 1). In order to control if participants' actually liked regulation strategy, they were asked to rate if the regulation was a) helpful or liked, b) unhelpful or not liked c) neither. Participants responded to this question after they listened to each regulation strategy (reported-preference). As a third factor, the dependent variable was measured first before regulation strategy and a second time after regulation strategy was presented. Regulation strategies were presented acoustically. Affective experiences were measured using subjective rating scales.

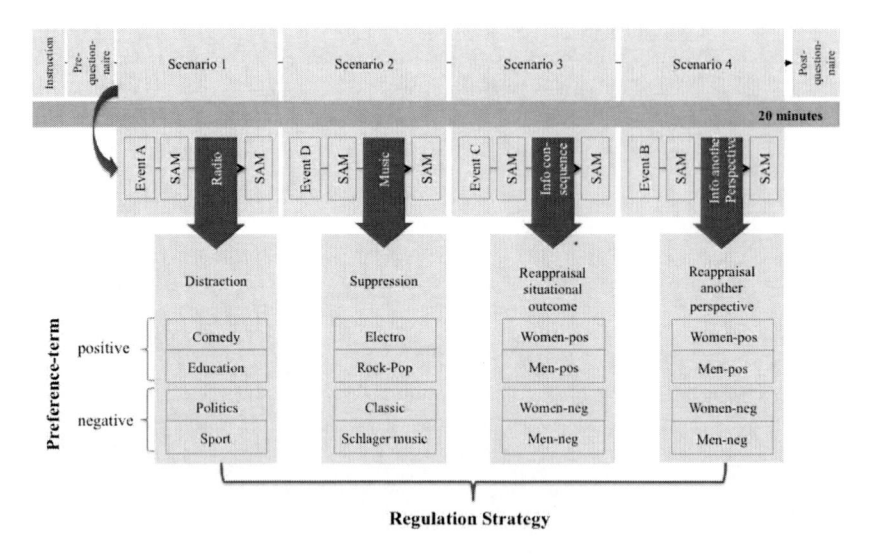

Figure 1. Sequence of scenarios, and regulation strategies taken across one example of experimental trial; Independent variables regulation strategy (distraction, suppression, reappraisal situational outcome, reappraisal another perspective), and Preference-term (positive = assumed to be liked; negative = assumed to be not liked)

In order to balance scenario effects each particular regulation strategy was assigned to different scenarios. Four experimental sets with different mappings of strategy-conditions to anger events were constructed, balancing all within-subject factorial combinations across events.

Measures

Self-report ratings of affect: The Self Assessment Manikin (SAM) scale is a reliable and valid method to measure affect via self-report introduced by Bradley and Lang (1994). This non-verbal pictorial assessment technique directly measures three affective dimensions: valence, arousal, and dominance, of which we measured ratings on the first two dimensions. Subjective affect ratings for valence and arousal were assessed on the screen after each anger scenario and regulation strategy. Participants indicated their current affective state online by a mouse click on a scale mark located below and between the figures, yielding a nine-point rating scale (1 = unhappy and low arousal; 9 = happy and high arousal). The two self-report scales were measured once for every scenario and regulation strategy, yielding 8 ratings on every scale (16 affect ratings per participant; see Figure 1).

Procedure

Once participants started the online survey a random function allocated participants to experimental settings. A short intro was followed by a pre-questionnaire, the main questionnaire, and a post-questionnaire. The pre- and post-questionnaire was the same for all participants. The pre-questionnaire asked for biographical data (age, gender, education, driving experience) as well as preferences in music and radio and preferred navigation/system voice. The main questionnaire included four parts each following this sequence: after participants read the anger-scenario and rated their imagined emotion in this driving situation (SAM-Rating t1) they listened to a regulation strategy and again rated their imagined emotion (SAM-Rating t2). Following each regulation strategy, they also rated if the regulation would be a) helpful or liked, b) unhelpful or not liked or c) neither in the prior described anger-scenario. The scenarios were presented as text-vignettes and were randomized. The regulation strategies were presented as audio-messages. For the questionnaire approximately 20 minutes were needed.

Material

Anger eliciting events: Drivers were asked to imagine that they are experiencing time pressure while driving in rush-hour/high traffic congestion. However, a slow driving car in front impeded their progress towards their destination and no escape.

Emotion regulation strategy: Following each anger vignette, participants heard one out of four regulation strategies and were asked to imagine themselves driving and listening to them. Concerning the distraction strategy, four radio broadcast genres were applied: comedy, education, politics, and sport. For the suppression strategy, different musical pieces were selected to modulate drivers' affective state: electro, rock-pop, classic, and Schlager music. For the reappraisal strategies the participants were asked to take a different perspective on the anger source. Either, information was provided about the situation of the person driving the vehicle in front (reappraisal by offering another perspective), or requested participants to rethink the consequences of the event (reappraisal by revealing situational outcome).

For *preference-term* of regulation, either positive regulation intended to mitigate emotion, or negative regulation, assumed to intensify emotion were varied (see Fig. 1). Thus, for distraction and suppression two radio and music genres were assumed to be liked, and two genres assumed to be disliked by participants. The assumption was made based on a pre-test in which a different test group was asked which music they liked and disliked. Accordingly, the reappraisal was presented as positive or negative information about the anger-eliciting event (information about another perspective) versus positive or negative information about the implications for the driver's situation (information about situational outcome). Since studies show, that the kind of computer voice has an impact on how helpful the interaction with a system is perceived (Lee et al., 2000), the information was either presented by a male or a female voice.

Results

As outlined above, two subjective ratings of affective feelings (valence and arousal) were analysed as dependent variables, which are indicators of affect. We report analyses separately for each research question. To investigate the effects of regulation strategy on emotion regulation, differences of emotion ratings SAM-t1 (assessed after participants read the anger event) to SAM-t2 (assessed after participants heard regulation strategy) were used.

Do different regulation strategies cause specific affective regulation?

All data were subjected to a univariate ANOVA, with repeated measurement on regulation strategy as within-subject factor and preference-term as between subject factor. If violations of shpericity occurred it was checked if conclusions change after Greenhouse-Geisser correction, which was not the case. Hence we only report the traditional ANOVA tests.

Table 1. Means and Standard Deviations (in Parentheses) of Dependent Variable controlled for preference-term

Strategy global	Distraction (Radio)		Suppression (Music)		Reappraisal (Info another perspective)		Reappraisal (Info situational outcome)	
Preference-term	pos	neg	pos	neg	pos	neg	pos	neg
Valence	2.19 (2.418)	1.01 (1.921)	1.60 (2.175)	0.89 (2.449)	1.41 (1.887)	-0.53 (1.820)	0.64 (1.954)	0.05 (2.082)
Arousal	-1.39 (2.162)	-.87 (1.989)	-0.70 (2.035)	-0.04 (2.820)	-1.30 (2.057)	1.14 (2.082)	-0.56 (1.949)	0.26 (1.928)

Note: N = 226 participants; N = 112 for preference-term positive; N = 114 for preference-term negative; Valence and arousal measured on a 9-point scale, differential mean values displayed

Following, we report ANOVA results in order to test for within-group differences of regulation strategy: distraction, suppression, reappraisal by another-perspective, and reappraisal by situational-outcome, and between-group differences of preference-term.

Valence: A significant main effect was found for regulation strategy ($F(3, 222) = 18,657$, $p<.000$) and a significant interaction effect for regulation strategy and preference-term ($F(3, 222) = 6,821$, $p<001$). A post-hoc t-test showed significant differences comparing distraction and reappraisal ($p<.000$) and suppression and reappraisal ($p<.000$). For preference-term-positive higher mean values were measured compared to preference-term-negative (see Table 1).

Arousal: A significant main effect was found for regulation strategy ($F(3, 222) = 12,349$, $p<.000$) and for interaction effect of regulation strategy and preference-term

($F(3,222) = 11,412$, $p<.000$). A conducted t-test showed significant differences comparing distraction and suppression ($p<.000$) and distraction with reappraisal ($p<.000$). For preference-term-positive lower mean values were measured compared to preference-term-negative for each factor of regulation strategy (see Table 1).

Is the affective impact of regulation strategies moderated by individual preferences?

For the analysis presented below, the data were divided with respect to whether participants actually preferred a regulation strategy. The division depended on their rating following each regulation strategy. Concerning this matter, the participants were asked whether they actually liked a specific regulation, considered it as helpful, or whether they had perceived the specific regulation as annoying and unhelpful (reported-preference). Linear regression analyses were conducted for valence and arousal as depended variables, with reported-preference as a predictor. Since intentional manipulation of preference-term (assumed to be liked vs. assumed to be not liked) did not yield differences in actual reported-preference, subsequently, results are reported in total for distraction and suppression (see Fig. 2).

Valence: Regression analyses yield a significant effect of preference for distraction ($\beta = 2.161$; $p<.000$), suppression ($\beta = 3.012$; $p<.000$) and reappraisal-pos (another-perspective-pos $\beta = 1.200$; $p=.007$; situational-outcome-pos $\beta = 1.364$; $p=.001$); Only reappraisal-neg is not significant (another-perspective-neg: $\beta = 0.299$; $p=.520$; situational-outcome-neg: $\beta = 0.739$; $p=.151$).

Arousal: Regression analyses yield a significant effect of preference for distraction ($\beta = -1.171$; $p<.000$), suppression ($\beta = -2.050$; $p<.000$) and reappraisal-pos (another-perspective-pos $\beta = -1.313$; $p=.006$; situational-outcome-pos $\beta = -1.146$; $p=.008$); Only reappraisal-neg is not significant (another-perspective-neg $\beta = 0.160$; $p=.789$; situational-outcome-neg: $\beta = -0.614$; $p=.208$).

Thus, except for reappraisal negative, affective responses are influenced by preference; yielding significant higher valence means and lower arousal means for regulations, which are liked or perceived as helpful versus regulations which are not liked or perceived as unhelpful (see Figure 2).

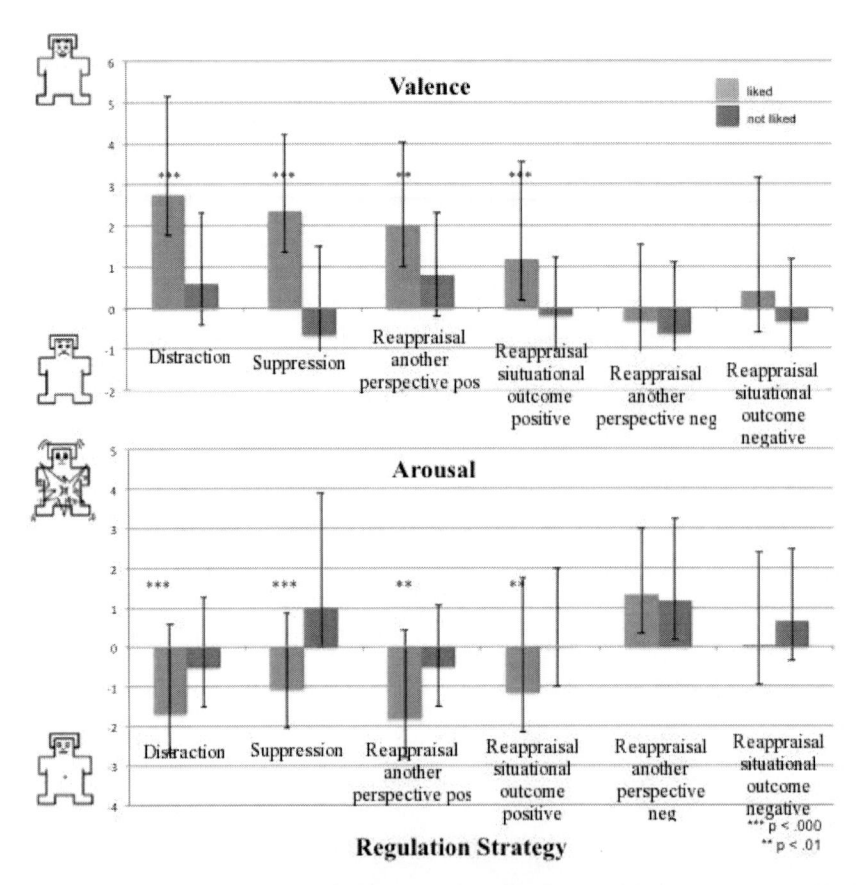

*Figure 2. Means of subjective ratings as a function of preference (liked interaction, not liked interaction); Error bars indicate standard deviation. Significant effect * p<.05; ** p<.01; *** p<.001*

Discussion

Driver-vehicle-interaction aimed to regulate drivers' emotional state may lead to decreased levels of negative emotion. Results indicate that distraction, suppression and under some conditions reappraisal modify emotion imagination, increase envisaged positive affect and decrease envisaged arousal. Furthermore, preference for regulation strategy has a crucial impact on emotion regulation, yielding higher valence and lower arousal ratings if a regulation strategy is liked (see Fig. 2). This applies to all regulation strategies, except for the reappraisal by information about another perspective. Even if information is rated as helpful, valence ratings are lower and arousal ratings are higher after a regulation strategy is conveyed. These findings apply at least for this hypothetical setting.

Distraction and Suppression (radio und music): Listening to radio broadcast or music while imagining to drive in an anger-eliciting situation may influence affective experience; elicit higher positive valence and lower arousal ratings. The results are in line with Van der Zwaag et al. (2011), who demonstrated that music mediates the state of anger in a driving simulation. Findings show, furthermore, modifying emotion in this positive way is even stronger if radio broadcast and music is liked. But, if music is not liked, maladjusted emotions responsible for risky driving behaviour are strengthened. Similarly, Wiesenthal et al, (2003) found that listening to owns favourite music while driving leads to lower occurrence of mild aggression. We are not aware of any study examining radio broadcast and emotion regulation. We assume, however, that radio broadcast could have similar mechanisms such as music while driving. In sum, if preference is considered, music and radio act in the desired direction and may reduce maladjusted emotions, which are under some circumstances responsible for risky and aggressive driving behaviour.

Reappraisal via information: Furthermore, provide information about the reason for a hold-up has an effect on imagined emotion regulation. Positive information about another perspective – offer comprehensible information why the driver in front goes slowly - is stronger in moderating negative affect than provide positive information about the situational outcome (minor actual delay).

Surprisingly, negative information about the situational outcome (ten minutes delay) leads to slightly higher valence ratings and lower arousal ratings. This could indicate that information itself (knowledge about what happens in a situation) has a regulative effect on participant's emotion. Noticeable, negative information about another perspective (driver in front talks on the phone) leads to decreased valence and higher arousal ratings. This result supports current appraisal research; anger is triggered if another person is to blame for a given situation (Roidl et al., 2010; Mesken et al., 2007). The effects are visible even when the information is deemed as useful.

The present study has methodological *limitations*. The present study was conducted online. Thus, the study permits no conclusion as to whether the emotion regulation would actually lead to safer driving. The simulation is purely hypothetical. To ask participants to read text-vignettes and visualise themselves as being in anger eliciting situations depends on their ability for vivid imagination. This ability may be strongly dependent on their real-life experiences. Furthermore, emotion regulation strategies lasted only for 12 (information) to 45 (radio and music) seconds. Thus, findings need to be validated in real driving or in a driving simulator. As mentioned above, the experience and regulation of emotion is highly dependent on individual differences as e.g. age and situational experience. Prospective, the test sample should represent more elderly and driving experienced participants.

In sum, in this study we compare several emotion regulation strategies, asking whether driver assistant systems can cause emotion regulation via certain interactions. Results indicate that at least in an emotion imagination setting it is possible to regulate emotions within a technical system such as a imagined car. In this study participants imagine themselves regulate their anger by induced distraction, suppression and, under some conditions, by reappraisal. Furthermore, individual preferences for these kinds of imagined regulation strategies have a significant impact on emotion regulation.

Acknowledgement

This research project was supported by a grant (F.A.-Nr. 2006.63) from the 'Arbeitsgemeinschaft Innovative Projekte' (AGIP) of the Ministry of Science and Culture, Lower Saxony, Germany.

References

Allen, K., & Blascovich, J. (1994). Effects of music on cardiovascular reactivity among surgeons. *Journal of the American Medical Association, 272,* 882-884.

Anderson, R.A., Baron, R.S., & Logan, H. (1991). Distraction, control, and dental stress. *Journal of Applied Social Psychology, 21,* 156-171.

Arnett, J., Offer, D., & Fine, M.A. (1997). Reckless driving in adolescence: 'state' and 'trait' factors. *Accident Analysis and Prevention, 29,* 57-63.

Bradley, M.M., & Lang, P.J. (1994). Measuring emotions: the self-assessment manikin and the semantic differential. *Journal of Behavior Therapy and Experimental Psychiatry, 25,* 49-59.

Bushman, B.J., Bonacci, A.M., Pedersen, W.C., Vasquez, E.A., & Miller, N. (2005). Chewing on it can chew you up: Effects of rumination on triggered displaced aggression. *Journal of Personality and Social Psychology, 88,* 969-983.

Campbell-Sills, L., & Barlow, D.H. (2009). Incorporating Emotion Regulation into Conceptualizations and Treatments of Anxiety and Mood Disorders. In J. Gross (Ed.), *Handbook of Emotion Regulation* (pp. 542-559). New York: Guilford.

Campos, J., Frankel, C., & Camras, L. (2004). On the nature of emotion regulation. *Child Development, 75,* 377-394.

Charles, S.T., & Carstensen, L.L. (2009). Emotion Regulation and Aging. In J. Gross (Ed.), *Handbook of Emotion Regulation* (pp. 307-327). New York: Guilford.

Deffenbacher, J.L. (2009). Angry Drivers: Characteristics and Clinical Interventions. *Revista Mexicana de Psicología, 26,* 5-16.

Deffenbacher, J.L. (2006). Evidence for effective treatment of anger-related disorders. In E.L. Feindler (Ed.), *Anger-related disorders* (pp. 43-69). New York: Springer.

Deffenbacher, J.L., Petrilli, R.T., Lynch, R.S., Oetting, E.R., & Swaim, R.C. (2003). The Driver's Angry Thoughts Questionnaire: A Measure of Angry Cognitions When Driving. *Cognitive Therapy and Research, 27,* 383-402.

Deffenbacher, J.L., Filetti, L.B., Lynch, R.S., Dahlen, E.R., & Oetting, E.R. (2002). Cognitive-behavioral treatment of high anger drivers. *Behaviour Research and Therapy, 40,* 895-910.

Deffenbacher, J.L., Huff, M.E., Lynch, R.S., Oetting, E.R., & Salvatore, N.F. (2000). Characteristics and treatment of high anger drivers. *Journal of Counseling Psychology, 47,* 5-17.

Gross, J.J., Carstensen, L.L., Pasupathi, M., Tsai, J.J., Skorpen, C.G., & Hsu, A.Y.C. (1997). Emotion and Aging: Experience, Expression, and Control. *Psychology and Aging, 12*, 590-599.

Gross, J.J., & Thompson, R.A. (2009). Emotion Regulation: Conceptual Foundations. In J.J. Gross (Ed.), *Handbook of Emotion Regulation* (pp. 3-24). New York: Guilford.

Gross, J.J. (2001). Emotion regulation in adulthood: Timing is everything. *Current Directions in Psychological Science, 10*, 214-219.

Hatfield, J., & Chamberlain, T. (2008). The effect of audio materials from a rear-seat audiovisual entertainment system or from radio on simulated driving. *Transportation Research Part F, 11*, 52-60.

John, O.P., & Gross, J.J. (2009). Individual Differences in Emotion Regulation. In J.J. Gross (Ed.), *Handbook of Emotion Regulation* (pp. 3-24). New York: Guilford.

Juslin, P.N., & Sloboda, J.A. (Eds.), (2010). *Handbook of music and emotion: Theory, research, applications.* New York: Oxford University Press.

Kunzmann, U., Kupperbusch, C.S., & Levenson R.W. (2005). Behavioral Inhibition and Amplification During Emotional Arousal: A Comparison of Two Age Groups. *Psychology and Aging, 20*, 144-158.

Lajunen, T., & Parker, D. (2001). Are aggressive people aggressive drivers? A study of the relationship between self-reported general aggressiveness, driver anger and aggressive driving. *Accident Analysis and Prevention, 33*, 243-255.

Lazarus, R.S. (2001). Relational meaning and discrete emotions. In K.R. Scherer, A. Schorr, and T. Johnstone (Eds.), *Appraisal processes in emotion: Theory, methods, research.* New York: Oxford University Press.

Lee, E., Nass, C., & Brave, S. (2000). Can computer-generated speech have gender? An experimental test of gender stereotype. *CHI'00 extended abstracts on Human factors in computing systems*, 289-290.

Martinez, R. (1997). Statement of the Honorable Recardo Martinez. M.D. of the National Highway Traffic Safety Administration. Sub-committee on Surface Transportation, Committee on Transportation and Infrastructure, US House of Representatives, Washington, DC.

Mesken J., Hagenzieker, M.P., Rothengatter, T., & De Waard, D. (2007). Frequency, determinants, and consequences of different drivers' emotions: An on-the-road study using self-reports, (observed) behaviour, and physiology. *Transportation Research Part F, 10*, 458-475.

Parker D., Lajunen T., & Summala H. (2002). Anger and aggression among drivers in three European countries. *Accident Analysis and Prevention, 34*, 229-235.

Parkinson, B. (2001). Anger on and off the road. *British Journal of Psychology, 92*, 507-526.

Ray, R.D., Wilhelm, F.H., & Gross, J.J. (2008). All in the Mind's Eye? Anger Rumination and Reappraisal. *Journal of Personality and Social Psychology, 94*, 133-145.

Rusting, C.L., & Nolen-Hoeksema, S. (1998). Regulating responses to anger: Effects of rumination and distraction on angry mood. *Journal of Personality and Social Psychology, 74*, 790-803.

Roidl, E., Höger, R., & Pfister, H.-R. (2011). Emotional States of Drivers and the Impact on Driving Behaviour A Simulator Study. In D. de Waard, N. Gérard, L. Onnasch, R. Wiczorek, and D. Manzey (Eds.), *Human Centred Automation* (pp. 171-182). Maastricht, the Netherlands: Shaker Publishing.

Scherer, K.R., Schorr, A., & Johnstone, T. (2001). *Appraisal processes in emotion: Theory, methods, research.* New York: Oxford University Press.

Shinar, D. (1998). Aggressive driving: the contribution of the drivers and the situation. *Transportation Research Part F, 1,* 137-160.

Sidorenko, V.N. (2000). Effects of medical resonance therapy music in the complex treatment of epileptic patients. *Integrative Physiological & Behavioral Science, 35,* 212-217.

Stratton, V.N., & Zalanowski, A.H. (1984). The relationship between music, degree of liking, and self-reported relaxation. *Journal of Music Therapy, 21,* 184-192.

Stephens, A.N., & Groeger, J.A. (2006). *Do emotional appraisals of traffic situations influence driver behaviour?* Paper presented at the Behavioural Studies Seminar, Bath, 3.04-4.04.2006.

Snyder, C.R., Crowson, J.J., Houston, B.K., Kurylo, M., & Poirier, J. (1997). Assessing hostile automatic thoughts: Development and validation of the HAT Scale. *Cognitive Therapy and Research, 21,* 477-492.

Thaut, M.H., & Wheeler, B.L. (2010). Music therapy. In P.N. Juslin and J.A. Sloboda (Eds.), *Handbook of Music and Emotion: Theory, Research, Applications* (pp. 819-848). New York: Oxford University Press.

Van der Zwaag, M.D., Fairclough, S.H., Spiridon, E., & Westerink, J.H.D.M. (2011). *The impact of music on affect during anger inducing drives.* Paper presented at Affective Computing and Intelligent Interaction (ACII), Memphis, USA.

Wheeler, B.L. (2005). *Music therapy research* (2nd ed). Gilsum, NH: Barcelona Publishers.

Wiesenthal, D.L., Hennessy, D.A., & Totten, B. (2000). The influence of music on driver stress. *Journal of applied social psychology, 30,* 1709-1719.

Wiesenthal, D.L., Hennessy, D.A., & Totten, B. (2003). The influence of music on mild drivers aggression. *Transportation Research Part F, 6,* 125-134.

Precision of congestion warnings: Do drivers really need warnings with precise information about the position of the congestion tail

Ingo Totzke, Frederik Naujoks, Dominik Mühlbacher, & Hans-Peter Krüger
Center for Traffic Sciences (IZVW), University of Würzburg,
Germany

Abstract

In a driving simulator with a motion system, warnings were provided to N = 16 participants (25-72 years) en route to the tail of a congestion. Two different kinds of congestion tails were simulated. In the first, the speed of the surrounding traffic was abruptly reduced before the tail of the congestion was reached and in the second, the speed of the surrounding traffic was gradually reduced. In the simulated runs, congestion warnings were given at different distances ("3.5 km" vs. "1.5 km" vs. "0.3 km" prior to the congestion tail) and the precision of the warnings ("precise warning": distance to the congestion is indicated and updated regularly vs. "imprecise warning": without a clear distance indication) was varied. Furthermore, the tail was approached without any warning. Drivers were asked to work on a secondary task (handling a menu system) during the entire run. Overall, precise warnings have greater effects on driving safety when approaching a congestion tail than imprecise warnings. Driving safety was at its lowest when drivers approached the congestion tail without any warning. Precise warnings given 1.5 km prior to the congestion tail show the highest driving safety and were preferred by drivers.

Introduction

Motorway accidents predominantly occur when speeds vary and traffic is dense. According to Lee, Saccomanno and Hellinga (2002), approximately 88% of accidents on US-highways are caused by drivers who abruptly decelerate or by non-moving vehicles. Similar results were reported by Oh, Oh, Ritchie and Chang (2001) as well as by Zheng, Ahn and Monsere (2010). In Germany, approximately 30% to 40% of motorway accidents occur because: (1) the geometry of the road ahead is badly visible and (2) slowly moving or stopped traffic (German Federal Statistical Office, 2010). The latter scenario is typical cause for traffic jams on motorways.

In the following, the term "congestion tail" is used to describe the transition between free and congested traffic (Treiber & Kesting, 2010). Kerner (1999) defines the congestion tail as the region in which traffic volume, traffic density, and mean velocity varies within a very short section of the motorway. Two types of congestion tails can be distinguished (following Buld, 2003; Kerner, 2004; Kim, 2002):

1. Congestion tails with synchronized traffic (so-called "soft congestion tails"): Transitions between synchronized and congested traffic with a smooth reduction of velocity from the drivers' perspective while approaching the congestion tail (range of velocities: 0 to 70 km/h)
2. Congestion tails with free traffic (so-called "hard congestion tails"): Transitions between free and congested traffic with a strong reduction of velocity (range of velocities: 0 to 110 km/h)

It is assumed that warning systems that alert the driver prior to critical driving situations (e.g. congestion tails) might be beneficial if drivers are made aware of a potentially dangerous situation and can thus react faster. However, only a few studies have been done that focus on congestion warnings or warnings given prior to objects on motorways.

Van Driel, Hoedemaker and Van Arem (2007) as well as Brookhuis, Van Driel, Hof, Van Arem and Hoedemaeker (2008) examined, for instance, a so-called "congestion assistant" in a driving simulator. In this study, information about congestion was given 5 km prior to the tail. Distance information was updated every 0.5 km. Subjects were asked to drive under normal and foggy conditions. Under normal conditions, subjects chose a lower speed with the "congestion assistant" when approaching the congestion tail than without the assistant. In foggy conditions, however, subjects drove more slowly, as a result no positive effect of the "congestion assistant" on velocity could be found.

Similarly, Popiv, Rommerskirchen, Bengler, Duschl and Rakic (2010) had subjects in a simulation study and demonstrated that a congestion warning (system) positively effects driving behaviour if given when approaching a congestion tail behind a curve: If drivers were warned of the congestion tail, they used the brake pedal earlier and decelerated less abruptly than if they weren't warned.

Finally, Alm and Nilsson (2000) examined the effects of precise or imprecise congestion warnings in a driving simulator study. Imprecise warnings were simulated by superimposing red signals on the roadside or by simply naming the kind of danger on a display (e.g. "congestion tail ahead"). Warnings were made more precise by including the distance to the danger (e.g. "congestion tail in 1 km ahead") and were given 1 km in advance. In each of these scenarios, drivers reduced their velocity approximately 800 m prior to the congestion tail.

The present study's aim was to examine possible effects of such a congestion warning depending on the precision as well as on the time it was given. It was assumed that the type of congestion tail is of major importance: The harder it is for the driver to anticipate the existence and position of a congestion tail, the stronger the effects of a congestion warning are. In summary, this paper explores the following questions:

- Does the type of a congestion tail moderate the effects of a congestion warning on driving behaviour when approaching the congestion tail?
- Does the precision of a congestion warning and the time it is given moderate the warning's effect?

Methods

Apparatus

In this study, the subjects drove in a driving simulator with a motion system (see Figure 1 on the left). The driving task consisted of a two-lane motorway that was 115 km and had varying traffic densities (see Figure 1 on the right); the speed limit was 130 km/h. The participants were instructed to comply with road traffic regulations (e.g. driving on the right lane) and to adhere to the speed limit. The simulator course took approximately 90 min.

Figure 1. Driving simulation with motion system of the Würzburg Institute for Traffic Sciences (WIVW GmbH, on the left). Screenshot of approaching one of the congestion tails (on the right).

Within the simulator course, the subjects approached a congestion tail 16 times. These segments were of 4.550 m long. In each of these segments, the simulated traffic in the right lane drove approximately 90 km/h. In the left lane, the simulated traffic travelled at a speed of up to 130 km/h (area: 0 m to 3.300 m). Two variations of the environmental traffic were integrated 1.200 m prior to the congestion tail:

1. "Soft congestion tail": Simulated traffic decelerated smoothly as it approached the congestion tail. Traffic density and velocity of the simulated vehicles in both lanes converge.
2. "Hard congestion tail": Simulated traffic decelerated abruptly as it approached the congestion tail. Traffic density and velocity of the vehicles varied significantly between the two lanes of the motorway.

Additionally, eight irrelevant driving situations were introduced in the simulator course. In each of these situations the simulated traffic behaved similarly to the above-mentioned situations. However, no congestion tail appeared that forced vehicles to slow down.

Furthermore, the following features of the congestion warning were varied in this study:

- Precision of warning: Precise or imprecise information about the congestion tail's position
- Warning distance: The time when of the first congestion warning was given

In the "precise warning" condition, information about the subsequent situation (i.e. "congestion tail ahead") as well as its distance was given on a display. The distance information was updated every 300m (in the 3.5km-warning, the first update was given after 500 m). In the "imprecise warning" condition, only information about the subsequent situation (without any distance information) was presented. Figure 2 illustrates these warnings (on the left: "precise warning", on the right: "imprecise warning"). The design of the warning displays was based on work done within the research project sim^{TD} ("Safe and Intelligent Mobility - Testfield Germany", www.simTD.de).

Figure 2. Screenshots of congestion warning: Precise warning (on the left) and imprecise warning (on the right)

The time when the congestion warning was first given ("warning distance") was varied:

- First warning 3.5 km prior to the congestion tail ("3.5 km-warning")
- First warning 1.5 km prior to the congestion tail ("1.5 km-warning")
- First warning 0.3 km prior to the congestion tail ("0.3 km-warning")

In a control condition, the subjects approached one hard and one soft congestion tail without any given warning. In these situations, "Kein Empfang" (in English: "No reception") was shown on the system's display. This information also was presented in four of the eight irrelevant situations in order to make congestion tails less predictable. Moreover, the drivers were asked not to work on the secondary task while approaching one hard and one soft congestion tail.

The congestion warnings were shown on a display in the upper middle console of the the simulator vehicle (see Figure 1 on the right). When a warning was displayed (as mentioned above), an acoustic tone was given. The warning stopped when the driver reached the congestion tail.

An additional task was introduced in this study: The subjects were instructed to work on tasks given in a menu system (e.g. "Please search for the radio station 'Bayern 3'.") by using a joystick (see Rauch, Totzke & Krüger, 2004). The menu system was displayed in the lower middle part of the console (see Figure 1 on the right). It was assumed that this secondary task would distract drivers and draw their attention away from driving, especially while approaching the congestion tails. This task did not have a systematic influence on the drivers' braking behaviour while approaching a congestion tail (Totzke, Naujoks, Mühlbacher & Krüger, 2011) and, therefore, the results are not considered in this paper.

This experiment was carried out in the driving simulator of the Wuerzburg Institute for Traffic Sciences (WIVW GmbH, for further information please visit www.wivw.de; see Figure 1 on the left). The simulator with a motion platform consists of a 180° front projection system and three rear projectors. The driver sits in the front part of a BMW. The vehicle is ensconced on a platform that is moved by a 6 degrees-of-freedom motion system. The simulator is controlled by personal computers. By means of a flexible scripting language developed by the WIVW, scenarios with numerous models for the behaviour of other traffic participants (including vehicles and pedestrians) can be defined easily. Based on the behaviour of the simulator's driver, these are recalculated during simulation and it is possible to adapt the behaviour of other traffic participants so that they dynamically interacted with the simulator car.

Design

In this study, a within-factors design with the factors "type of congestion tail" (hard vs. soft), "precision of warning" (precise vs. imprecise) and "warning distance" (3.5 km vs. 1.5 km vs. 0.3 km) was realized. In order to control possible sequence effects, four different sequences of congestion tails were construed.

Parameters and strategy of analysis

Objective and subjective data were considered in this study to determine the effects of the congestion warning. In order to determine the drivers' reaction to the congestion warnings, the following parameters were recorded (data were sampled with a frequency of 100 Hz):

- Point of time of readiness to brake [s]
- Point of time of onset of braking [s]

The first-mentioned parameter describes the point of time when the driver approached the brake pedal with the right foot without touching it. This parameter was measured by an infrared-sensor. The last-mentioned parameter refers to the point of time when the brake pedal was pressed by the driver for the first time. For both parameters threshold values were identified after extensive preliminary tests. As soon as the respective threshold value was exceeded, the point of time of this event was identified and used for further analyses.

Immediately after stopping at the congestion tail, the subjects were questioned via a microphone and had to answer the following (in closed-ended answering format) using a 16-point scale (1 = "very weak" until 15 = "very strong" and additionally with 0 = "not at all", Heller, 1985):

- "How safe was it approaching the congestion tail?"
- "How surprising was the position of the congestion tail?"

Moreover, a detailed inquiry with questions and statements (in open- or closed-ended answering format) was conducted at the end of each session. In this paper, the following statements will be addressed:

- "Driving with a congestion warning system was more – less strenuous than driving without."
- "I had to make more – fewer quick decisions while driving with the congestion warning system than without."
- "With the congestion warning system I estimated the position of the congestion tail worse – better than without."

To evaluate these statements, so-called "visual-analogue scales" were used: These scales consist of a horizontal line (length: 16 cm) with verbal labels on both ends of the line (e.g. "Driving with congestion warning system was…than driving without", left pole: "more strenuous", right pole: "less strenuous"). The centre of the line was verbally anchored by the label "comparable to driving without warning system". The subjects' task was to position a cross on this horizontal line to rate the impact of the precise and the imprecise warnings in comparison with driving without a congestion warning. Additionally, questions in an open-ended answering format were used. In this paper the following questions will be cited:

- "Please rank: Which of the congestion warnings (i.e. precise or imprecise warnings) did you prefer?"
- "Please rank: Which of the warning distances (i.e. 3.5 km, 1.5 km or 0.3 km) did you prefer?"

The inferential analysis for the above-mentioned objective data was conducted in several steps with the factors "type of congestion tail", "precision of warning" and "warning distance":

1. Exploratory analysis by calculating means and 95%-confidence intervals for each factor level
2. Comparison of driving with and without the congestion warning system using t-tests for dependant samples (separated for hard and soft congestion tails)
3. Evaluation of the impacts of the type of congestion warning ("precision of warning" and "warning distance") and the type of congestion by means of full-factorial profile analysis (MANOVA)

For the subjective data a similar procedure was used:

1. Exploratory analysis by calculating means and 95%-confidence intervals
2. Comparison of driving with and without the congestion warning system using t-tests for dependant samples (separated for precise und imprecise warnings)

The subjects' answers to the questions in open-ended answering format were only descriptively analysed, taking into account the qualitative nature of the data.

Procedure

The experimental session can be divided into three main parts: (1) Instructions and exercises, (2) simulator rides, and (3) inquiries. The whole session took approximately 3.5 hours.

In the first part "instructions and exercises", the subjects were welcomed and informed about the aim of the study. Additionally, they were given instructions on how to handle the menu system. In order to become familiar with the handling of the menu system, the subjects were given 45 tasks in the menu system to complete. After this, short simulator rides were conducted in order to familiarize the subjects with driving in a simulator with a motion system and operating the menu system while driving. In one of these runs, the congestion warnings were also demonstrated. Prior to this ride, the subjects were given a short written description of the congestion warning system. In all, this part of the session took approximately 90 min.

In the second part "simulator rides", two rides that lasted 45 min each were conducted. In each of these rides, the subjects approached eight congestion tails and four irrelevant driving situations. After stopping at a congestion tail, a short inquiry was conducted using a microphone. This part of the session lasted approximately 90 min.

In the third part "inquiries", the subjects were to answer several questions or evaluate statements in a paper-pencil questionnaire (closed-ended answering format) as well as in a face-to-face interview (open-ended answering format), respectively. This part of the session took approximately 30 min.

Sample

A total of N = 16 subjects between 25 and 72 years of age (m = 47.7, sd = 17.3) participated in the study. All subjects had a minimum driving average of 10,000 km per year. Prior to the experiment, all subjects had taken part in an extended simulator training that lasted approximately three hours (Hoffmann & Buld, 2006). The subjects were granted an expense allowance for their participation in this study.

Results

In the first step, it was assumed that the subjects would be ready to brake and start braking with a larger distance to the congestion tail due to the congestion warning. As shown in Figure 3, this assumption can only be partially confirmed: When approaching a hard congestion tail, the subjects only profit from precise warnings if compared to the control condition. Imprecise warnings do not have beneficial effects on the subjects' braking behaviour when approaching hard congestion tails. When approaching soft congestion tails, neither precise nor imprecise warnings lead to an earlier braking behaviour compared to the control condition (see Table 1). Figure 3 also shows that the warning distance influences the point of time when braking begins: The subjects start braking at greater distances with the 1.5 km-warning compared to the 0.3 km- and 3.5 km-warning. The warning distance does not have any impact on the subjects' readiness to brake (see Table 2).

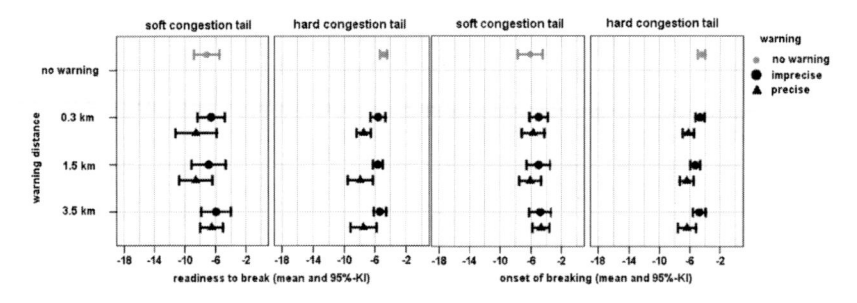

Figure 3. Point of time of readiness to brake (on the left) and point of time of onset of braking (on the right). Pictured is the mean with 95%-confidence interval

Table 1. Results of t-Tests for dependant samples comparing with vs. without congestion warning for "readiness to brake" and "onset of braking"

Congestion tail	Warning precision	Variable					
		Readiness to brake			Onset of braking		
		t	df	p	t	df	p
soft	precise	-0.67	15	.516	0.81	15	.431
	imprecise	0.63	15	.539	1.37	15	.191
hard	precise	-4.95	15	.000	-5.07	15	.000
	imprecise	-2.01	15	.063	-1.52	15	.149

Table 2. Results of MANOVA for "readiness to brake" and "onset of braking". Velocities of leading vehicle while approaching soft congestion tail (on the left) and hard congestion tail (on the right)

Variable	Effect	F	df_1	df_2	p	Partial η^2
Readiness to brake	Congestion tail	2.24	1	15	.155	0.13
	Warning distance	2.94	2	14	.086	0.30
	Warning precision	7.80	1	15	.014	0.34
	Congestion tail x Warning distance	1.52	2	14	.253	0.18
	Congestion tail x Warning precision	0.74	1	15	.405	0.05
	Warning distance x Warning precision	0.29	2	14	.754	0.04
	Congestion tail x Warning distance x Warning precision	0.33	2	14	.724	0.05
Brake Onset	Congestion tail	5.78	1	15	.030	0.28
	Warning distance	7.39	2	14	.006	0.51
	Warning precision	7.74	1	15	.014	0.34
	Congestion tail x Warning distance	1.76	2	14	.208	0.20
	Congestion tail x Warning precision	7.51	1	15	.015	0.33
	Warning distance x Warning precision	0.27	2	14	.770	0.04
	Congestion tail x Warning distance x Warning precision	1.26	2	14	.313	0.15

The subjects evaluate the congestion warning for both hard and soft congestion tails in a positive way: They report that approaching the congestion tail with congestion warning is safer than without (see Figure 4 on the left and Table 3). Perceived safety is higher for precise warnings than for imprecise warnings. The warning distance is largely irrelevant in this point. Furthermore, the subjects say they were less surprised by the position of the congestion tail when precise warnings were given than when imprecise warnings were given (see Figure 4 on the right and Table 4). This effect is greater for hard congestion tails than for soft congestion tails. 1.5 km- and 3.5 km-warnings are judged as superior to 0.3 km-warnings when talking about perceived surprise.

Table 3. Results of t-Tests for dependant samples comparing with vs. without congestion warning for "perceived safety" and "perceived surprise"

Congestion tail	Warning precision	Variable Perceived safety			Perceived surprise		
		t	df	p	t	df	p
soft	precise	6.32	15	.000	-4.97	15	.000
	imprecise	2.75	15	.015	-1.90	15	.077
hard	precise	5.78	15	.000	-5.40	15	.000
	imprecise	2.79	15	.014	-1.19	15	.253

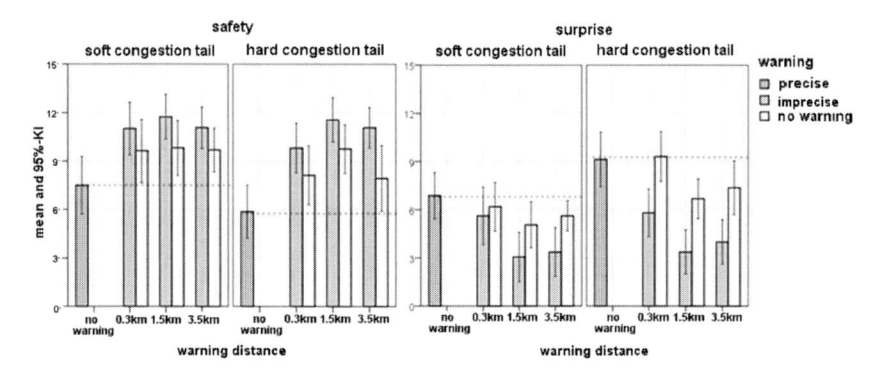

Figure 4. Judgements about perceived safety (on the left) and perceived surprise (on the right) while approaching a congestion tail. Judgements were given on 16-point scales. Means are displayed with 95%-confidence intervals

Table 4. Results of MANOVA for "perceived safety" and "perceived surprise"

Variable	Effect	F	df_1	df_2	p	partial η^2
Perceived safety	Congestion tail	1.75	1	15	.206	0.10
	Warning distance	2.03	2	14	.168	0.23
	Warning precision	21.6	1	15	.000	0.59
	Congestion tail x Warning distance	1.47	2	14	.264	0.17
	Congestion tail x Warning precision	0.56	1	15	.466	0.04
	Warning distance x Warning precision	0.41	2	14	.670	0.06
	Congestion tail x Warning distance x Warning precision	0.37	2	14	.696	0.05

Perceived surprise	Congestion tail	1.00	1	15	.334	0.06
	Warning distance	8.73	2	14	.003	0.56
	Warning precision	24.8	1	15	.000	0.62
	Congestion tail x Warning distance	0.22	2	14	.806	0.03
	Congestion tail x Warning precision	7.74	1	15	.014	0.34
	Warning distance x Warning precision	0.40	2	14	.681	0.05
	Congestion tail x Warning distance x Warning precision	1.44	2	14	.271	0.17

Comparing the congestion warning to driving without any warning at the end of the session, the drivers favour precise warnings (see Figure 5): With precise warnings, the subjects report that approaching the congestion tail is less strenuous, less spontaneous decisions are necessary, and the position of the congestion tail can be estimated more precisely. Nevertheless, imprecise warnings are preferred over driving without any congestion warnings. Overall, the subjects accept the introduction of a congestion warning system (see Table 5).

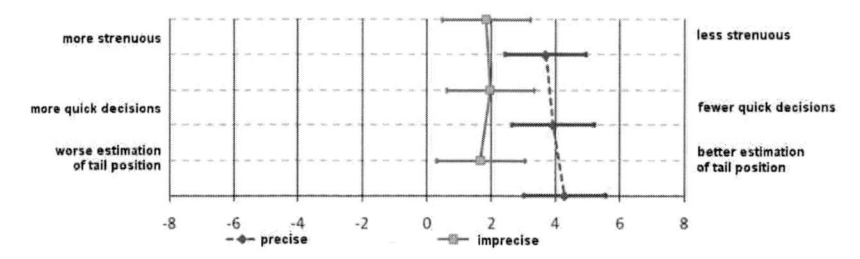

Figure 5. Judgements in selected questions of the inquiry at the end of the session. Positive values imply a positive evaluation of the congestion warning compared to the control condition. Means are displayed with 95%-confidence intervals

Finally, the subjects were asked explicitly to rank which type of congestion warnings is the best. 12 of the 16 subjects prefer the precise congestion warning with the following reasons (examples):

- "You are prepared and you drive more carefully when approaching the congestion tail."
- "If you know the precise distance to the congestion tail, you might have the possibility to choose another route."
- "You can warn subsequent traffic."

Table 5. Results of t-Tests for dependant samples comparing with vs. without congestion warning for different questions of the inquiry at the end of the session

Item	Warning precision	t	df	p
less – more strenuous	imprecise	2.37	15	.032
	precise	4.47	15	.000
more – fewer quick decisions while driving	imprecise	2.23	15	.042
	precise	5.37	15	.000
estimated the position of the congestion tail worse – better	imprecise	2.37	15	.032
	precise	5.88	15	.000

Only 4 of the 16 subjects favour the imprecise warning for everyday driving. They make the following points:

- "You are warned by the system and must react on your own. Therefore, the distance of the congestion tail is not necessary."
- "You pay more attention when driving with the imprecise warning than with the precise warning. Therefore, imprecise warnings are better."
- "With precise warnings you are tempted to rely on the given distance given unduly."

Independent from the precision of the congestion warning the subjects prefer the 1.5 km-warning: 11 of the 16 subjects agree that this congestion warning as the best alternative, 5 of the 16 subjects rate it as the second best alternative. Accordingly, the 0.3 km- and 3.5 km-warnings are not evaluated as positive as the 1.5 km-warning.

Discussion

This study's aim was to examine the effects of congestion warnings on drivers' behaviour while approaching a congestion tail. For this purpose, soft and hard congestion tails were simulated in a driving simulator with motion a system. The congestion warnings differed regarding the precision of the warning and the distance at which the warning was given. Firstly, the results of this study confirm published results for congestion warnings (e.g. Alm & Nilsson, 2000; Brookhuis et al., 2008; Popiv et al., 2010; Van Driel et al., 2007): Congestion warnings result in adapted driving behaviour when approaching a congestion tail. Secondly, the relevance of the warning's precision and of the time it is first given has not yet been considered. Therefore, this study broadens the understanding of the effects of congestion warnings on driving behaviour.

When approaching hard congestion tails, precise warnings have the most beneficial effects on driving: Drivers are prepared to brake and start braking at a greater distance than when driving without a congestion warning. Imprecise warnings do not have any impact on the drivers' braking behaviour. Moreover, the warning distance influences the braking behaviour because drivers start braking earlier with the 1.5 km-warning than with the 3.5 km- and 0.3 km-warning. The point of time when drivers are ready to brake is not changed by the warning distance. When approaching soft congestion tails, no systematic beneficial effects for any of the congestion warnings were found.

Likewise, the drivers in our study reported of benefiting more by the precise congestion warnings than by the imprecise warnings: Driving with precise congestion warnings is safer and less strenuous; as a result, less spontaneous decisions are necessary and the position of the congestion tail can be estimated more precisely. This, in turn, eliminates the element of surprise for drivers. In all, drivers prefer precise congestion warnings. However, imprecise warnings are still favoured over driving without any congestion warning at all. The warning distance is of minor importance for drivers' overall evaluation of the congestion warning, though 3.5 km-

and 1.5 km-warnings reduce the perceived surprise when approaching the congestion tail if compared to warnings given at 0.3 km.

To sum up, significant advantages result for precise congestion warnings compared to imprecise warnings. The beneficial effects on driving behaviour are most pronounced when drivers approach hard congestion tails. Positive effects of the congestion warnings when approaching soft congestion tails can only be seen if the drivers' judgements are taking into consideration, but not in driving behaviour. This implies that congestion warnings will have most impact on driving safety when drivers approach non-predictable congestion tails (i.e. hard congestion tails). Therefore, the implementation of congestion warnings might be particularly useful when approaching hard congestion tails. Considering that congestion warnings do not positively influence driving when approaching soft congestion tails, warnings in this situation might make drivers dislike congestion warnings after extensive use.

The present study was conducted in a driving simulator with a motion system; as a result, the empirical results of this study cannot be generalized to everyday driving in real traffic without some amount of limitation. Therefore, the identified results for the drivers' braking behaviour (i.e. point of time of readiness to braking and onset of braking) should not be taken in absolute values, but only as a comparison between different experimental conditions (e.g. gain of time in braking behaviour while approaching a congestion tail by congestion warnings compared to driving without a congestion warning). Moreover, the drivers in this study were asked to work on a secondary task (i.e. solving given tasks within a menu system) while driving. This task was introduced so that the drivers were not able to fully concentrate on driving when approaching a congestion tail. Totzke et al. (2011) demonstrated that this task did not have a major impact on driving behaviour when approaching a congestion tail in the present study. In conclusion, it is necessary to pose the question whether or not similar results could be found in a simulator ride without integrating a secondary task. Further studies should explicitly consider this factor.

Acknowledgement

This study was funded by the Federal Highway Research Institute (BASt, Germany).

References

Alm, H. & Nilsson, L. (2000). Incident warning systems and traffic safety: A comparison between the PORTICO and MELYSSA test site systems. *Transportation Human Factors, 2*, 77-93.

Brookhuis, K.A., Van Driel, C.J.G., Hof, T., Van Arem, B., & Hoedemaeker, M. (2008). Driving with a congestion assistant: Mental workload and acceptance. *Applied Ergonomics, 40*, 1019–1025.

Buld, S. (2003, Juli). *INVENT - Das Learnability-Lab bei der Untersuchung von Fahrerassistenzsystemen*. Informations- und Assistenzsysteme im Automobil - Erlernbarkeit als Beitrag zur Fahrsicherheit. Bergisch-Gladbach: Bundesanstalt für Straßenwesen.

Heller, O. (1985). Hörfeldaudiometrie mit dem Verfahren der Kategorienunterteilung (KU). *Psychologische Beiträge, 27*, 478-493.

Hoffmann, S., Krüger, H.-P., & Buld, S. (2003). Vermeidung von Simulator Sickness anhand eines Trainings zur Gewöhnung an die Fahrsimulation. In VDI-Gesellschaft Fahrzeug- und Verkehrstechnik (Hrsg.), *Simulation und Simulatoren - Mobilität virtuell gestalten* (VDI-Berichte, Nr. 1745, S. 385-404). Düsseldorf: VDI-Verlag.

Kerner, B.S. (1999). Theory of congested traffic flow. In A. Ceder (Ed.), *Transportation and Traffic Theory: Proceedings of the 14th International Symposium on Transportation and Traffic Theory* (pp. 147-171). Amsterdam: Elsevier.

Kerner, B.S. (2004). *The physics of traffic: Empirical freeway pattern features, engineering applications, and theory*. Heidelberg: Springer.

Kim, Y. (2002). *Online traffic flow model applying dynamic flow-density relations*. Dissertation. München: Technische Universität München.

Lee, C., Saccomanno, F., & Hellinga, B. (2002). Analysis of crash precursors on instrumented freeways. *Transportation Research Record, 1784*, 1-8.

Oh, C., Oh, J.S., Ritchie, S.G., & Chang, M., (2001, January). *Real-time estimation of freeway accident likelihood*. Presented at the Annual Meeting of the Transportation Research Board, Washington D.C.

Popiv, D., Rommerskirchen, C., Bengler, B., Duschl, M., & Rakic, M. (2010). Effects of assistance of anticipatory driving on drivers' behaviour during deceleration phases. In J. Krems, T. Petzold & Henning, M. (Eds.), *European Conference on Human Centred Design for Intelligent Transport Systems* (pp. 133-145). Lyon: HUMANIST Publications.

Rauch, N., Totzke, I., & Krüger, H.-P. (2004). Kompetenzerwerb für Fahrerinformationssysteme: Bedeutung von Bedienkontext und Menüstruktur. In VDI-Gesellschaft Fahrzeug- und Verkehrstechnik (Hrsg.), *Integrierte Sicherheit und Fahrerassistenzsysteme* (VDI-Berichte, Nr. 1864, S. 303-322). Düsseldorf: VDI-Verlag.

Statistisches Bundesamt (2010). *Verkehrsunfälle 2009* (Fachserie 8, Reihe 7). Wiesbaden: Statistisches Bundesamt.

Totzke, I., Naujoks, F., Mühlbacher, D., & Krüger, H.-P. (2011). Stauendewarnungen im Fahrzeug: Eine (un)geeignete Unterstützung für ältere Fahrer? In S. Schmid, M. Elepfandt, J. Adenauer & A. Lichtenstein (Hrsg.), *Reflexionen und Visionen der Mensch-Maschine-Interaktion - Aus der Vergangenheit lernen, Zukunft gestalten"* (Fortschritt-Berichte VDI Reihe 22, Nr. 33, S. 12-13). Düsseldorf: VDI-Verlag.

Treiber, M. & Kesting, A. (2010). Datengestützte Analyse der Stauentstehung und -ausbreitung auf Autobahnen. *Straßenverkehrstechnik, 1*, 5-12.

Van Driel, C.J.G., Hoedemaeker, M., & Van Arem, B (2007). Impacts of a Congestion Assistant on driving behaviour and acceptance using a driving simulator. *Transportation Research Part F, 10*, 139–152.

Zheng, Z., Ahn, S., & Monsere, C.M. (2010). Impact of traffic oscillations on freeway crash occurrences. *Accident Analysis and Prevention, 42*, 626-636.

The existence and impact of the Psychological Refractory Period effect in the driving environment

Daryl Hibberd, Samantha Jamson, & Oliver Carsten
Institute for Transport Studies, University of Leeds
UK

Abstract

Driver distraction from in-vehicle tasks can have negative impacts on longitudinal and lateral vehicle control and brake reaction time. The distraction problem is well-established in the literature, and is increasing due to advances in the functionality, availability, and number of in-vehicle systems. One approach to a solution is managing in-vehicle task presentation to reduce associated distraction. This paper reports a driving simulator experiment, designed to investigate the existence of the Psychological Refractory Period (PRP) in the driving context and its effect on driver performance. The PRP effect is observed when a surrogate in-vehicle task is presented in close temporal proximity to a lead vehicle braking event. Brake responses are subject to an increasing delay as the interval to an in-vehicle task is decreased. In-vehicle task modality modulates this effect. The impact of the PRP effect on driving performance is quantified and recommendations are made for reducing the driver distraction problem through the management of in-vehicle task timing and modality. The potential impact of these results on driver safety is discussed.

Introduction

Driver distraction

Driver distraction due to interaction with systems inside the vehicle has been the subject of research for nearly half a century (Brown, 1965) and has recently become a topical issue at both the academic and governmental level. Driver distraction is estimated to play a contributory role in approximately 25% of vehicle crashes (Stutts et al., 2001), with in-vehicle systems proving to be a prominent source of distraction (Klauer et al., 2006, Neale et al., 2005). The exact contribution of driver distraction to unsafe driving behaviours is difficult to quantify given widespread inconsistencies in the precise definition of the construct (Regan et al., 2011) and the reporting protocols used following vehicle accidents. Furthermore, the safety costs of distracted driving are likely to be under-estimated in the current statistics (Stutts et al., 2001). However, the negative impact of in-vehicle driver distraction on driving performance is indisputable, with examples of degradation of longitudinal and lateral control (Horrey & Lesch, 2009, Lansdown et al., 2004), reduced event detection (Horrey et al., 2008, McKnight & McKnight, 1993) and slower braking responses (Alm & Nilsson, 1995) prevalent in the literature. In addition, the magnitude of the problem is likely to increase in future vehicles, as the level of technological advancement, availability and uptake of in-vehicle systems continues to rise (Damiani et al., 2009, Young & Regan, 2007). If this is considered alongside demonstrations of poor driver awareness of distracting activities and the associated negative effects (Lerner, 2005), it would suggest that management of in-vehicle tasks is required to reduce the problems that could occur.

Psychological Refractory Period

Two tasks that are presented in close temporal proximity have been shown to impact on each others' performance; termed dual-task interference. The Psychological Refractory Period (PRP) effect (Telford, 1931, Welford, 1952) is an example of this, and results from a fundamental limitation in human task performance, whereby two tasks cannot be processed entirely in parallel (Pashler, 1990, Welford, 1952), regardless of their simplicity. The second of two tasks presented in quick succession is subjected to a delay in its processing due to an immediately preceding task occupying the processing resources that it also requires. This is postulated to occur due to limited capacity processing resources at the response selection stage of task processing (termed the Central Bottleneck), which prevents people selecting responses to two tasks at the exact same moment (perceptual and response execution processes are hypothesized to proceed in parallel, Pashler, 1984, Smith, 1967, Welford, 1952). The enforced 'queuing' of one task at this stage of processing ensures that it experiences a concomitant increase in reaction time. The delay in the performance of the second task (relative to its performance in isolation) varies in a manner that is dependent on the interval or stimulus onset asynchrony (SOA) between the two tasks (see Figure 1). The effect manifests itself as an increase in the reaction time to Task 2 with decreasing SOA between Task 1 and Task 2.

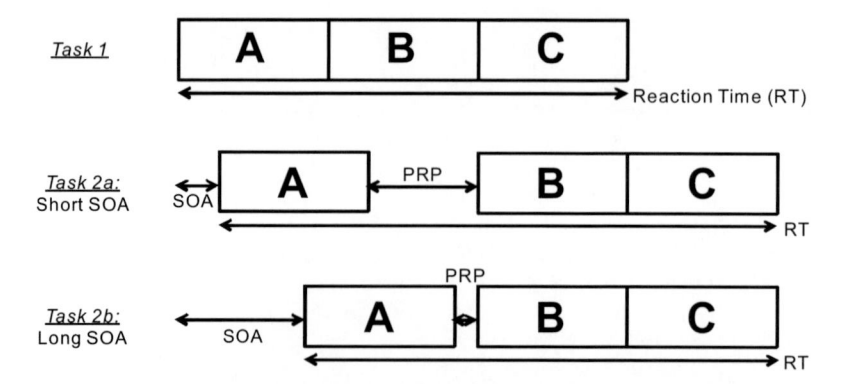

Figure 1. Illustration of the PRP effect. Task processing consists of three serial stages (A = perception, B = response selection, C = response execution). The response selection stages of two tasks cannot overlap hence the longer delay to reaction time when the second of two tasks is presented at short stimulus onset asynchrony (SOA) (Task 1 and 2a) compared to long SOA (Task 1 vs. 2b)

This interference effect is typically observed in simple laboratory studies for SOA up to 350-500 milliseconds (Pashler, 1994, Van Selst et al., 1999), and is resistant to variations in task modalities (Brebner, 1977, Pashler, 1990), task difficulties (Glass et al., 2000, Hein & Schubert, 2004, Karlin & Kestenbaum, 1968) and extent of prior task practice (Tombu & Jolicoeur, 2004). Outside of this range, there is little impact of a preceding task on subsequent task performance.

The need for vehicle drivers to multi-task during driving means that it is necessary to investigate this form of dual-task interference in the driving context. There has been a prior investigation of the existence of the PRP effect in the driving environment. Levy et al. (2006) found that brake reaction time slowed with decreasing interval to a preceding surrogate in-vehicle task (presented in the visual or auditory modality). The greatest delay to braking performance was observed when the distracter task was presented simultaneously with the brake lights of a lead vehicle. This study considers the impact of an in-vehicle task on performance of the driving task. Estimating the delay to braking performance caused by the PRP effect could offer potential methods for minimizing the slowing effects of in-vehicle tasks on brake reaction time.

Multiple Resource Theory

Dual-task interference effects can depend on the similarity of two concurrent tasks, in addition to their temporal proximity. It has been proposed that humans possess distinct processing resource channels in the brain, which are specialized to deal with particular types of information, and are each capacity-limited. The theory contends that dichotomous resource supplies exist at the perceptual, central, and response execution stages, meaning that tasks that do not share common processing modalities or codes provide less competition for each others' resources (Wickens, 1984, 2008). Multiple Resource Theory predicts that two visual stimulus tasks will interfere more and cause greater delays to the others' processing than one visual stimulus and one auditory stimulus task. Two differing tasks are less likely to produce a processing resource demand that exceeds the supply available from any single channel. This theory is relevant to the study of interference effects in the driving context due to the largely visual nature of the driving task, and the need to explore alternative presentation methods in the search to reduce the interference effects from in-vehicle tasks.

Current study

The literature identifies driver distraction from in-vehicle tasks as a key causal factor in the degradation of driving performance and the increase in crash risk, both currently and in the future. One approach to mitigating the effects of driver distraction is to ensure that in-vehicle tasks are not presented in a way that can impair performance of safety-critical aspects of the driving task. This study considers the possible methods to manage in-vehicle task presentation to minimize the chance of negative distraction-related effects. The impact of an in-vehicle task on the braking response is assessed, with the interval between the two tasks being varied systematically.

Method

Participants

Participants were recruited at the University of Leeds. 48 participants were tested (30 males, 18 females). Their mean age was 27.5 (SD = 8.2) and mean time since passing a driving test was 7 years 3 months (Min. = 6 months). All participants took part in a single 80-minute testing session and received £10 honorarium. Participants with difficulty detecting the visual, auditory and tactile stimuli selected for the experiment were not considered for further study.

Materials

The study was conducted on a desktop computer driving simulator. All elements of the simulation, including the vehicle dynamics model, the graphical subsystem and the presentation of the various stimuli were provided by a Dual-Core Toshiba laptop with an nVidia workstation-class graphics card. The simulator software consisted primarily of freely available OpenSceneGraph for the rendering process and programs developed by staff at the University of Leeds. The laptop was connected to an Acer 19" flat-panel display 1.0m in front of the driver. A real-time, fully textured and anti-aliased, 3-D graphical scene of the virtual world was displayed. The display was a single 1280x1024 channel with a horizontal field of view of 50° and a vertical field of view of 39°. The simulator was equipped with a Logitech G25 force-feedback steering wheel and spring-loaded foot pedals (accelerator, brake and clutch). There was no gear lever. The steering wheel provided force feedback to simulate the aligning torque of the wheel. Manual response paddles were located on the upper rear-side of the wheel. Vocal responses were recorded using an Olympus WS-321M Digital Voice Recorder attached to a Griffin Lapel Microphone. Vocal reaction times were manually measured using Praat; spectral analysis software.

Simulator environment

Participants drove on a single-carriageway, straight, rural road (maximum length = 1km). The road was centrally-divided with a dashed white line. In addition to the participant vehicle, there was one vehicle present in the driving scene (Figure 2). This vehicle was a black Mitsubishi Shogun, which drove with a fixed speed (40mph) and headway (1500ms) in front of the participant vehicle. Participants were required to operate the simulator using their right foot on the accelerator and brake pedal only. The accelerator pedal activated a controller system that maintained the speed of the participant vehicle. The participant had no control over vehicle speed, but was required to depress the accelerator pedal (>50%) to ensure that all braking responses involved a foot movement between the two pedals. The participant had full lateral control. Throughout the experiment, background vehicle engine noise was presented via the laptop speakers. A simulated vehicle dashboard was visible to the participants – including functional speedometer and tachometer.

Figure 2. Simulator screenshot showing the participants forward field of view. Speedometer and tachometer functioned realistically

Tasks

Braking task
Participants completed a simple, car-following task. The lead vehicle braked with fixed deceleration (-5 ms^{-2}) and duration (3 seconds) on random trials (57.1% of total trials). The braking event occurred after a variable foreperiod (Range = 8-23s). The braking event involved illumination of the two lead vehicle rear side-lights and a centre high-mounted stop light (CHMSL). The correct response involved immediate depression of the brake pedal to stop the vehicle. The braking task was selected for its precisely measurable onset and the high incidence, economic and human cost of rear-end collisions (McIntyre, 2008).

In-Vehicle task
Participants were presented with a two-choice, speeded response task, acting as a surrogate in-vehicle task. Participants were randomly allocated to one of six groups defined by the stimulus modality and response modality of the surrogate in-vehicle tasks (see Table 1).

Table 1. The stimulus modality and response modality combinations for the surrogate in-vehicle task used

Group	Stimulus Modality	Response Modality
1	Visual	Manual
2	Visual	Vocal
3	Auditory	Manual
4	Auditory	Vocal
5	Haptic	Manual
6	Haptic	Vocal

All tasks involved a discrimination decision followed by a simple, single-action response (International Organisation for Standardisation, 2002). The ease of discrimination was confirmed via pilot work and the necessary procedural checks demonstrated no difference in reaction time to the six types of in-vehicle task. Stimulus duration was short and response actions were distinct and common in the driving environment (see Table 2). Stimulus-response relationships were trained before the experimental phase of the study.

Table 2. Summary of the in-vehicle task stimulus parameters and the responses required

Stimulus Modality	Stimulus Presented
Visual	One of two colour rectangles (blue or yellow) presented centrally on the simulated dashboard for 400ms (128 x 235 pixels)
Auditory	One of two sawtooth wav files (300 or 900Hz) presented for 200ms at 75dB
Haptic	One of two steering wheel vibrations (0.8 or 0.4Nm amplitude) presented for 200ms with fixed period (100ms)
Response Modality	**Response Required**
Manual	Single press of a manual response paddle: left or right
Vocal	Single word vocalization: 'one' or 'two'

Design

Participants were asked to perform the braking and in-vehicle tasks under single-task and dual-task conditions (Table 3). In the dual-task condition, the in-vehicle task was presented before (or at the same time) as the braking event. On such trials, the SOA between the onsets of the two tasks was varied. A mixed ANOVA, three-factorial design (8x3x2) was used. The first factor was manipulated within-subjects and represented the eight levels of SOA that could be presented on dual-task trials (0, 50, 150, 250, 350, 450, 850 and 1000ms). This range was selected based on prior experimental work (Levy et al., 2006, Van Selst et al., 1999). The levels were spread unevenly within the range to ensure high data collection around the intervals likely to elucidate information about the duration of the PRP effect (300-500ms) (Van Selst et al., 1999, Pashler, 1990, Levy et al., 2006, Allen et al., 1998). The second and third factors were between-subjects factors relating to the stimulus modality and response modality of the in-vehicle task. These variables had three (visual, auditory and haptic) and two levels (manual and vocal) respectively.

Table 3. Summary of the four possible trial types, randomly selected on each trial

Trial Type	Description
Dual-task	Surrogate in-vehicle task followed by lead vehicle braking task – SOA counterbalanced
Single-task (braking)	Lead vehicle braking task only
Single-task (in-vehicle)	Surrogate in-vehicle task only
Catch task	No task to perform; included to ensure that a response was not required on all trials, which could foster artificially high driver vigilance.

Dependent variables

Data was collected from both the braking task and the in-vehicle task. Braking response data included the brake reaction time (measured from the onset of the lead vehicle braking event until brake pedal depression), maximum deceleration, minimum time-to-collision, and minimum distance headway. Analysis focused on brake reaction time due to possible confounding of the remaining measures by the instruction to brake as harshly as possible. Each trial was tagged with the trial type and the SOA used (dual-task trials only). All parameters were measured and recorded by the simulator system at a sampling rate of 60 Hz. In-vehicle task performance was assessed through the collection of task reaction time and response accuracy measures.

Procedure

Participants were permitted a short practice session to familiarize themselves with the simulator controls and its operation. The surrogate in-vehicle task was demonstrated and then repeated until ceiling level performance was reached (12 consecutive correct responses). Braking responses to the lead vehicle braking event were practiced both in isolation and in combination with the selected in-vehicle task. One participant was omitted from the study at this point due to difficulties controlling the simulator vehicle.

In the experimental phase, participants were presented with 112 trials. A four-identical block design was used to allow rest periods to reduce fatigue effects. Participants experienced 32 versions of each trial type (except catch trials). The dual-task trial was presented four times for each SOA level. The order of presentation of SOA levels was partially counterbalanced across participants to reduce potential order effects. Trial type and task onset were randomized on each trial (max. trial duration = 29.5s). One task or task combination was presented on each trial. Inter-trial interval was participant-controlled and was accompanied by a black screen on the simulator.

Instructions

Participants were requested to follow the lead vehicle and maintain their position in the left-hand lane. All tasks required an urgent response, and response performance was requested in the order that the tasks were presented. The braking event was described as the lead vehicle approaching a traffic jam, and collision avoidance was emphasized. Constraints were placed on participant behaviours to minimize confounding of manual response times by individual differences in movement speed (hold response paddles throughout) and to prevent lack of detection of haptic stimuli (maintain a loose grip on the steering wheel).

Results

Reaction time data for both tasks was collected for up to 3500ms after task presentation. Dual-task brake reaction time data was analyzed from correct in-vehicle task response trials only. Brake reaction times were accepted for further analysis if they exceeded 200 milliseconds. 93.8% of the data was included in the statistical analyses. All data were subjected to the Kolmogorov-Smirnov test for normally-distributed data and the Levene's Test of Equality of Error Variances. Brake reaction time data produced significant results in both cases and as such the analyses were performed on reciprocal-transformed data. A split-half analysis of possible trial exposure effects on brake reaction time showed no significant effect thus confirming that brake reaction time data from the experimental phase could be pooled across all trials.

Braking task

Consideration of the impact of SOA on brake reaction time data involved analysis of dual-task data only. Mean brake reaction time data was subjected to mixed ANOVA with SOA as a within-subjects variable and in-vehicle task stimulus modality and response modality as between-subject variables. When violations of sphericity were present in the data, the Greenhouse-Geisser correction was applied to the degrees of freedom used in the ANOVA. The effect of SOA on brake reaction time was highly significant, [$F(5.187, 217.857) = 51.239$, $p = .000$, $\mu^2 = .550$]. The plot of brake reaction time vs. SOA shows the typical PRP curve, with increasing brake reaction time as SOA decreases, for SOA within the 0-350ms range (see Figure 3). The longest brake reaction time (1096ms) was observed with coincident presentation of both tasks (0ms SOA). For SOA above 350ms, there is a plateau on the graph, with brake reaction time remaining relatively constant. Post-hoc Bonferroni-corrected pairwise comparisons support these trends, with a pattern of significant effects between brake reaction time for short-short SOA and short-long SOA comparisons, but not long-long SOA comparisons.

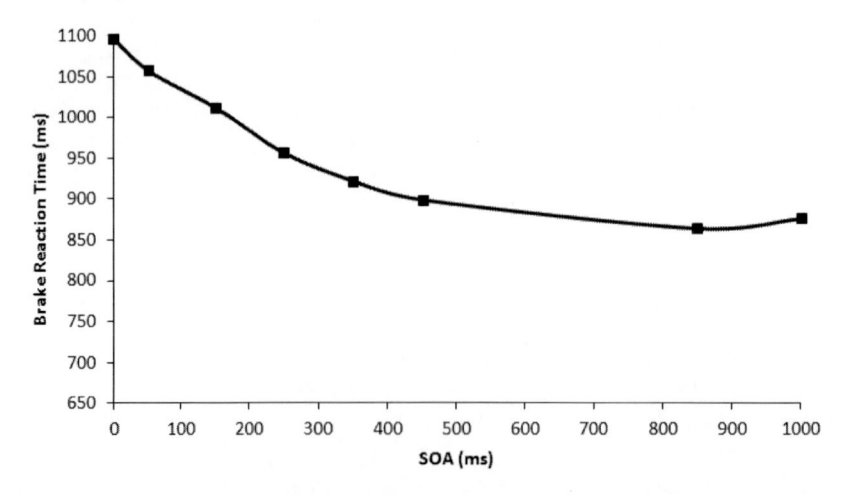

Figure 3. Plot of brake reaction time (dual-task trials only) vs. stimulus onset asynchrony (SOA)

There was a main effect of preceding in-vehicle task stimulus modality on brake reaction time, [$F(2,42) = 6.070$, $p = .005$]. Braking performance was significantly faster after an auditory [M=921ms, SE=350] or haptic in-vehicle task [M=903ms, SE=350] than after a visual task [M=1052ms, SE=350]. A main effect of preceding task response modality was also observed [F(1,42)=6.890, p=.012], with faster braking responses after a vocal response task [M=909ms, SE=280] than after a manual response task [M=1004ms, SE=280]. Neither the interaction of SOA with stimulus modality or response modality reached significance. In both cases, the significant main effect of modality was present across the entire SOA range.

Performing a similar mixed ANOVA with SOA as a within-subjects variable but with a single between-subjects factor of stimulus-response modality combination, yields a significant main effect of modality on brake reaction time, [F(5,42)=3.906, p=.317]. Post-hoc pairwise comparisons show that braking performance is faster following an auditory stimulus-vocal response or haptic stimulus-vocal response task compared to a visual stimulus-manual response task (see Figure 4). A two-way, between-subjects ANOVA using in-vehicle task reaction time data from single-task trials showed neither a significant main effect of stimulus modality or response modality on task reaction time. This finding allows conclusions about task modality effects on brake reaction time to be made without the risk of confounding by differences in the reaction time to each task.

Mean Brake Reaction Time (milliseconds)

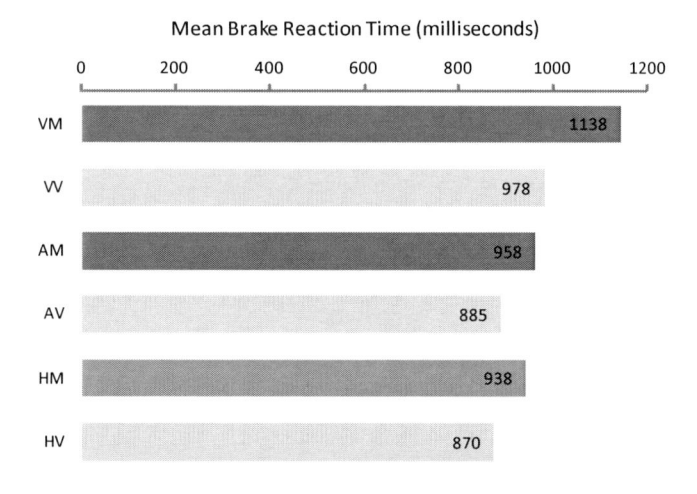

Task	BRT
VM	1138
VV	978
AM	958
AV	885
HM	938
HV	870

Figure 4. Bar chart shows mean brake reaction time (BRT) on dual-task trials, split by in-vehicle task type (VM = visual stimulus-manual response, VV = visual-vocal, AM = auditory-manual, AV = auditory-vocal, HM = haptic-manual, HV = haptic-vocal)

In-Vehicle task

An 8 (SOA) x 3 (stimulus modality) x 2 (response modality) mixed ANOVA was run on the in-vehicle task reaction time data from correct response trials only. A significant main effect of SOA on reaction time was observed, [F(7,294)=4.557, p=.000, μ^2 = .098] (see Figure 5).

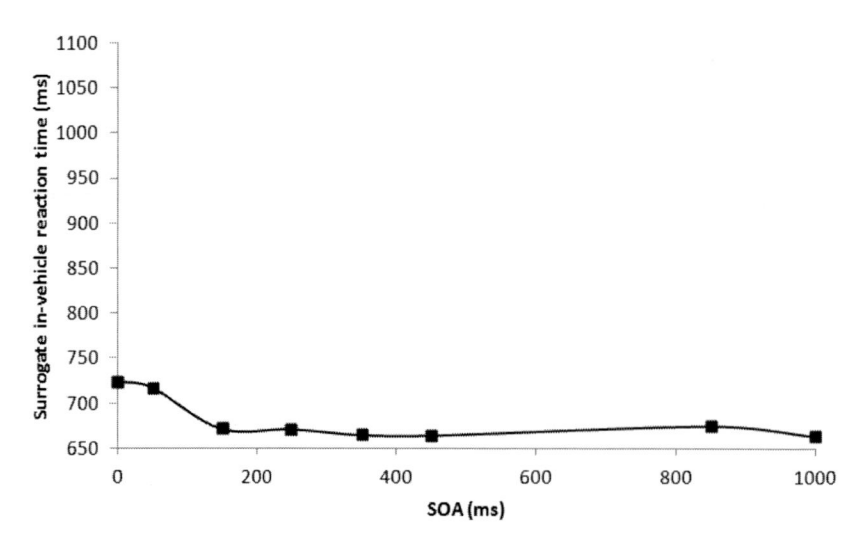

[Figure: Plot showing surrogate in-vehicle reaction time (ms) on y-axis ranging from 650 to 1100, against SOA (ms) on x-axis from 0 to 1000. Values start near 725ms at low SOA and drop to approximately 660-675ms from 200ms onwards.]

Figure 5. Plot of in-vehicle stimulus reaction time against SOA (dual-task data only)

Discussion

The PRP effect has been demonstrated in the driving environment. This study has shown that an in-vehicle task interferes with the performance of a subsequent braking event in a way that is dependent on the interval between the two tasks. For stimulus onset asynchronies (SOA) in the 0-350ms range, decreasing temporal spacing of the two tasks causes a slowing of the braking response. For SOAs outside of this range, the effect of task temporal separation on braking performance diminishes.

This study extends the work of Levy et al. (2006) by using post-hoc pairwise comparison analysis – rather than inspection of the brake RT graph – to identify the SOA at which an in-vehicle task ceases to impair a braking response. The outcome is similar to previous work in that the PRP effect seems to exist for SOA in the range 0-350ms. Furthermore, the delay to the braking response across this range is identical in this and the aforementioned study (174ms). To quantify this effect, for a vehicle travelling at 70mph, a 174ms increase in brake reaction time would equate to an increase in stopping distance of approximately 5.45 metres. It would seem reasonable to suggest that this effect could have a noticeable impact on driver safety, either in increasing the likelihood of a collision with the lead vehicle, or increasing the severity of a collision that is unavoidable. These figures lead systematically to the conclusion that there are potential safety benefits to be obtained through the prevention of concurrent presentation of an in-vehicle task stimulus with lead vehicle braking, with a gradual decrease in this advantage until an inter-task interval in excess of 350ms.

The modality of a preceding in-vehicle task affects dual-task brake reaction time and modulates the magnitude of the PRP effect (see Table 4). Unlike the prior study of the PRP in the driving domain (Levy et al., 2006), there are significant main effects of both in-vehicle task stimulus and response modality on subsequent braking performance. There was no interaction of either modality with SOA, however, the mean trends show that the delay caused by the PRP effect is longer for tasks that share greater stimulus or response modality overlap with the braking task. This fits with Multiple Resource Theory predictions (Wickens, 1984, 2008) showing increased dual-task interference between similar tasks. However, it should be noted that while the main effect of SOA was present within each in-vehicle task modality group, the pattern of pairwise comparisons did not show a strong PRP effect for either the auditory-vocal or haptic-vocal groups.

Table 4. The impact of preceding in-vehicle task modality on the brake reaction time delay caused by the PRP effect. The PRP effect delay is calculated by subtracting mean brake reaction time on 0ms SOA trials from the same variable on 350ms SOA trials.

In-Vehicle Task Modality	PRP Effect Delay (ms)	+ Stopping Distance at 70mph (m)
Visual-Manual	291	9.106
Visual-Vocal	223	6.978
Auditory-Manual	257	8.042
Auditory-Vocal	96	3.004
Haptic-Manual	163	5.101
Haptic-Vocal	95	2.973

A surprising result was found for braking performance on long SOA dual-task trials (450-1000ms), where braking responses were more rapid than single-task braking responses. This suggests a beneficial effect of an additional in-vehicle task on driver safety. It could be that the presence of an in-vehicle stimulus primes for faster responses to subsequent tasks. However, it is likely that this effect is an artifact of the methodology employed. The frequency of an in-vehicle task-braking task co-occurrence is much greater in this experimental study than would be expected in real-world driving, and as such, the provision of an in-vehicle task stimulus may have been associated with a subsequent braking event, thus priming for a rapid response. This learned association would not be possible in the more unpredictable on-road driving environment, and the authors would predict that this effect would therefore not be observed.

The main effect of SOA on in-vehicle task reaction time of dual-task trials is a surprising result. Prior studies of the PRP effect tend to show forward interference effects on the second task, but little impact of the second task on speed of performance of the first (Levy et al., 2006, Van Selst et al., 1999). The effect of SOA would suggest that the first task is not always winning the race to gain access to the limited-capacity central processing resources. However, post-hoc analysis revealed that the only significant pairwise comparisons involved the 0ms SOA condition. This trial type might be expected to produce slower in-vehicle task performance because the braking task is presented simultaneously, and thus may be perceived as the first task to arrive, subsequently delaying the 'second' in-vehicle task. This does not suggest a backwards interference effect on the in-vehicle task or the PRP effect for the first task presented in the dual-tasking scenario.

It is interesting to note that a comparison of brake reaction time on single-task and dual-task trials (no division by SOA level) produced a non-significant result. Brake response was as fast on dual-task trials as single-task trials, due to an improvement in braking speed at long SOAs relatively to single-task performance. The authors would suggest that this effect could be the result of a high frequency of in-vehicle task/braking task combinations (28.6% of all trials), allowing the in-vehicle task to be a relatively accurate predictor of a subsequent braking event (50% hit rate). This could produce pre-emptive brake responses; a response strategy that would not be possible in more realistic driving scenarios, where an in-vehicle task stimulus would not be as intrinsically linked with a following braking stimulus.

The application of these results is currently limited by the inability to predict the exact onset timing of lead vehicle braking events. However, a conservative application of these findings still offers potential improvements to driver safety. For example, braking events at certain road geometry features (motorway off-ramps, intersections, traffic control signals) or in heavy congestion could be approximately predicted using forward sensing technology and GPS data. Furthermore, a 1998 study (Koter, 1998) showed that harsh vehicle braking manoeuvres tend to be preceded by a specific accelerator release profile. The communication of this information between leading and following vehicles could allow the management of in-vehicle task presentation in following vehicles within the SOA range identified in this study.

Further work should be conducted to determine whether the PRP effect is observed across a range of braking scenarios, with more accurate in-vehicle task simulations. Real in-vehicle task stimuli would be beneficial. Also, this study considers brake reaction time to repeated, highly-expected braking events. Expectancy is a variable that has a significant impact on braking performance (Engström et al., 2010), and therefore recommendations about in-vehicle task presentation would be more reliable if considered with more realistic levels of braking task expectancy. Presentation guidelines may also need to be tailored to driver age, due to the generalized increase in brake reaction with age (Glass et al., 2000, Hein & Schubert, 2004).

References

Allen, P.A., Smith, A.F., Vires-Collins, H., & Sperry, S. (1998). The Psychological Refractory Period: Evidence for Age Differences in Attentional Time-Sharing. *Psychology & Aging, 13*, 218-229.

Alm, H. & Nilsson, L. (1995). The effects of a mobile telephone task on driver behaviour in a car following situation. *Accident Analysis & Prevention, 27*, 707-715.

Brebner, J. (1977). The Search for Exceptions to the Psychological Refractory Period. In S. Dornic (Ed.) *Attention & Performance VI, Proceedings of the Sixth International Symposium on Attention and Performance.* Hillsdale, New Jersey: Lawrence Erlbaum Associates, Publishers.

Brown, I.D. (1965). Effect of a car radio on driving in traffic. *Ergonomics, 8*, 475 - 479.

Damiani, S., Deregibus, E., & Andreone, L. (2009). Driver-vehicle interfaces and interaction: where are they going? *European Transport Research Review, 1*, 87-96.

Engström, J., Aust, M.L., & Viström, M. (2010). Effects of Working Memory Load and Repeated Scenario Exposure on Emergency Braking Performance. *Human Factors, 52*, 551-559.

Glass, J.M., Schumacher, E.H., Lauber, E.J., Zurbriggen, E.L., Gmeindl, L., Kieras, D.E., & Meyer, D.E. (2000). Aging and the Psychological Refractory Period: Task-Coordination Strategies in Young and Old Adults. *Psychology & Aging, 15*, 571-595.

Hein, G. & Schubert, T. (2004). Aging and Input Processing in Dual-Task Situations. *Psychology & Aging, 19*, 416-432.

Horrey, W.J. & Lesch, M.F. (2009). Driver-initiated distractions: Examining strategic adaptation for in-vehicle task initiation. *Accident Analysis & Prevention, 41*, 115-122.

Horrey, W.J., Lesch, M.F., & Garabet, A. (2008). Assessing the awareness of performance decrements in distracted drivers. *Accident Analysis & Prevention, 40*, 675-682.

International Organisation for Standardisation (2002). Road vehicles – Ergonomic aspects of transport information and control systems – Dialogue management principles and compliance procedures (ISO 15005). Geneva, Switzerland.

Karlin, L. & Kestenbaum, R. (1968). Effects of number of alternatives on the psychological refractory period. *Quarterly Journal of Experimental Psychology, 20*, 167 - 178.

Klauer, S.G., Dingus, T.A., Neale, V.L., Sudweeks, J.D., & Ramsey, D. J. (2006). *The impact of driver inattention on near-crash/crash risk: An analysis using the 100-Car Naturalistic Driving Study data.* Washington DC: National Highway Traffic Safety Administration.

Koter, R. (1998). Advanced Indication of Braking: A practical Safety Measure for Improvement of Decision-Reaction Time for Avoidance of Rear-End Collisions. In R.K. Jurgen (Ed.) *Object Detection, Collision Warning and Avoidance Systems.* PT-70 ed. Warrendale, Pennsylvania, USA: Society of Automotive Engineers, Inc.

Lansdown, T.C., Brook-Carter, N., & Kersloot, T. (2004). Distraction from multiple in-vehicle secondary tasks: vehicle performance and mental workload implications. *Ergonomics, 47*, 91-104.

Lerner, N. D. (2005). *Driver Strategies for Engaging in Distracting Tasks Using In-Vehicle Technologies: Final Report.* Report under Contract DTNH22-99-D-07005. Washington, DC: National Highway Traffic Safety Administration.

Levy, J., Pashler, H., & Boer, E. (2006). Central Interference in Driving: Is There Any Stopping the Psychological Refractory Period? *Psychological Science, 17*, 228-235.

Mcintyre, S.E. (2008). Capturing attention to brake lamps. *Accident Analysis & Prevention, 40*, 691-696.

Mcknight, A.J. & Mcknight, A.S. (1993). The effect of cellular phone use upon driver attention. *Accident Analysis & Prevention, 25*, 259-265.

Neale, V.L., Dingus, T.A., Klauer, S.G., Sudweeks, J., & Goodman, M.J. (2005). *An Overview of the 100-Car Naturalistic Study and Findings*. In International technical conference on the enhanced safety of vehicles (CD-ROM). Washington DC: National Highway Traffic Safety Administration.

Pashler, H. (1984). Processing stages in overlapping tasks: Evidence for a central bottleneck. *Journal of Experimental Psychology: Human Perception & Performance, 10*, 358-377.

Pashler, H. (1990). Do Response Modality Effects Support Multiprocessor Models of Divided Attention? *Journal of Experimental Psychology: Human Perception and Performance, 16*, 826-842.

Pashler, H. (1994). Dual-Task Interference in Simple Tasks: Data and Theory. *Psychological Bulletin, 116*, 220-244.

Regan, M.A., Hallett, C., & Gordon, C.P. (2011). Driver distraction and driver inattention: Definition, relationship and taxonomy. *Accident Analysis and Prevention, 43*, 1771-1781.

Smith, M.C. (1967). Theories of the Psychological Refractory Period. *Psychological Bulletin, 67*, 202-213.

Stutts, J.C., Reinfurt, D.W., Staplin, L., & Rodgman, E.A. (2001). *The Role of Driver Distraction in Traffic Crashes*. Washington DC: AAA Foundation for Traffic Safety.

Telford, C.W. (1931). The refractory phase of voluntary and associative responses. *Journal of Experimental Psychology: General, 14*, 1-36.

Tombu, M. & Jolicoeur, P. (2004). Virtually No Evidence for Virtually Perfect Time-Sharing. *Journal of Experimental Psychology: Human Perception & Performance, 30*, 795-810.

Van Selst, M., Ruthruff, E., & Johnston, J.C. (1999). Can Practice Eliminate the Psychological Refractory Period Effect? *Journal of Experimental Psychology: Human Perception & Performance, 25*, 1268-1283.

Welford, A.T. (1952). The "psychological refractory period" and the timing of high-speed performance - A review and a theory. *British Journal of Psychology, 43*, 2-19.

Wickens, C.D. (1984). Processing Resources in Attention. In R. Parasuraman and D.R. Davies (Eds.) *Varieties of Attention*. London: Academic Press.

Wickens, C.D. (2008). Multiple Resources and Mental Workload. *Human Factors: The Journal of the Human Factors and Ergonomics Society, 50*, 449-455.

Young, K. & Regan, M. (2007). Driver distraction: A review of the literature. In I.J. Faulks, M. Regan, M. Stevenson, J. Brown, A. Porter, and J.D. Irwin (Eds.) *Distracted driving*. Sydney, NSW: Australasian College of Road Safety.

Vulnerable Road Users

In D. de Waard, N. Merat, A.H. Jamson, Y. Barnard, and O.M.J. Carsten (Eds.) (2012). *Human Factors of Systems and Technology* (pp. 263). Maastricht, the Netherlands: Shaker Publishing.

Drivers' visual behaviour at cycle crossings

Carmen Kettwich[1] & Carina Fors[2]
[1]Karlsruhe Institute of Technology (KIT), Germany
[2]Swedish National Road and Transport Research Institute (VTI), Sweden

Abstract

Two field studies on drivers' ability to detect cyclists and cycle crossings have been carried out. In the first study, the night-time visibility distances of cyclists and of cycle crossings along straight road stretches were investigated. It was found that cyclists were detected at a significantly longer distance than cycle crossing markings. In wet weather, the visibility distance of cycle crossing markings was significantly shorter than in dry weather. No such effect was seen for the visibility distance of cyclists. In the second study, drivers' gaze behaviour at cycle crossings located immediately after right turns was investigated. A cyclist approached the cycle crossing from the same direction as the participant. The visibility of cycle crossings during night-time was rated lower compared to daytime. The cyclist was detected by all participants. Gaze parameters such as dwell time, maximum duration length was higher and the number of glances was lower at night. No statistically significant differences in the gaze parameters were found between the daytime and night-time condition.

Introduction

More than 90 % of the information required for driving is obtained through the visual system (Fastenmeier, 1994). At night, visibility can be severely degraded. Nevertheless most car drivers are not aware of their visual impairment at night (Owens, 1999). Thus risky situations may arise, for example when crossing pedestrians or cyclists are not detected in sufficient time.

Visual processes can be divided into central and peripheral vision. Central vision is mainly used for target detection and identification, whereas the peripheral vision is associated with lane keeping and lateral control (Mourant, 1972; Summala, 1996). Some main functions of target detection, like the perception of contrast, distance and depth, are severely restricted at night, whereas peripheral vision is less affected (Leibowitz, 1977). Amongst other things, gaze behaviour depends on velocity (Weise, 1997), traffic volume, road geometry and surface as well as time of day (Diem, 2004; Gut, 2011).

In urban areas, street lighting is often present, which partly compensates for the absent daylight. But even on roads with street lighting, visual conditions can be poor, either because of insufficient street lighting (Lundkvist & Nygårdhs, 2007) or because of other light sources that cause glare or make the environment visually

In D. de Waard, N. Merat, A.H. Jamson, Y. Barnard, and O.M.J. Carsten (Eds.) (2012). *Human Factors of Systems and Technology* (pp. 265 - 276). Maastricht, the Netherlands: Shaker Publishing.

cluttered (Murray et al., 1998; Sayer & Mefford, 2004). Furthermore, the use of retroreflective tags and/or lighting among vulnerable road users is rather low (Thulin & Kronberg, 1998; Lindahl & Stenbäck, 1999), which reduces the chances for drivers to detect them.

Cyclists tend to have a higher risk of having an accident during the hours of darkness than in daylight. In Swedish urban areas the estimated cyclist accident risk is increased by 42% in darkness compared to in daylight (Johansson et al., 2009). In a previous focus group study, motorists as well as cyclists thought that motorists often have problems detecting cyclists at night (Fors & Nygårdhs, 2010). Poor visibility may have serious consequences when motorists and cyclists share the same space, such as at cycle crossings. In order to investigate visibility at cycle crossings, two field studies have been conducted, which are presented below. The objectives of the two studies were to:

- Investigate the night-time visibility distance of cycle crossings and of cyclists at cycle crossings, in dry and in wet weather
- Investigate driver gaze behaviour at cycle crossings located after right turns, in daylight and at night

The studies have been reported in the VTI report series (Nygårdhs et al., 2010; Kettwich & Fors, 2011) and in a previous paper (Nygårdhs & Fors, 2010).

Cycle crossings in Sweden

The Swedish Transport Administration states that a cycle crossing can be marked on a place meant to be used by cyclists to cross a road (Trafikverket, 2004). The cycle crossing should be designed with white road markings in the form of squares with each side 0.5 m long. Cycle crossings are often combined with pedestrian crossings ("zebra crossings").

When a cycle path crosses a straight stretch of the road, cyclists should give way to vehicles on the road. However, if the cycle crossing is located immediately after a turn, drivers of motor vehicles should give *cyclists an opportunity to pass*, according to Swedish traffic rules.

Study 1

Method

Twelve drivers (six females) between 26–57 years of age participated in the study. They all had a valid driving licence and self-reported normal or corrected vision. A route of approximately 10 km was used as a test route. The test route was located in an urban area and included nine cycle crossings and two separate starting positions. The posted speed limit was 50 km/h along the entire route. All nine cycle crossings were combined with pedestrian crossings and located on straight road stretches. Cycle crossings at roundabouts and cycle crossings immediately after turns were not included. The experiment consisted of two parts. In the first part, the visibility distance of cyclist dummies standing at the cycle crossings waiting to cross the road was investigated. In the second part, the visibility distance of the cycle crossing markings was determined. All participants participated in both parts. For half of the participants, the road surface was dry, while for the other half, the road surface was wet. The wet condition was obtained by conducting the experiment immediately following a rainfall.

The participants sat in the passenger seats in an instrumented car. There were two participants in the rear seat and one in the front seat. Those in the rear seat were instructed to lean their heads forward in order to see the road. The car was driven along the test route at a constant speed by a test leader. Four noiseless push-buttons were linked to a computer in the boot of the car. The computer registered the exact time for each push-button being pushed down. The driver (i.e. the test leader) pushed his button each time a cycle crossing was passed. The participants pushed their buttons when they detected a cyclist dummy (part 1) or a cycle crossing (part 2). Vehicle speed was continuously logged and thereby the distance travelled between the participants' and the driver's buttons being pushed down could be computed.

Figure 1. Cyclist dummy at a cycle crossing

Before the first part of the experiment started, the participants were given written and oral instructions about the task and they were also shown a photo of the cyclist dummy. They were instructed to push the push-button only when they were certain that there was a cyclist dummy. When the first part was complete, the cyclist dummies were removed and the participants were instructed about the second part. They were shown a picture of cycle crossing markings and were told to push the button when they were certain that they saw a cycle crossing – correctly marked with white squares – on the road on which they were travelling. They were also informed that all of the cycle crossings along the test route were combined with pedestrian crossing markings but that not all of the pedestrian crossings were combined with cycle crossings.

The cyclist dummies in the first part of the experiment were placed at exactly the same cycle crossings used in the second part, but this was not known by the participants. To minimize the expectations, the two parts began at different starting positions, i.e. the order of the crossings was not the same in part 1 as in part 2. The cyclist dummies consisted of a real cycle and dummy of an outdoor-dressed person, painted in matt grey. The cycle had no lighting or reflection tags. The cyclist dummies were always placed at a cycle crossing and at the right hand side of the road, standing as if they were about to cross the road, Figure 1.

Visibility distance was analysed using a split plot design ANOVA of group (wet and dry road surface) x target (cycle crossing and cyclist dummy) x crossing (nine crossings). If the assumption of sphericity was violated, the degrees of freedom were corrected using the Greenhouse-Geisser correction. A p-value of <0.05 was considered significant.

Results

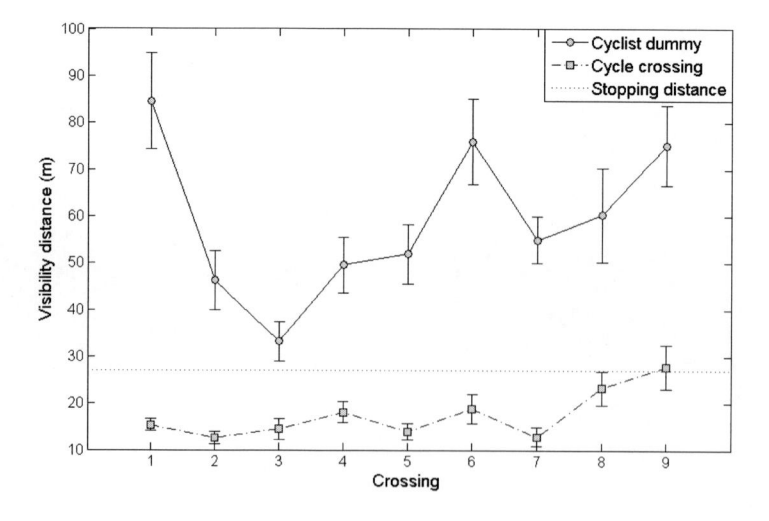

Figure 2. Visibility distance of cycle crossings and cyclist dummies as well as the stopping distance when driving 50 km/h assuming a reaction time of 1 s followed by full brake

The statistical analysis showed main effects of target, $F(1,10)=59.12$, $p<0.0001$ and crossing, $F(8,80)=12.05$, $p<0.0001$, and there was also a significant interaction between target and crossing, $F(8,80)=8.76$, $p<0.0001$. Figure 2 shows the visibility distances for cyclist dummies and cycle crossing markings at each cycle crossing. The average visibility distance of cyclist dummies was 59.1±2.9 m (mean ± standard error) while the visibility distance of cycle crossing markings was 17.5±1.0 m. This can be compared to the approximate stopping distance which is 26–28 m when driving in 50 km/h and assuming a reaction time of 1 s followed by full brake (shown in Figure 2).

There was no interaction effect between group (road surface condition) and target. However, separate analyses of road surface condition showed that cycle crossing markings were detected at a significantly longer distance ($p<0.05$) on dry roads (20.9±1.5) than on wet roads (14.1±1.2). No such effect was observed for cyclist dummies ($p=0.797$). Figure 3 and 4 show the mean visibility distance of cycle crossing markings and of cyclist dummies, respectively.

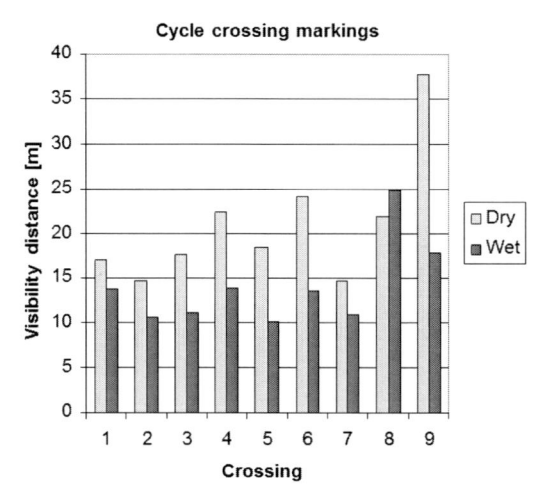

Figure 3. Mean visibility distance of cycle crossing markings on dry and on wet road surface

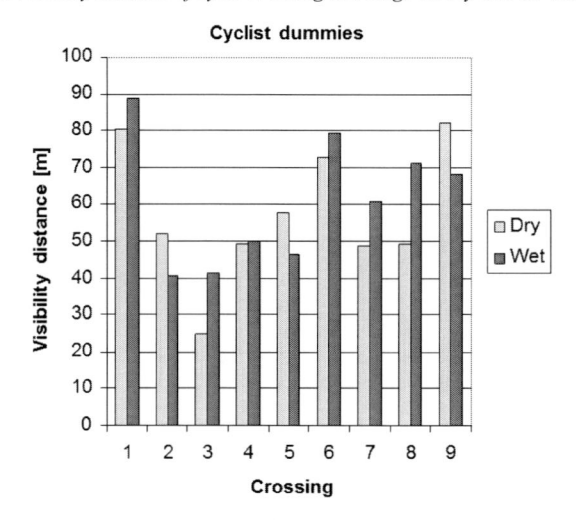

Figure 4. Mean visibility distance of cyclist dummies on dry and on wet road surface

Study 2

Method

A Volvo V70 of the year 2009 served as test vehicle. The car had a manual gear box. Speed, pedal positions and activation of direction indicator were obtained from the CAN network and recorded by a VBOX data logger (Racelogic, UK). GPS position and GPS based speed was provided through the VBOX itself. In order to study the gaze behaviour a head-mounted eye tracking system called SMI iView X (SensoMotoric Instrument, Germany) was used. The eye tracking systems consists of two main parts: a headset including tracking and scene camera as well as an eye tracking computer.

The test track was 19 km long. It comprised mainly urban roads. The average time for one test drive amounted to 40 minutes. There were five cycle crossings along the route. The first four were not marked, whereas the last one was a correctly labelled cycle crossing. All the cycle crossings were located immediately after a right turn. The speed limit was either 50 km/h (cycle crossings 1, 2, 3, 5) or 70 km/h (cycle crossing 4).

Twenty-one participants between 25 and 46 years of age took part in the study. They had a valid driver licence for at least six years, drove at least once a week and were not required to wear glasses while driving. Only non-professional drivers were included in the study. Each participant had to drive two times along the pre-defined route: once in daylight and once at night. The order of the sessions was counterbalanced. In the second driving session a cyclist approached the last crossing from the same direction as the participant. This means that half of all participants experienced the cyclist in the daylight condition and half of them at night. In order to avoid rush-hours all daytime sessions were conducted in the late morning or the early afternoon. Night-time sessions took part between 8:30 p.m. and midnight. Test drives were only conducted on dry road.

Participants were told to drive the test route with help of a GPS navigator without knowing the actual purpose of the study. They were urged to drive as they are used to. After the written and oral introduction of the driving task, the eye-tracking system was mounted and calibrated. During the first ten minutes of the test drive the participants had time to get used to the test vehicle, the GPS instructions as well as the head-mounted eye tracking system. Thereafter the actual investigation started. During the driving session the test leader did not talk to the participant, unless GPS instructions were unclear or the driver took the wrong way. After the first test drive, participants filled out a questionnaire with questions comprising their background (like age, gender, driving licence years), familiarity of the test route, experience following the GPS instructions and experience of the eye tracking system.

The procedure was repeated at the second occasion, with the exception that a cyclist approached the last cycle crossing. The cyclist was dressed in dark clothes and at night-time sessions the bicycle lighting was activated. The test leader and the cyclist communicated without saying anything with a mobile phone. The cyclist was positioned about 60 m from the crossing and started cycling when the test vehicle was about 200 m from the crossing. Thus the cyclist and the car reached the crossing approximately at the same time. The cyclist turned right at the crossing in order to avoid a collision. After the second driving session participants were asked to fill out another questionnaire, including questions about the visibility of cycle and pedestrian crossings and about traffic rules concerning vulnerable road users.

Gaze and driving behaviour were analysed for two zones. Zone 1 started approximately 125 m before the crossing and was about 115-120 m long. Zone 2 started where zone 1 ended, that means 5-10 m before the crossing and ended just after the right turn. In order to find the same starting point of zone 1 and 2 well-defined objects were needed. Thus the length of zone 1 and 2 varied somewhat between the investigated crossings.

Results

The results from the questionnaires showed that 57 % of all participants did not know that cyclists have to give way to car drivers at uncontrolled cycle crossings on straight roads. Figure 5 shows the ratings of pedestrians and cycle crossings visibility in general. The visibility during night-time is rated lower compared to daytime. Furthermore visibility of cycle crossings is rated lower than the visibility of pedestrian crossings.

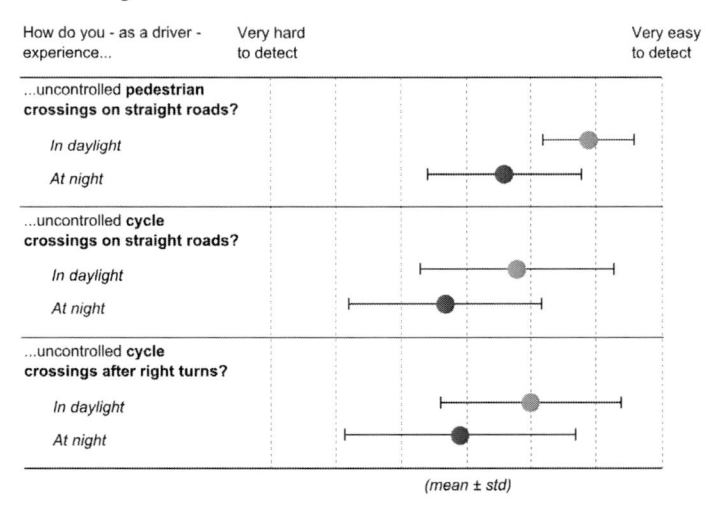

Figure 5. Ratings of the visibility of pedestrian and cycle crossings in general

More than half of all participants thought that the visibility of uncontrolled pedestrian and cycle crossings should be improved, particularly the visibility at cycle crossings. Figure 6 shows the participants' views on treatments for improved visibility at cycle crossings after right turns. Treatments getting the highest ratings were intense street lighting, reflecting material on sign posts, road sign with lights that flash when a cyclist is about to cross and pavement markers with reflective material.

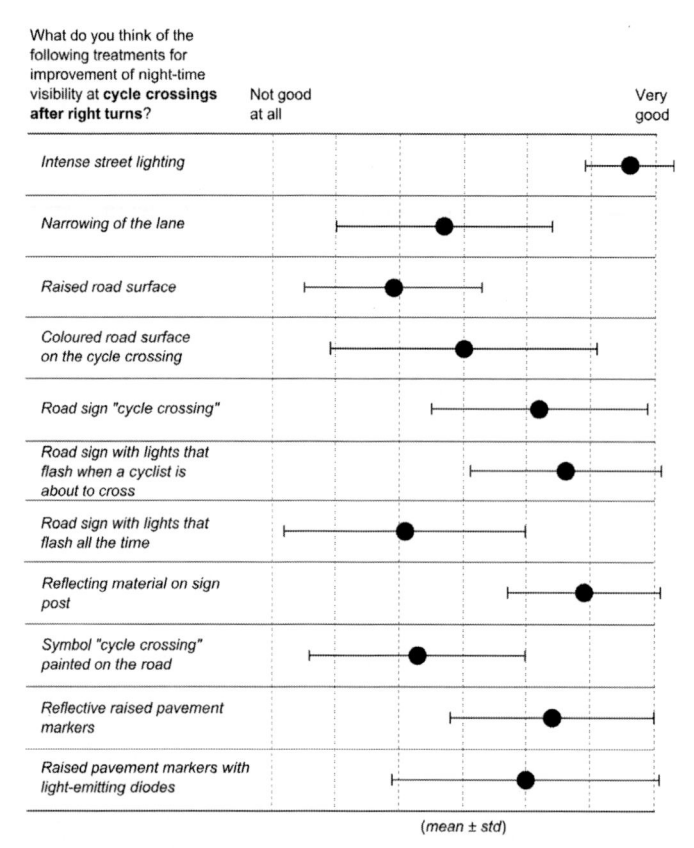

Figure 6. Participants' views on treatments for improved visibility at cycle crossings after right turns

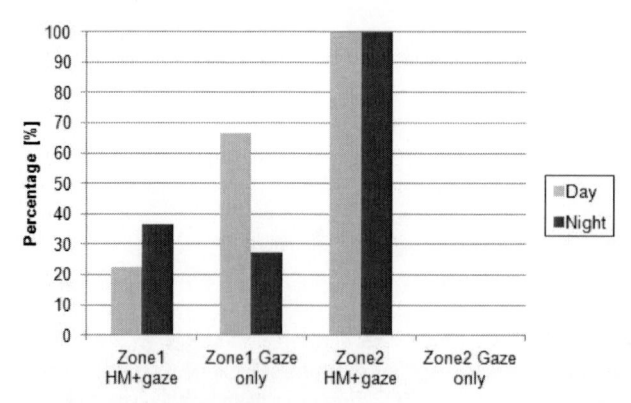

Figure 7. Percentage of participants who looked at the cyclist in zone 1 and zone 2 (HM+gaze: head movement and gaze towards the cycle path; gaze only: only gaze towards the cycle path)

drivers' visual behaviour at cycle crossings 273

Gaze behaviour was analysed for all cycle crossings. Results of cycle crossing 1-4, where no cyclist was present, are presented in Kettwich & Fors (2011). The results presented here only refer to cycle crossing 5. Nine participants had a cyclist in the daylight condition and eleven in the night-time condition next to crossing 5.

The percentage of all participants who gazed at the cyclist depicts figure 7. During daytime eight out of nine test persons detected the cyclist in zone 1. Whereas seven out of eleven participants gazed at the cyclist in zone 1 at night, all participants looked at the cyclist in zone 2.

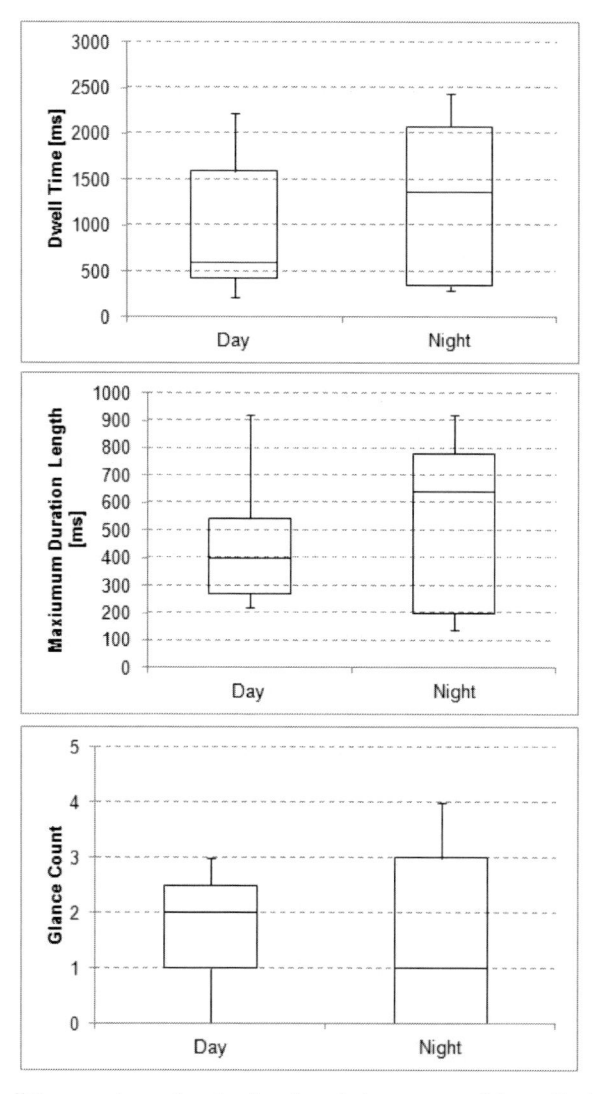

Figure 8. Dwell time, maximum duration length and glance count of the cyclist in zone 1

Regardless whether a cyclist was present or not, all participants looked towards the cycle path either in zone 1 or in zone 2 or in both zones at night as well as under the daylight condition. In case of a cyclist, dwell time, maximum duration length and glance count were analysed in zone 1 (figure 8). At night the dwell times and maximum duration lengths are higher and the average number of glances is lower compared to the daylight condition. The standard deviation of the above mentioned gaze parameters are higher during the night-time condition. No statistically significant differences were found between the daytime and night-time condition.

Discussion

Two studies on visibility and gaze behaviour at cycle crossings have been conducted. The first study showed that cyclist dummies waiting to cross the road at cycle crossings were detected at a significantly longer distance than the cycle crossing itself at night-time. The results imply that the cycle crossing markings do not help drivers to become observant of the cycle crossing until the vehicle is very close to the crossing, where the driver might not be able to slow down or stop if a cyclist suddenly appears.

A wet road surface resulted in significantly shorter visibility distances of cycle crossing markings than a dry road surface. No such effect was found for the visibility distance of cyclist dummies. A reasonable explanation is that since the cyclist dummies were located on the side of the road, the background – which is strongly related to visibility – was most often made up by lawn, buildings etc. which are not much affected by wet weather. A wet road surface, on the other hand, often provides worse visibility conditions than a dry road (at night), because of specular reflections in the road surface.

Both visibility studies were carried out on a public road with real cycle crossings. There are, however, some shortcomings with the method used in the first study. First, the participants could pay their full attention to the visibility task, since they were passengers and not drivers, and they were also told to look for either cyclist dummies or cycle crossing markings. This is not a very realistic situation and hence, visibility distances can be expected to be shorter in reality. Second, the cyclist dummies were standing still. In the real situation they are expected to be moving – sometimes pretty fast – which changes the way they are perceived.

In the second study, another method was used, where the participants themselves were driving along a public road without being told to look for certain objects. The drivers' natural gaze behaviour was measured at cycle crossings located after right turns. In the present paper, only results from the last cycle crossing, where a cyclist was approaching, is presented. There were no significant differences in gaze behaviour between the daytime and night-time conditions, however, a larger percentage of the drivers looked towards the cyclist in daylight than at night, in zone 1.

The visibility at the cycle crossing with the cyclist is fairly good. There are no obscuring trees or shrubberies, or any conspicuous buildings or signs around the crossing. The illuminance of the cycle path is relatively high compared to other cycle crossings. The fact that all participants detected the cyclist and that there were no significant differences in gaze behaviour between the day and the night group indicates that it is possible to obtain good visual conditions also at night. But since only one cycle crossing with a cyclist was included in the study, no general conclusions about gaze behaviour and visibility can be drawn. The cyclist event, which was expected to be a bit tricky to carry out, was however successful and therefore, it should be possible to repeat the study with more cyclists.

In the first study, the visibility of cycle crossings was found to be rather poor. This is in agreement with the answers from the questionnaire in the second study. However, improved visibility does not necessarily imply that the safety will increase. According to Swedish traffic regulations, cyclists should give way to motor vehicles at cycle crossings on straight stretches of road, and as long as these regulations apply, improved visibility of the cycle crossings may cause confusion and hesitation about who should give way. The traffic regulations for cyclists have however been under discussion for some years. At the time of the first study, the Swedish government had proposed new rules, which suggested that motorists should give way to cyclists at cycle crossings. If the proposal had passed, (which it did not since it was withdrawn), it would probably have been necessary to improve the visibility of cycle crossings in order to make it possible for drivers to obey the rules. Changed rules in combination with improved visibility may seem beneficial from a safety point of view, but there is also a risk that such treatments will lull cyclists into a false sense of security which, in turn, may lead to an unfavourable behaviour. For example, it has been shown that blue pavement markings on cycle lanes resulted in fewer cyclists turning their heads to scan for traffic or using hand signals (Hunter et al., 2000) and that a painted cycle crossing resulted in somewhat higher cyclist speeds (Räsänen et al., 1998). Future studies on the design and visibility of cycle crossings should thus include investigations of the behaviour of drivers as well as of cyclists.

References

Diem, C. (2004). *Blickverhalten von Kraftfahrern im dynamischen Straßenverkehr [Gaze behaviour of car drivers in dynamic road traffic]*. Darmstadt, Germany: Herbert Utz Verlag.

Fastenmeier, W. (1994). *Verkehrstechnische und verhaltensbezogene Merkmale von Fahrstrecken – Entwicklung und Erprobung einer Typologie von Straßenverkehrssituationen [Traffic and behavioural characteristics of traffic routes – development and testing of a typology of traffic situations]*. PhD thesis, München Technische Universität, München, Germany.

Fors, C. & Nygårdhs, S. (2010). *Trafikanters upplevda behov och problem i mörkertrafik i tätort – En fokusgruppsstudie med cyklister, äldre bilförare och äldre fotgängare [Road users' experienced problems and needs in night-time traffic in urban areas]*. (Report VTI notat 5-2010). Linköping, Sweden: VTI.

Gut, C. (2011). *Untersuchung des Blickverhaltens von Kraftfahrzeugführern in Kurven bei Nacht [Gaze behaviour of car drivers in curves during night-time driving]*. Thesis, Karlsruher Institute of Technology, Karlsruhe, Germany.

Hunter, W.W., Harkey, D.L., Stewart, J.R. & Birk, M.L. (2000). *Evaluation of blue bike-lane treatment in Portland, Oregon*. Transportation Research Record, 1705, 107-115.

Johansson, Ö., Wanvik, P.O., & Elvik, R. (2009). A new method for assessing the risk of accident associated with darkness. *Accident Analysis and Prevention, 41*, 809-815.

Kettwich, C. & Fors, C. (2011). *Driver gaze behaviour at cycle crossings in daylight and at night. (Report VTI report 733a)*. Linköping, Sweden: VTI.

Leibowitz, H.W. &Owens,D.A. (1977). Night-time driving accidents and selective visual degradation. *Science, 197*, 422-423.

Lindahl, E. & Stenbäck, I. (1999*). Provundersökning av cyklisters synbarhet i tätort oktober – november 1998 [Investigation of cyclists' visibility in urban areas between October and November in 1998]*. Borlänge, Sweden: Trafikverket (The Swedish Transport Administration).

Lundkvist, S.-O. & Nygårdhs, S. (2007). *Upptäckbarhet av fotgängare i mörker vid övergångsställen [Night-time visibility of pedestrians at zebra crossings]. (Report VTI notat 5-2007)*. Linköping, Sweden: VTI.

Mourant, R.R. & Rockwell,T.H. (1972). Strategies of visual search by novice and experienced drivers.*Human Factors,14*, 325-335.

Murray, I.J., Plainis, S., Chauhan, K., & Charman, W.N. (1998). Road traffic accidents: The impact of lighting. *The Lighting Journal, 63*, 42-46.

Nygårdhs, S. & Fors, C. (2010). Field test on visibility at cycle crossings at night. *European Transport Research Review, 2*, 139-145.

Nygårdhs, S., Fors, C., Eriksson, L., & Nilsson, L. (2010). *Field test on visibility at cycle crossings at night. (Report VTI report 691a)*. Linköping, Sweden: VTI.

Owens, D.A. &Tyrrell, R.A. (1999). Effects of luminance, blur, and age on nighttime visual guidance: A test of the selective degradation hypothesis. *Journal of Experimental Psychology: Applied,5*, 115-128.

Räsänen, M., Summala, H., & Pasanen, E. (1998). The safety effect of sight obstacles and roadmarkings at bicycle crossings. *Traffic Engineering and Control, 39*, 98-102.

Sayer, J.R. & Mefford, M.L. (2004*). The roles of retroreflective arm treatments and arm motion in nighttime pedestrian conspicuity. (Report 2004-21)*. Ann Arbor, MI, USA: The University of Michigan Transportation Research Institute.

Summala, H., Nieminen,T., & Punto, M. (1996). Maintaining lane position with peripheral vision during in-vehicle tasks. *Human Factors, 38*, 442-451.

Thulin, H. & Kronberg, H. (1998*). Gåenderesor och cykelresor i olika trafikmiljöer [Travelling of pedestrians and cyclists in different traffic environments]. (Report VTI notat 47-1998)*. Linköping, Sweden: The Swedish National Road and Transport Research Institute.

Trafikverket (2004). *Vägar och gators utformning [Road and street design]*. Borlänge, Sweden: Trafikverket (The Swedish Transport Administration).

Weise, G. & Durth,W. (1997). *Straßenbau, -planung und Entwurf [Road construction, -planning and design]*. Berlin, Germany, Verlag Bauwesen.

Towards understanding hazard perception abilities among child-pedestrians

Anat Meir, Tal Oron-Gilad, Avinoam Borowsky, & Yisrael Parmet
Ben-Gurion University of the Negev
Israel

Abstract

The present study aimed to address child and adult pedestrians' perception of hazards through a traffic-scene categorization task. Twenty young-children (6-8 years-old), twenty-two older-children (9-12 years-old) and twenty-one adults (24-28 years-old) were requested to observe 12 traffic-scene still photos taken from a pedestrian's perspective and to categorize them according to similarities in their hazardousness. Results have shown that experienced adult pedestrians tended to be more aware of potential hazards (i.e., obscured field of view (FOV) from where a hazard instigator might appear) than both younger and older child-pedestrians. Consistent with expectations, child pedestrians categorized the photos on the basis of a single criterion (e.g., a hazard instigator such as the presence of a vehicle) while adult pedestrians established a categorization criterion based on a combination of aspects derived from the traffic environment (e.g., hazard instigator and traffic environment). The present study used an innovative paradigm to investigate child pedestrians' conceptions regarding road crossing situations. Understanding child-pedestrians shortcomings in accurately assessing the traffic situation may help in creating intervention techniques which may increase child-pedestrians' awareness to potential and hidden hazards and help in reducing their over-involvement in traffic crashes. Conclusions and implication for further studies are discussed.

Introduction

Pedestrian road crashes pose one of the most serious threats to contemporary life. They are amongst the most substantial causes of death, injury and long-term disability among children, particularly among those in the age range of 5-to 9-years (e.g., Whitebread & Neilson, 2000; Tabibi & Pfeffer, 2003), who endure four times the injury rate of adults, in spite of their lower levels of exposure to traffic (Thomson et al., 2005). Negotiating traffic requires a variety of cognitive and perceptual skills (e.g., Tabibi & Pfeffer, 2003; Te Velde, Van der Kamp & Savelsbergh, 2003; Thomson, Tolmie, Foot, & McLaren, 1996). When a pedestrian's skills are not properly developed, his or her road-related decisions will probably be inadequate (Thomson et al. 1996). Indeed, young children are more involved in traffic crashes, hence leading to the conclusion that they are less competent in traffic than adults (e.g., Tabibi & Pfeffer, 2003; Hill, Lewis & Dunbar, 2000). One might have thought that prohibiting children's crossing the road alone until the age of 9 would be enough to reduce their over-involvement in pedestrian crashes. However, it was shown that elementary-school children do tend to cross the road without adults' accompaniment, especially when coming back from school (e.g., Van der Molen, 1981; Macpherson, Roberts & Pless, 1998). Clearly, in order to reduce road crashes amongst child-pedestrians it is insufficient to assume that young children will avoid crossing roads by themselves. Rather, there is a need to examine and assess the skills and knowledge necessary for children in order to behave safely when coping with the traffic environment, so as to provide them with means for increasing those abilities (Hill et al., 2000).

Prior research has suggested that rather than attitudinal shortcomings, young children suffer from poor pedestrian skills and visual search strategies (Whitebread & Neilson, 2000; Tolmie, Thomson, Foot, McLaren, & Whelan, 1998) as well as other limitations in identifying the factors which compose of dangerous road-crossing sites (Ampofo-Boateng & Thomson, 1991). Thomson et al. (1996) have suggested that pedestrians require several underlying abilities in order to interact safely with traffic: (a) Making judgments about whether crossing places are 'safe' or 'dangerous'; (b) Detecting the presence of traffic that could be a source of danger; and (c) Integrating information from different parts of the relevant traffic environment including different directions (Whitebread & Neilson, 2000). Recent studies have indicated that the ability to identify safe and dangerous road-crossing sites increases with age (e.g., Tabibi & Pfeffer, 2003). Moreover, it was suggested that the ability to resist interference from irrelevant stimuli increases with age (e.g., Tabibi & Pfeffer, 2003; Tolmie et al., 1998). Lastly, examining children's ability of integrating information from different parts of the relevant traffic environment, Underwood, Dillon, Farnsworth and Twiner (2007) have shown that younger children exhibit an idiosyncratic perspective of the road, as compared to the older children, who are able to observe the road from a global perspective (Underwood et al., 2007).

Inspecting the abilities suggested by Thomson et al. (1996) as vital for inducing safe road behaviour, a common denominator is revealed. All of these abilities are hazard-perception-related. Hazard perception (HP) can be defined as the process of evaluating the hazardousness of a traffic situation (Benda & Hoyos, 1983). It can also be described as the ability to 'read the road' and anticipate forthcoming events; a situation awareness for hazardous situations in the traffic environment (Horswill & McKenna, 2004). Situation awareness (SA) was described by Endsley (1995) as a state of knowledge which enables a holistic perception of the environment. Achieving this state involves a three-level process which includes perception of elements in the environment, comprehension of their integrated meaning in order to create a holistic appreciation of the current situation and projection of their status in order to predict near future events (Endsley, 1995). Early research has indicated that part of the reason for the higher accident rate with young children may be their relative inability to perceive hazards correctly (Martin & Heimstra, 1973). More recent studies have also indicated that an important ability for child-pedestrians is the ability to 'read the road' (e.g., Foot et al., 2006). According to Hill et al. (2000) there is evidence that young children are poor at identifying unsafe situations. These researchers argue that the high accident rate for child-pedestrians is a result of the failure to recognize a potential danger when unprompted, thus, they suggested that young children's understanding of danger is not robust (Hill et al., 2000).

Similarities between child-pedestrians and young-novice drivers

Another population of novices whose lack of experience is highly correlated with high crash rate is the young novice drivers' population. Indeed, the literature on young-inexperienced drivers shows that some of the major causes for their over involvement in traffic crashes can be attributed to their lack of driving experience and poor hazard awareness. These impediments entail high resemblance to those reported previously with regard to child-pedestrians. For example, Armsby, Boyle & Wright (1989), asked novice and experienced drivers to rate traffic scene photographs in terms of level of hazardousness (i.e., more or less hazardous). Novices rated a photograph of a pedestrian crossing the road as more hazardous than experienced drivers, but rated a photograph of fog as less hazardous than experienced drivers. Hence, as in the case of child-pedestrians (e.g., Ampofo-Boateng & Thomson, 1991), the lack of experience prevented novice drivers from considering potential hazards unless they were salient and physically present in the environment. Borowsky, Shinar, and Oron-Gilad (2010) have shown that novices, as opposed to experienced drivers, rarely fixate on merging roads when driving a car. As in the case of child-pedestrians (who tend to focus on the most salient factor, see Ampofo-Boateng & Thomson, 1991; Demetre & Gaffin, 1994; Foot, Tolmie, Thomson, McLAren & Whelan, 1999), these young-novice drivers' ability of integrating information from different parts of the relevant traffic environment, is poor (Benda & Hoyos, 1983).

Since there is a tight link between the behaviour of young-novice drivers in their ability to detect hazards and the task requirements of young pedestrians crossing a road, the methods and sophistication developed for assessing HP among drivers, such as using categorization task (e.g., Borowsky, Oron-Gilad & Parmet, 2009; Borowsky, Meir, Oron-Gilad, Shinar & Parmet, 2010), might be helpful in providing in-depth understanding of prominent road crossing deficits among child pedestrians and in creating a platform for developing the HP abilities. Indeed, examining the driving domain, several studies have focused on measuring the degree of perceived hazardousness associated with a situation, frequently by utilizing still-picture traffic-scene or movies rating tasks (e.g., Benda & Hoyos, 1983; Finn & Bragg, 1986; Armsby et al., 1989). Benda and Hoyos (1983) for example, requested participants to categorize a mixture of traffic situations into an arbitrary number of groups regarding similar hazardous situations. Young-novices tended to categorize movies according to similarities in hazard instigator characteristics, where experienced drivers tended to categorize them according to similarities in traffic environment characteristics. It was concluded that the more driving experience a driver has, the more he or she is able to assess hazardousness as a holistic attribute of the traffic situation and to integrate multiple different aspects of a situation (Benda & Hoyos, 1983). In a later study, Armsby et al. (1989) found that while young-novices were more inclined to rate fog situation as less hazardous than a pedestrian crossing a road, experienced drivers tended to rate the fog situation as more hazardous than the pedestrian crossing. Borowsky et al. (2009) had examined the traffic-scenes categorization techniques of drivers varying in age and driving experience. Young-novice, experienced, and older-experienced drivers were asked to categorize traffic-scene movies after being asked to actively detect hazards featured in them first. Experienced drivers tended to demonstrate a coherent, holistic attitude and to categorize traffic-scenes movies according to general traffic environment characteristics whereas young-novice drivers tended to relate to the actual events' instigators as the categorization criterion (e.g., to group movies together since a pedestrian appeared in all of them). In order to establish understanding of children traffic behaviour patterns (e.g., when and where do children cross the road?), and following Borowsky et al. (2009; 2010b) results, the present study utilize a similar procedure to examine hazard-categorization task patterns between experienced and young-inexperienced pedestrians as means for evaluating their road-crossing HP abilities.

Implications for the current study

The HP-based traffic-scenes database was created according to taxonomy of factors aiming to differentiate between pedestrians at different age and experience levels. Previous research had suggested that the majority of children are injured on non-arterial roads, particularly in residential areas (e.g., Roberts, Keall & Frith, 1994). Past research had also suggested that pedestrians' traffic crashes which occur in inhabited areas tend to take place in locations where visibility is restricted, e.g., at curves, or near parked cars (Ampofo-Boateng & Thomson, 1990). On-street vehicle parking presents a particular risk for child-pedestrians since parked cars interfere with their ability to detect oncoming vehicles, while also obstructing the motor vehicle drivers' vision, preventing them from noticing child-pedestrian who may be masked by stationary vehicles along the road (e.g., Aoki and Moore, 1996; Petch & Henson, 2000). Studies have suggested that there are lower pedestrians road crashes rates on one-way streets as compared with two-way streets (Zegeer, 1991). However, Summersgill & Layfield (1998) have indicated that there is no difference in the level of pedestrians' crash risk between one-way and two-way roads with the same cross-section. According to Ampofo-Boateng & Thomson (1991), the younger child-pedestrians are, the more they are likely to base their evaluation of a crossing site as safe or dangerous on a single factor- the presence or absence of cars on the road. Children aged 5 and 7 were found to determine the safety of a site purely on whether they can see cars on the road, where the presence of a vehicle anywhere (even remotely in the vicinity of the location) was correlated with these younger children judgment of the situation as dangerous. Another critical factor is familiarity. Using a virtual road crossing environment, Johnston & Peace (2007) investigated the road crossing behaviour of pedestrians in familiar and unfamiliar environments (familiarity was manipulated using traffic direction). Results have indicated that pedestrians had a lower safety ratio in crossing when presented with an unfamiliar direction, indicating of an unsafe crossing behaviour and a higher risk for traffic crashes compared to familiar environments.

In summary, three prominent different factors found to influence pedestrians and child-pedestrians' road crossing behaviour were manipulated within the present experimental setup: (1) Traffic density (high vs. low), (2) Field of view (restricted vs. unrestricted), and (3) Familiarity (familiar vs. unfamiliar scenery). High density settings included moving vehicles, while low density included none. Field of View (FOV) limitation was achieved by presence of a parked vehicle. Familiarity was gained by presenting school children with scenery from the proximity of their elementary school while non-familiarity was gained by presenting them with scenery from the area near the Ben-Gurion University (BGU) and vice versa for the BGU students. Each scenario displayed a specific road-crossing environment and was presented from a pedestrian's point of view, as if standing on one side of the pavement intending to cross over to the other side.

Lastly, the general agreement among traffic safety professionals is that children under the age-range of 9-10 should not cross roads alone (e.g., Percer, 2009). Indeed, it was argued that children under the age-range of 9-10 are not at the appropriate stage of development allowing them to master the complex skills involved in the pedestrian task (Vinje, 1981; Sandel, 1975). Taken together with the findings indicating that child-pedestrians' crashes are a significant cause of injury mainly in the age range of 5-to 9-years (e.g., Whitebread & Neilson, 2000; Tabibi & Pfeffer, 2003), the current research focused on primary-school children at the age range of 6-to-12. The age-group allocation refers to young children, under the age of 9 (which has been suggested should not engage in road-crossing alone) and children within the age-range of 9-12.

Method

Participants

Sixty participants, twenty 6 to 8 year olds (mean age=6.9, SD=0.7), twenty two 9 to 12yearolds (mean age=11.3, SD=0.6), twenty one 24-28-year-olds experienced pedestrians (mean age=26.2, SD=1.1) completed this experiment as volunteers. All participants had normal vision, with uncorrected Snellen static acuity of 6/12 or better and normal contrast sensitivity. Participants were all requested to sign an informed consent form, approving their participation in the experiment. In addition, parental consent was obtained for participants under the age of 18.

Apparatus

Classification hard copies. As aforesaid, several different factors were found to influence pedestrians and child-pedestrians' road crossing behaviour. Each of the twelve (13x18 cm) still pictures (see Figure 1) depicting traffic-scenes taken from a pedestrian's perspective was portrayed as a different combination of values given to each of the factors: (1) Traffic density (high vs. low density), (2) Field of view (restricted vs. unrestricted FOV), and (3) Familiarity (familiar vs. unfamiliar scenery).

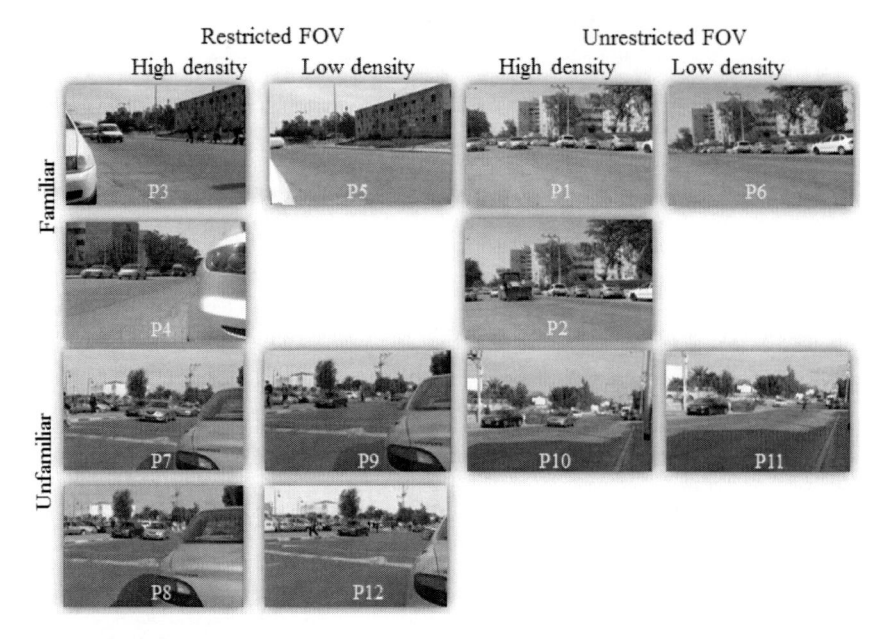

Figure 1. Traffic-scenes database. Each traffic-scene was portrayed as a different combination of values given to the factors: (1) Traffic density, Field of view, and Familiarity. Still pictures' numbers are cited on the bottom (colour image available from http://extras.hfes-europe.org)

Statistical Analysis Tool. Decision Tree (DT), a machine learning algorithm was used, utilizing the 'Clementine' software. DTs provide a hierarchy of if-then rules for dividing records into predefined groups or predicting possible outcomes of large data sets based on particular target variables' values (Larose, 2005). It is constructed through a process called "induction", an economical process which requires a small number of passes through the training data, which generally involves growing the DT and pruning it (Kulkarni et al., 2005). Based on the specific decision rules, the root-node splits into two or more child nodes. Selecting a variable to a node is usually based on examining the entropy of each variable and preferring the one that best differentiates between the output variables. Once a node splits, the same process is repeated for each of its child nodes, with the dataset further subdivided based on the split criteria at the parent node. Once the DT is constructed, it is necessary to prune it in order to avoid over-fitting the dataset. Before deploying it, it is required to study its accuracy and generalization abilities by passing a test dataset throughout it and examining the outcome. Then, as the DT is ready for deployment, it may be utilized for predicting a new case (Kulkarni et al., 2005).

Hazard definition. Participants were all given the same definition of "Hazard" based on Haworth, Symmons and Kowadlo's (2001, p. 3).

Procedure

In a separate room in their elementary school, one at a time, each of the child-participants observed 12 still pictures depicting traffic-scenes taken from a pedestrian's perspective as hard copies. Students performed the same task in an experimental laboratory at the University. Participants were asked to categorize these hard-copy pictures by dividing them into an arbitrary number of non-overlapping sets according to similarities in their level of hazardousness (this procedure resembled the one used in Borowsky et al., 2009). After the classification procedure had ended, participants were asked to suggest an appropriate title for each of the groups.

Results

Preliminary analysis

Analysis started by referring to the number of verbal descriptions made by participants to each one of the factors (see Figure 2). Conducting one-way ANOVA revealed that the age groups responded differently to the Traffic density factor ($F_{(2,60)}=5.45$, $p<0.01$). LSD post hoc pair-wise comparisons revealed that 6-8-year-olds (Mean number of verbal references=1.10, SD=1.02) tended to relate significantly ($p<0.001$) less to the Traffic density factor than both the experienced-adult pedestrians (1.95, 1.43) and the 9-12-year-olds (2.23, 0.92). However, from examining the verbal descriptions given by participants to each of the non-overlapping sets they created, it could be seen that the number of sets created by participants rose significantly ($F_{(2,60)}=21.42$, $p<0.001$) with age.

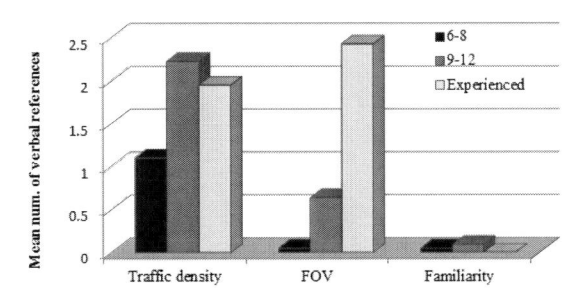

Figure 2. Distribution of pedestrians' verbal references to each traffic-related factor, according to age-group

While 6-8-year-olds created on average 2.25 groups, 9-12-year-olds created on average 2.86 groups and experienced-adults created 3.81 groups. LSD revealed that the differences between each of the age groups were significant as well (6-8 to 9-12: $p<0.05$; the rest- $p<0.001$). Thus, it may be that the growth in the number of verbal descriptions of Traffic density mentioned by participants stemmed from the increased number of sets referred to by each age-group (which rose with age). Moreover, it was revealed that the age groups responded differently to the Field of view factor ($F_{(2,60)}=35.91$, $p<0.001$). LSD revealed that experienced-adult pedestrians tended to be significantly ($p<0.001$) more aware of the potential hazardousness of restricted FOV (2.43, 1.21) than both younger (0.05, 0.22) and older child-pedestrians (0.64, 1.05). Lastly, the age groups did not respond differently to the familiarity factor ($F_{(2,60)}=0.96$, N.S.).

Decision tree analysis

Analysis began by listing all the different picture set-arrangements and categorization patterns displayed by the participants. A picture set-arrangement was defined as the unification of several HP still pictures to a specific set. Each was marked according to the pictures it had integrated (e.g., categorization of pictures 2 and 6 together, as members of the same set, was marked "P_2_6"). Altogether, participants categorized the hard-copy pictures into 119 different set-arrangements. Results indicated (see Table 1) that experienced pedestrians tended to classify still pictures 6 and 11 together, as members of the same set (P_6_11) (6 of its members used this set out of the 6 participants who chose to use this combination, whereas none of the 6-8-year-olds' and the 9-12-year-olds' ensembles used this categorization). Applying Fisher exact test suggested that experienced-adults were significantly more likely to categorize the traffic-scenes according to P_6_11 (Experienced (E) vs. 6-8-year-olds: p=0.01; E vs. 9-12: p<0.01). While 6-8-year-olds were not significantly different than the 9-12-year-olds (N.S.). Experienced pedestrians' tendency to classify movies 6 and 11 together may indicate upon these pedestrians' higher sensitivity to the potential hazardousness of the FOV dimension compared to the 6-8-year-olds' ensemble. Experienced-adult pedestrians were also inclined to use the categorization P_1_2_10. Applying Fisher exact test suggested that experienced-adults were significantly more likely to categorize the traffic-scenes according to P_1_2_10 (Experienced (E) vs. 6-8-year-olds: p<0.05). However, 9-12-year-olds were not significantly different than the 6-8-year-olds or the Experienced-adults (p=N.S.). Experienced-adult pedestrians' tendency towards categorization P_1_2_10 indicates upon their higher sensitivity to the potential hazardousness of the field of view dimension compared to the 9-12-year-olds. Moreover, using variables P_6_11 and P_1_2_10 separately may indicate that the experienced-adult pedestrians' tend to relate to different aspects of the same situation and to categorize the still pictures on the basis of more than one characteristics (i.e., relate to both the Traffic density and the Field of view dimensions).

Table 1. Distribution of participants in the set-arrangements

Categorization	6-8-year-olds	9-12-year-olds	Experienced-adults
P_6_11	(0, 0, 0)	(0, 0, 0)	(6, 1, 0.29)
P_1_2_10	(0, 0, 0)	(1, 0.17, 0.05)	(5, 0.83, 0.24)
P_5_6_11	(0, 0, 0)	(4, 0.44, 0.18)	(5, 0.56, 0.24)
P_5_6_9_11_12	(2, 0.33, 0.67)	(4, 0.67, 0.44)	(0, 0, 0)

*Presented are the number and proportion of participants from each age-group (relative to the total number of (a) participants utilizing the set-arrangement and (b) participants in the age-group, respectively).

Indeed, the DT algorithm revealed that the variables P_6_11, P_1_2_10 and P_5_6_11 separated the experienced pedestrians from the young-novices (see Figure 3). The best split among all possible splits and over all variables was variable

P_6_11, which separated the experienced pedestrians (who tended to make use of it) from the 6-8-year-olds (who refrained from doing so). The next split between the age groups was variable P_1_2_10. Participants who tended to make use of the P_1_2_10 categorization, were inclined to be experienced pedestrians, whereas a pedestrians who refrained from doing so tended to belong to the 6-8-year-old' ensemble. The final split between the groups was variable P_5_6_11. Participants who tended to make use of the P_5_6_11 categorization, were inclined to be experienced pedestrians, whereas a pedestrian who refrained from doing so tended to belong to the 6-8-year-old' ensemble. A test set model was run to compare the model's predictive ability against the training set. Results showed that twenty of the twenty 6-8-year-olds were identified correctly by the model (a total accuracy of 100%) and 13 of the 21 experienced-adults were identified correctly by the model (61.9%), however, none of the twenty-two 9-12-year-olds were identified correctly by the model. Overall, the model presented a medium accuracy level (52.4%) in recognizing participants' age-group. The DT model separated between experienced-adults and 6-8-year-olds according to their categorization patterns via three cardinal variables. However, no specific variable or group arrangement was found to differentiate 9-12-year-olds from either one of the other age-groups. None of these twenty-two 9-12-year-olds were identified correctly by the model. Thus, it can be suggested that this group of pedestrians lacked idiosyncratic characteristics compared to the other age-groups. Notably, examining the model identification trials, it was found that 18 out of the 22 9-12 year olds (81.8%) were labelled "6-8-year-olds" by the model, while the rest (18.2%) were labelled "experienced-adults". These findings may indicate that 9-12-year-olds' characteristics resemble those of the 6-8-year-olds more than those of the experienced-adults. The model's inability to separate the 9-12-year-olds' group from the other age-groups may stem from its resemblance to each one of them. Indeed, being situated in the midst, both age-wise and development-wise, the 9-12-year-olds shared characteristics with both age-groups thus made it difficult for the model to separate this age group from the others.

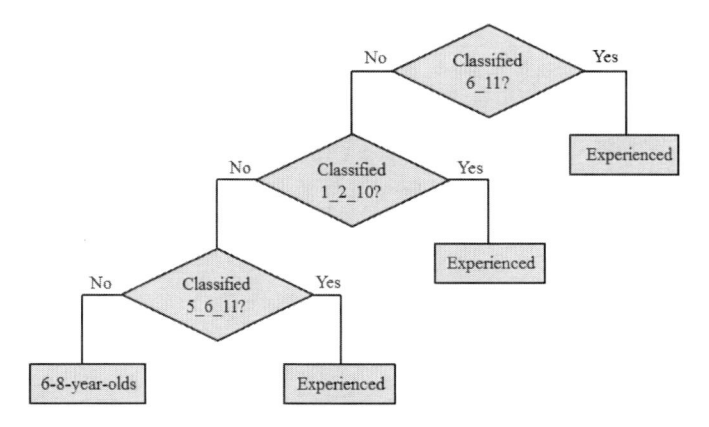

Figure 3. Decision Tree algorithm

In the hope of enhancing the model's accuracy level and effectively identifying 9-12-year-olds, an attempt was made to refine the data set. Due to the large number of group arrangements and in order to increase the reliability level of a future DT model by screening distracters, the number of participants referring to each of the 119 set-arrangements was examined. Indeed, it was found that many of the set-arrangements were created and used by a single participant, thus representing only this particular individual and not contributing to the differentiation between age-groups. Hence, a participant whose set-arrangement was unique (i.e., he or she were the only one categorizing according to this arrangement) and his or her full categorization pattern suggested no "Important" group arrangement (i.e., which differentiates effectively between participants' age groups) was excluded from the DT analysis. This dataset reduction led to a total of 43 set-arrangements, categorized by three 6-8-year-olds (creating a non-representative sample), nine 9-12-year-olds and thirteen adults. Fisher exact test suggested that experienced-adults were significantly less susceptible to categorize the traffic-scenes according to $P\_5\_6\_9\_11\_12$ (Experienced (E) vs. 6-8-year-olds: $p<0.05$; E vs. 9-12: $p<0.05$). 6-8-year-olds were not significantly different than the 9-12-year-olds p=N.S.). Indeed, DT's best split was variable $P\_5\_6\_9\_11\_12$, which separated the 9-12-year-olds (who tended to make use of it), from the experienced pedestrians (who refrained from doing so). Thus, 9-12-year-olds' tendency to combine pictures 5, 9 and 12 (all illustrating restricted field of view) with pictures 6 and 11 (both illustrating unrestricted field of view) highlights their obliviousness to the Field of view factor, a characteristic which was evident in the preliminary analysis as well. Running a test set model for comparing the modified model's predictive ability against the training set showed that (as was expected due to their low number in the sample) none of the three 6-8-year-olds were identified correctly by the modified model- a total accuracy of 0%. Moreover, four out of the nine 9-12-year-olds were identified correctly by the model- a total accuracy of 44.5%. However, thirteen out of the thirteen experienced-adults were identified correctly by the model. Overall, the model presented an accuracy level of 68% in recognizing both the experienced and young-novices pedestrians.

Discussion

The study used an innovative paradigm to investigate child-pedestrians' conceptions regarding road crossing situations. This paradigm has been used by us previously for analysis of comparing HP performance of drivers varying in age and experience-level (see Borowsky et al. 2009) and for the analysis of young novice drivers' HP training program (see Borowsky et al., 2010b).

Results are indicative of the differences among pedestrians from various levels of experience with regard to their HP abilities. Classification analysis with the Clementine- DT model revealed that experienced-adult pedestrians referred more to the Field of view element than both child groups. Moreover, as was hypothesized and consistent with Benda and Hoyos's (1983) findings, it was found that with age pedestrians were more inclined to relate to potential hazards (either by their verbal description or by their categorization patterns), as well as actual hazard, and to categorize traffic-scenes more holistically- on the basis of a combination of aspects and dimensions (e.g., Traffic density and Field of view), as opposed to doing so on the basis of a single criterion (primarily Traffic density).

Understanding child-pedestrians shortcomings in accurately assessing the traffic situation may help in creating intervention techniques aimed to increase child-pedestrians' awareness of potential and hidden hazards and help in reducing their over involvement in traffic crashes. Indeed, evidence from the road safety domain demonstrated the success of training a variety of road crossing skills to young children (e.g., Thomson et al., 1996; Demetre et al., 1992; Rothengatter, 1981). It was important to note that the familiarity element, mentioned previously in the literature as a contributor to pedestrians' behaviour was not perceived as a critical element in hazard classification and had much lower importance than the field of view and the presence of vehicles on the road. This finding should be further examined in future studies.

This study also had its limitations. It utilized traffic-scenes categorization for differentiating between HP abilities among pedestrians and laid the foundation of utilizing DT as an additional supporting tool. However, due to the limited sample size, the DT model accuracy was marginal at best. Future research should broaden the sample size. Moreover, the utilization of still traffic-scenes did not account for dynamics factors (e.g., moving vehicles, time and distance), did not examine the differences between pedestrians' performance in crossing decision tasks, and did not examine participants' eye-movements' patterns. These need to be addressed in future studies.

Acknowledgements

This research was supported in part by the Ran Naor Foundation, Dr. Tzipi Lotan, Technical Monitor.

References

Ampofo-Boateng, K., & Thomson, J. A. (1990). Child pedestrian accidents: A case for preventative medicine. *Health Education Research: Theory and Practice, 5*, 265–274.

Ampofo-Boateng, K., & Thomson, J.A. (1991). Children's perception of safety and danger on the road. *British Journal of Psychology, 82*, 487-505.

Aoki. M., & Moore. L. (1996). KIDSAFE: A Young Pedestrian Safety Study. *ITE Journal*, September, 36-45.

Armsby, P., Boyle, A.J., & Wright, C.C. (1989). Methods for assessing drivers' perception of specific hazards on the road. *Accident Analysis & Prevention, 21*, 45-60.

Benda, H.V., & Hoyos, C.G. (1983). Estimating hazards in traffic situations. *Accident Analysis & Prevention, 15*, 1-9.

Borowsky, A., Oron-Gilad, T., & Parmet, Y. (2009). Age and skill differences in classifying hazardous traffic scenes, *Transportation Research part F, 12*, 277-287.

Borowsky, A., Shinar, D., & Oron-Gilad, T. (2010a). Age, Skill, and Hazard Perception in Driving. *Accident Analysis and Prevention, 42*, 1240-1249.

Borowsky A., Meir A., Oron-Gilad T., Shinar D., & Parmet Y. (2010b). The effect of hazard perception training on traffic-scene movies categorization, *Proceedings of the Human Factors and Ergonomics Society Annual Meeting* September 2010, vol. 54 (24), 2101-2105.

Demetre, J.D. & S. Gaffin, S. (1994). The salience of occluding vehicles to child pedestrians. *British Journal of Educational Psychology, 64*, 243–251.

Demetre, J.D., Lee D.N., Pitcairn, T.K., Grieve, R., Thomson, J.A., & Ampofo-Boateng, K. (1992). Errors in young children's decisions about traffic gaps: Experiments with roadside simulations. *British Journal of Psychology, 83*, 189–202.

Endsley, M.R. (1995). Toward a theory of situation awareness in dynamic systems, *Human Factors, 37*, 32-64.

Finn, P. & Bragg, B. (1986). Perception of the risk of an accident by young and older drivers, *Accident Analysis and Prevention, 18*, 289-298.

Foot, H.C., Tolmie, A., Thomson, J.A., McLaren, B., & Whelan, K. (1999). Recognising the danger. *The Psychologist, 12*, 400–402.

Foot, H. C., Thomson, J. A., Tolmie, A., Whelan, K., Morrison, S., & Sarvary, P. (2006) Children's understanding of drivers' intentions. *British Journal of Developmental Psychology. 24*, 681–700.

Hill, R., Lewis, V. & Dunbar, G. (2000). Young children's concepts of danger. *British Journal of Developmental Psychology, 18*, 103–120.

Horswill, M.S., & McKenna, F.P. (2004). Drivers' hazard perception ability: Situation awareness on the road. In S. Banbury and S. Tremblay (Eds.). *A Cognitive Approach to Situation Awareness* (pp.155-175). Aldershot, UK: Ashgate.

Johnston, L. & Peace, V. (2007). Where did that car come from?: Crossing the road when the traffic comes from an unfamiliar direction. *Accident Analysis and Prevention. 39*, 886-893.

Larose, D.T. (2005). *Frontmatter, in Discovering Knowledge in Date: An introduction to Data Mining.* Hoboken, NJ, USA: John Wiley & Sons.

Macpherson A., Roberts I., & Pless I.B. (1998). Children's exposure to traffic and pedestrian injuries. *Journal of Public Health, 88*, 1840–1843

Martin, G.L., & Heimstra, N.W. (1973). The perception of hazard by young children. *Journal of Safety Research , 5*, 238-246.

Percer, J. (2009). *Child pedestrian safety education: Applying learning and developmental theories to develop safe street-crossing behaviors*. Washington, DC: National Highway Traffic Safety Administration.

Petch R.O. & Henson R.R. (2000). Child Road Safety in the Urban Environment. *Journal of Transport Geography, 8*, 197-211.

Roberts I.G., Keall M.D., & Frith W.J. (1994). Pedestrian exposure and the risk of child pedestrian injury. *Journal of Pediatric Child Health, 30*, 220–223.

Rothengatter, J.A. (1981). *Traffic safety education for young children*. Lisse, The Netherlands: Swets and Zeitlinger.

Sandels, S. (1975). *Children in traffic*. London: Elek.

Summersgill, I. & Layfield, R. (1998). *Non-junction accidents on urban single-carriageway roads*. TRL Report TRL183. Crowthorne, UK: Transport Research Laboratory.

Tabibi, Z. and Pfeffer, K. (2003). Choosing a safe place to cross the road: the relationship between attention and identification of safe and dangerous road crossing sites. *Child: Care, Health and Development, 29*, 237-244.

Thomson, J.A., Tolmie, A., Foot, H.C., & McLaren, B. (1996). *Child Development and the Aims of Road Safety Education*. Road Safety Research Report No. 1. London: HMSO.

Thomson, J.A., Tolmie, A, Foot, H.C., Whelan, K.M., Sarvary, P., & Morrison, S. (2005). *Journal of Experimental Psychology: Applied, 11*, 175-186.

Tolmie, A., Thomson, J. A., Foot, H.C., McLaren, B., & Whelan, K.M. (1998). *Problems of attention and visual search in the context of child pedestrian behavior*. Road Safety Research Report No. 8. London: Department of Transport, Environment & the Regions.

Underwood, J., Dillon, G., Farnsworth, B & Twiner, A. (2007). Reading the road: the influence of age and sex on child pedestrians' perceptions of road risk. *British Journal of Psychology, 98*, 93-110.

Van der Molen, H.H. (1981). Child Pedestrian's Exposure, Accidents and Behavior. *Accident Analysis and Prevention, 13*, 193-224.

Vinjé, M.P. (1981). Children as pedestrians: Abilities and limitations. *Accident Analysis & Prevention, 13*, 225-240.

Whitebread, D., & Neilson, K. (2000). The contribution of visual search strategies to the development of pedestrian skills by 4-11 year-old children. *British Journal of Educational Psychology, 70*, 539-557.

Zegeer C.V. (1991). *Synthesis of safety research- pedestrians*. Highway Safety Research Center, University of North Carolina, HSRC-TR90.

Skills and Remote Control

In D. de Waard, N. Merat, A.H. Jamson, Y. Barnard, and O.M.J. Carsten (Eds.) (2012). *Human Factors of Systems and Technology* (pp. 291). Maastricht, the Netherlands: Shaker Publishing.

On-shore supervision of off-shore gas production - Human Factors challenges

Ruud N. Pikaar[1], Renske B. Landman[1], Niels de Groot[1], & Leen de Graaf[2]
[1] ErgoS Engineering & Ergonomics
[2] UReason
The Netherlands

Abstract

Technology enables remote process control of off-shore gas production assets, thus reducing off-shore manpower. The human factors in control centre engineering include operator consoles, information presentation, interaction, alarm management, and job content. The human factors are all related to each other. Moving off-shore tasks to on-shore control centres requires a human factors approach, which includes an operator task analysis. For natural gas production, some new control room tasks appear, such as contract management and related production volume control.

Two cases of Human Factors engineering of *a move to shore* are presented. At the first case, a hierarchical task analysis was performed to get insight in the operator tasks. This enabled determination of the number and size of workplaces and revealed the importance of contextual off-shore platform information. Several years later, increased data transmission capacity between on- and off-shore, led to the implementation of an advanced alarm management philosophy, including an optimal visualisation of (grouped) alarms. The second case also concerned the design of an on-shore control centre for over 40 off-shore gas production assets. A major effort concerned the redesign and standardization of process graphics, in order to enable on-shore operators to supervise all processes adequately.

Human Factors Engineering

The aim of Human Factors (HF)/Ergonomics is to optimize the work system. Ergonomics can be defined as *user-centred design*, or user-centred engineering. This definition expresses a focus, both on the human being and design. In general terms, this requires an approach including both social and technical aspects of the system. Job design, operator workload, control centre layout, workplace layout, instrumentation, information display, environment, and many more topics have to be addressed. The HF professional may not have much background in process control or other engineering sciences. Therefore, he relies on a systematic analysis and design approach (ISO 11064, 1998). He tries to get insight in the relationships between relevant human factors, such as operator workload and job design, or the number of screens on a console and the measurements of the workplace. In addition, HF may fill the gap between technical engineering disciplines and users. Of course,

In D. de Waard, N. Merat, A.H. Jamson, Y. Barnard, and O.M.J. Carsten (Eds.) (2012). *Human Factors of Systems and Technology* (pp. 293 - 306). Maastricht, the Netherlands: Shaker Publishing.

a close cooperation between HF professional and technical engineering disciplines will be needed. The aim of this paper is to show the impact of a structured HF involvement in control centre design projects.

De Looze and Pikaar (2006) assume a gap between the work of HF scientists and the needs of HFE professionals. Closing the gap between practitioners and researchers is a challenge. Two steps should be taken: 1) organize access to the best practices developed in the field and 2) organize research programmes with potential societal and market value. Scientist are missing one important item: case material. Professional ergonomists have a tremendous amount of case material. Related to process control and control centre design, several systematic case reports have been published by Kragt (1992). Ten control centre case studies have been compiled by Pikaar (2007). Also, several cases from process industries are discussed by Rijnsdorp (1991) and Pikaar et al. (1997). These publications have in common a structured approach to present cases. For each case, the same topics are discussed with an emphasis rather on methodology, than describing the final design result. It should be noted that the system ergonomics approach to engineering projects has been the same for many of the cases mentioned by Kragt, Rijnsdorp and Pikaar. Another approach to use case material related to process control is presented by Henderson (2002). Anecdotal material of many cases results in an overview of important human factors issues to be considered when changing the degree of remote operation.

Practitioners, evaluating work and designing or implementing solutions, may develop good or even best practices. Publishing a report on a successful, or perhaps an unsuccessful project, seldom is part of the work contract. It is not a standard line of business if one is not affiliated to scientific research. In addition, getting a project report published may easily fail, because this type of work is not commonly accepted in the international journals. A project is never carried out twice (with or without ergonomics) to find out whether ergonomics makes a difference. The authors believe that HF experiences in industrial settings should be reported in literature notwithstanding the methodological problem of N=1. Anyhow, this paper is based on case studies.

Control centre ergonomics

Technology enables remote process control of off-shore gas production assets, thus reducing off-shore manpower. Moving off-shore tasks to on-shore control centres requires a human factors approach, which includes an operator task analysis and a reallocation of operator tasks. Some new control room tasks may appear, such as contract management and related production volume control.

The following related topics need to be addressed in control centre design projects (EEMUA, 2002; Pikaar et al., 1998): 1) job content and operator workload, 2) workplace design – operator console, 3) process graphics, 4) interaction design – navigation and control, and 5) alarm management. Each topic may be a (large) project on its own: the design of a work organization, control room layout and workplace design, the development of process control graphics, and so on.

Henderson et al. (2002) conclude that communication is of utmost importance for remote control. This can also be illustrated by the Esso Flexicoker Project, amongst others reported in Rijnsdorp (1991) and in Kragt (1992). For a major refinery extension, a decision had to be made between a new control room for the new process units, away from the old control room of the old process units, versus one new integrated control room for all units. The latter solution, although being far more expensive, was selected on the basis of communication issues. Moving an off-shore control room to shore is not different from other control centre design projects from a HF point of view, which will be illustrated by case material. In both situations, process units are remote operated.

Ergonomics engineering steps

Usually, an engineering project passes through several phases, starting with a feasibility study, via several design steps, to detailed engineering and implementation, as shown in figure 1 (Pikaar, 2007). Highlights of the HF engineering steps are discussed below. The HF professional needs knowledge of the actual operator tasks. Based on this knowledge, an accurate estimate of the new control room situation can be made (functional analysis). The main issue will be to what extent operator tasks change, when moving an off-shore control room to shore.

Step 1. Feasibility
Step 1 typically includes a review of the project owners' HF assumptions regarding work load, level of automation, and capabilities of operators. For the HF professional, it is important to be aware of such assumptions, and if needed, give feedback on a general level. For example, one could temper a too optimistic view on the number of operators needed.

Step 2. Problem definition
This step starts with a general description of the project and the purpose of the system to be designed. The outline of the design steps have to be negotiated with project management, including design constraints.

Step 3. Situation analysis
The aim of the situation analysis is to gain insight in existing and future tasks. It includes collecting formal documents and drawings of the existing system, analyzing work tasks by observations and interviews, and gathering knowledge on the new system (to be designed).

Step 4. Functional Design Specification
The functional design specification concerns the allocation of system tasks. An allocation procedure includes a discussion on the level of automation, job requirements, and the design of a local work organization. Topics are 1) the allocation of tasks to workplaces, 2) the lay out of a system, 3) shape and size of workstations and instruments, and 4) environmental requirements.

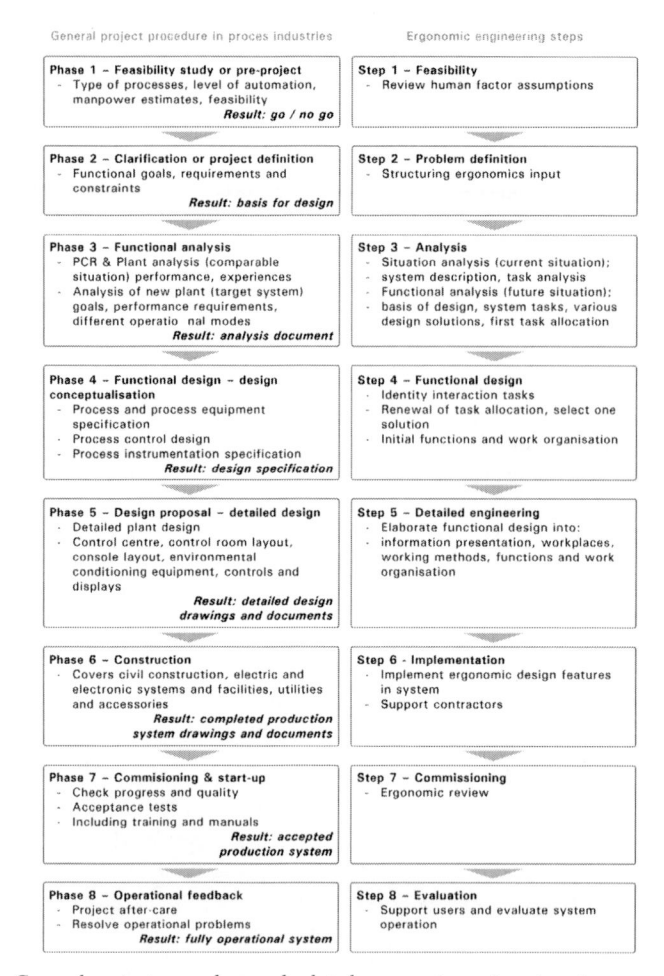

General project procedure in process industries	Ergonomic engineering steps
Phase 1 – Feasibility study or pre-project - Type of processes, level of automation, manpower estimates, feasibility *Result: go / no go*	**Step 1 – Feasibility** - Review human factor assumptions
Phase 2 – Clarification or project definition - Functional goals, requirements and constraints *Result: basis for design*	**Step 2 – Problem definition** - Structuring ergonomics input
Phase 3 – Functional analysis - PCR & Plant analysis (comparable situation) performance, experiences - Analysis of new plant (target system) goals, performance requirements, different operational modes *Result: analysis document*	**Step 3 – Analysis** - Situation analysis (current situation); system description, task analysis - Functional analysis (future situation); basis of design, system tasks, various design solutions, first task allocation
Phase 4 – Functional design – design conceptualisation - Process and process equipment specification - Process control design - Process instrumentation specification *Result: design specification*	**Step 4 – Functional design** - Identity interaction tasks - Renewal of task allocation, select one solution - Initial functions and work organisation
Phase 5 – Design proposal – detailed design - Detailed plant design - Control centre, control room layout, console layout, environmental conditioning equipment, controls and displays *Result: detailed design drawings and documents*	**Step 5 – Detailed engineering** - Elaborate functional design into: information presentation, workplaces, working methods, functions and work organisation
Phase 6 – Construction - Covers civil construction, electric and electronic systems and facilities, utilities and accessories *Result: completed production system drawings and documents*	**Step 6 – Implementation** - Implement ergonomic design features in system - Support contractors
Phase 7 – Commisioning & start-up - Check progress and quality - Acceptance tests - Including training and manuals *Result: accepted production system*	**Step 7 – Commissioning** - Ergonomic review
Phase 8 – Operational feedback - Project after-care - Resolve operational problems *Result: fully operational system*	**Step 8 – Evaluation** - Support users and evaluate system operation

Figure 1. General project procedure and related ergonomic engineering steps

Step 5. Detailed Design/Engineering
On the basis of functional design requirements, various design solutions can be developed. Choices have to be made, which implies weighing all aspects involved, including ergonomics. Tools to illustrate the results may be 3D-drawings, mock-up evaluations, or prototyping of graphics.

Step 6. Implementation (building the system)
Typically, the construction phase starts with the production of workshop drawings and building site drawings. A HF contribution is needed to avoid some typical errors. For example, an operator console may have been specified with two supporting legs. The workshop engineer decides that a third leg is needed for stability. He locates the additional leg in the middle of the console, which happens to be the central work position of the operator, thus reducing his leg room.

Step 7. Commissioning & step 8. Evaluation
Once finished, the formal commissioning of a working system is organized. Typically, the HF professional should review workplaces, information display and GUI's. Ideally, after a year, an evaluation of the running system should be organized, for example resulting in feedback on design and engineering of the project.

Case studies – general context

Over the years, the authors have been involved in several cases of moving operator tasks from North Sea natural gas production facilities to land based control centres. Several companies are active in this area, each operating several dozens of platforms. Satellite platforms produce onto larger platforms, which have recovery units for glycol and ethanol. Larger platforms are manned and have a local control room for remote control of process units on the platform and satellite platforms. Piping connects the platforms to a main entry point for shore going sales gas. At main platforms, usually a 24/7 manned control room can be found.

In the 90's, the authors redesigned their first on-shore control room for off-shore production. The off-shore control rooms were equipped with cctv-cameras, surveying the displays panels. Thus, off-shore operators could go to sleep, while colleagues watched their safety. In case of an alarm, a wake up call was placed.

Recently, the authors have been involved in two projects of moving a control room to shore. The main projects of case 1 case concern: 1) control centre and workplace layout, 2) central process overview graphic, and 3) alarm management. The main projects of case 2 are: 1) control centre and workplace layout and 2) process graphics redesign. The company of the second project was aware of the earlier findings at the first company. They visited this companies' operational control centre and copied several findings. The following sections give some highlights of both projects, however they are no full account of HF contributions.

Case 1A – Control room design

Starting point was a small on-shore control room for land based gas production assets and the off-shore gas receiving station. After selling the on-shore assets, the control room was moved to another location, tasks to be extended to supervise approximately 25 off-shore assets. Process supervision was based on <10% of the off-shore process control variables. The HF contribution to this project can be summarized by some key factors (more details can be found in Pikaar, 2007):

– project scope	upgrading and moving of an existing control room to another location
– investment	€ 200.000 exclusive of instrumentation and communication systems
– % HF engineering	10% of total investment / 200 hours
– management	project owners' engineering department

– project team	HF engineer, architect, and instrument engineering contractor
– main topics	room layout, workplaces, detailed design, large screen overview graphic
– workplaces	one double operator console, office desk, social area
– role HF professional	project management, ergonomic design

The project was organized along the system ergonomics engineering steps, as described earlier. A situation analysis was carried out in the existing on-shore control centre (observations, semi-structured interviews). Functional analysis concerned the expected new situation: daytime process control by local off-shore operators, night time process supervision on-shore. Of particular interest was the outcome of the functional analysis: an estimated 1.5 operator needed in the control room, which can only be realized by two operators. Hence, additional (office-type) tasks were added to realize a balanced work load. As a consequence, the control room design was based on a combination of an office desk and a double console, both having an easy access to a shared process overview. This also dictated the functional workplace design with one row of process screens (no tiled screens), in order to be able to look over the screens (see figure 2). Design tools the HF professional used were 3D-drawings and prototyping of a graphic overview display.

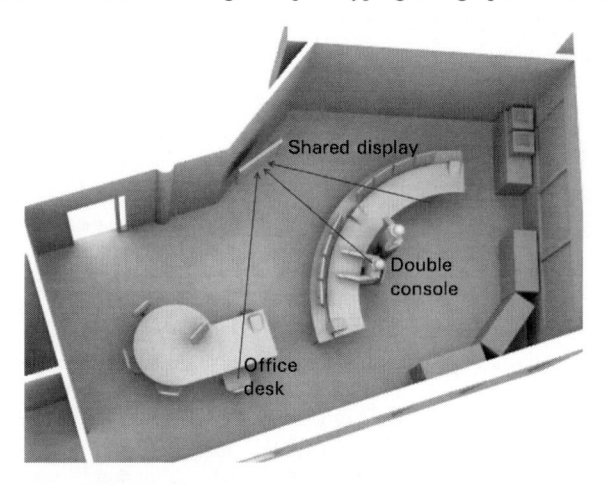

Figure 2. Control room layout – case 1

Two years later, the control room was moved to another location. Again, tasks were analyzed and a gradual change from supervision to dispatching and production volume control was found. Process control had become more important, due to changes in contracting (many small contracts instead of one large customer). This change required production flow control at platform level, however from an overall point of view. A new problem arose: it wasn't easy to control at platform level, because only 10% of the process data was available on-shore. Therefore, operators mainly acted upon off-normal messages (alarms). Improving off-normal messaging became the starting point of an extensive alarm rationalization project.

In addition, the company was advised to adapt the existing process operation philosophy to the new role of the on-shore control room. Of particular interest would be the division of tasks and responsibilities, including communication protocols. Work on the development of an operation philosophy is in progress.

Case 1B – Alarm management project

The alarm project can be considered a mix of HF and process control engineering. First, there was a need to get more knowledge on the characteristics of off-normal messages and the following up actions. Therefore, a detailed hierarchical task analysis (HTA) was carried out, using walk-through, talk-through discussions with experienced on-shore and off-shore operators. It showed that alarms may be initiated by process events, as well as by local activities or situations. In order to be able to understand an alarm message, an on-shore operator would need contextual information (you need to ask the local operator). In scientific research this topic is addressed as situation (or situational) awareness, and focuses primarily on interaction design (Van Erp, 1999).

Alarm philosophy

Parallel to the hierarchical task analysis, the project team developed an alarm philosophy, a strategy towards the effective handling of non-normal process situations. One of the dilemmas' faced: the more 'local' an operator is located, the better will be the quality of his context information. However, it is also more likely that messages are missed because the operator is not always in the local control room. On the other hand, operators in the central control room do not have much context information, but the control room is always manned.

Alarm management site survey

Next, a site survey on one platforms took place, to benchmark the current situation. The following Alarm Key Performance Indicators (KPI) were used:

1. Long term average; average number of alarms per hour, an indication of operator workload .
2. Alarm rate variation; does the average number of alarms/hour change much over time?
3. Frequent alarms; contribution of the most frequent alarms to the total alarm load.
4. Fleeting alarms; contribution of the most frequent fleeting alarms (active for a short period of time, up to 1 minute) to the total alarm load.
5. Number of alarms following upsets; plant upsets are periods of time where the load on the operator is particularly high.
6. Standing alarms; number of alarms active for a long period (>12 hours).

The results of the site survey have been presented in a spider chart (figure 3). The centre of the chart indicates a good score for the criteria, the outer sides indicate poor performance. Spider charts were used to set priorities for the alarm improvement project.

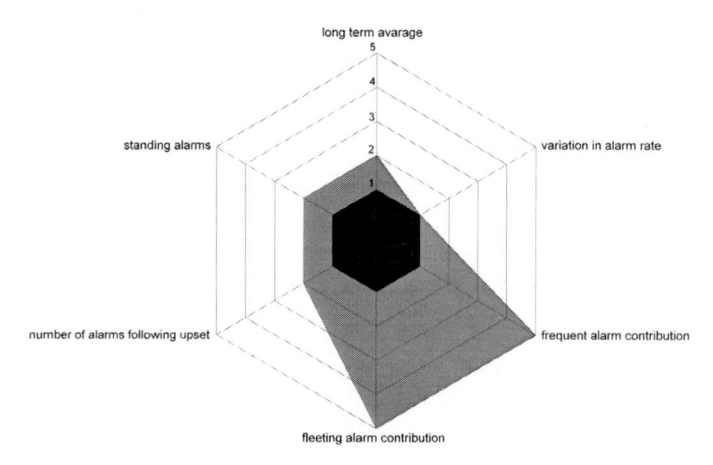

Figure 3. Spider chart of six Alarm key performance indicators

Alarm reporting and rationalization

Next step has been gathering data on actual off-normal messages of individual units and platforms. An online alarm reporting environment was installed, to determine and improve bad alarm actors and thereby reduce the alarm load. Next to a weekly alarm report, all alarms were systematically compared to the criteria for alarms set forth in the alarm philosophy. Refer to EEMUA (1999) for usable criteria. Although one expected that all platforms would be more or less the same, a major effort consisted of defining the same alarms and alarm levels at all units, thus improving consistency in process control.

Literature (EEMUA, 1999) suggests many solutions to reduce the number of alarms an operator faces. Dynamic alarm grouping proved to be very effective. Only one off-normal is presented of a group of related messages, though details are always literally at the fingertip of the operator. Also effective proved to be incident prediction, by using (a combination of) early indicators to detect abnormal process and/or equipment conditions.

Finally, safety alarms (such as Anti Collision, Man Over Board, Fire & Gas) should be presented and treated different from process alarms. Signals can only be judged adequately if the operator has access to context information. The operator cannot see whether an alarm is the result of a test (frequently done) or a real problem. Communication and responding to the alarm message is always needed. Safety devices (life jackets, man overboard alarm) are always tested before use. Testing takes place on site. Questions to be solved, for example by procedures, are whether the on-shore control room should be able to acknowledge test alarms (or not), should be aware of local testing (or not), and so on. Another example concerns fire & gas alarms. Fire and gas detectors may be sensitive to strong winds. Again, knowledge of the local situation is essential to make the right decision on-shore.

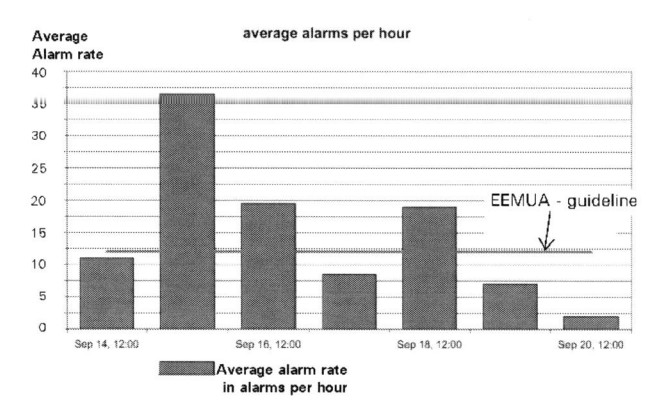

Figure 4. Impression of alarm reporting graphs – average alarms

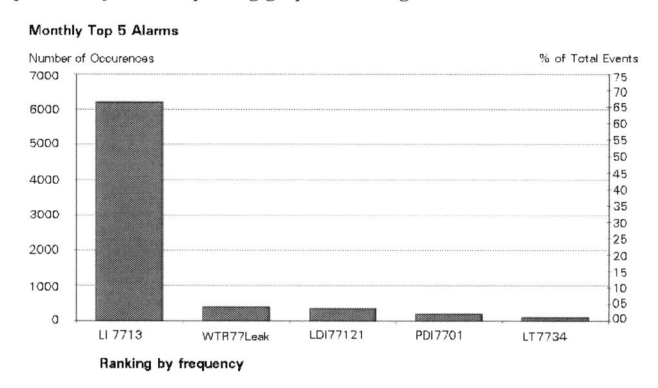

Figure 5. Impression of alarm reporting graphs – monthly top 5 alarms

Case 2A – Centralize off-shore control rooms and move to shore

Case 2 concerned moving supervision of gas production of approximately 30 off-shore platforms to shore, with the aim to improve gas contract handling and reduce decentralized control room manpower off-shore. The project team visited the Case 1 control centre and learned about system ergonomics. They decided that their own situation would be comparable and therefore a new task analysis was not considered necessary. While in Case 1 operators were available on-shore, here they were not. As a consequence, it is far more difficult to organize a task analysis at off-shore facilities (air transport, safety courses for the HF professional, costs). A step by step move to shore was decided for, in order to be able to cope with unexpected outcomes. First step, to reduce 24/7 staffing of off-shore control rooms to 12/7. This concerned three control room operator positions. Hence, by night the control room tasks are carried out on-shore. By daytime there is choice to be made: process control and supervision (also) on-shore, off-shore (at the off-shore central control room), or partially on- and off-shore. In any case, gas contract handling is taken over by the CCR onshore. The company expected that the local operator crew would become more focused on process control of their own plants thus expected to improve maintenance, reliability and consequently availability.

The control room design project included the design of a central control room in an existing office building. HF engineers developed a sketch design for several possibly locations. Next, the functional design of three workplaces took place. Functional design looked a lot like the control room of Case 1. Again, it was decided to develop a shared overview display wall combined with a triple operator console. Unless Case 1, an architect was introduced to make the final interior design.

Off-shore instrumentation needed to be upgraded in order to transfer process control to the new control room. The total count of existing graphics at over 30 platforms was 1800, in three different instrumentation systems. It was considered important to give the on-shore operators consistent and easy recognizable graphics. Therefore, a major HF project started to develop improved graphics, reducing the number of graphics considerably.

Case 2B – Process graphics design

A project team was composed of several experienced operators, a project leader, the head of instrumentation engineering, and two HF engineers. First, the project team got an introduction in ergonomic design guidelines for interaction design. Symbols, colours, and text size/format were defined by the project team. One large platform was chosen as a pilot for graphics redesign. The leading design principle for graphics redesign is: simplify (Pikaar, 2012; Bullemer et al., 2008). This can be done on the level of symbols (valve, pump), but also on the level of units (compressor, glycol recovery unit, furnace). Easy recognition of typical process units can be enhanced by applying a consistent layout.

Of course, navigation through many graphics can simply be simplified by reducing the number of graphics. The pilot graphics were thoroughly discussed by the project team. After consent, the rules to design graphics were compiled in a Human Computer Interactions Conventions document, amongst others to be used as a communication tool with the instrument (DCS) vendor. This document gives standards on colours, text size, symbols, arrangement of process values, and should give insight in why the graphics are designed as they are.

After the pilot phase, a selection of 136 graphics of the main production processes of 11 production facilities was made. It was argued that detailed graphics of utilities (and the like) would solely be used by local operators. Therefore, it was decided not to upgrade these graphics and have them still running at the local control rooms. With help of P&ID's and an experienced operator the graphics were designed as accurate as possible. Sometimes the operator needed the assistance of his off-shore colleagues to verify details and P&Ids that apparently were not all up to date (as build). What in fact happened, was a detailed tasks analysis on operator control tasks.

It was expected that the production platforms would be much alike. Designing a series of graphics would be simple: just copy. Though this approach would ensure easy recognition on a process unit level, it might be difficult to find out what platform you are looking at. In order to avoid mix-ups and keeping consistency in mind, some theoretical solutions were put forward:

- emphasize differences, if there are any
- use a watermark (graphic or textual) on each graphic
- use platform names in tooltips and title bars.

Later, it became clear that most of the platforms differed a lot from each other, no special solutions needed. Platform safety proved to be a very important issue. Questions were raised, whether the off-shore operators could trust that the on-shore control room has a full awareness of which platforms are being manned.

The selection of 136 graphics was redesigned by HF engineers, frequently consulting the experts: operators. This resulted in 25 new graphics, or a substantial reduction of 80%, which is in line with earlier findings of Pikaar (2012). Reduction of graphics was accomplished by simplifying symbols, omitting redundant or unimportant information (i.e., for on-shore supervision), and smart graphical solutions. A large contribution to this reduction occurred by using a standardized table for the line up of wellhead valves (figure 7). A typical example of a wellhead graphic is shown in figure 6. Three well heads are shown, each consisting from left to right of three valves in a row, a choke and two parallel valves for either gas to the production manifold or to the testing manifold.

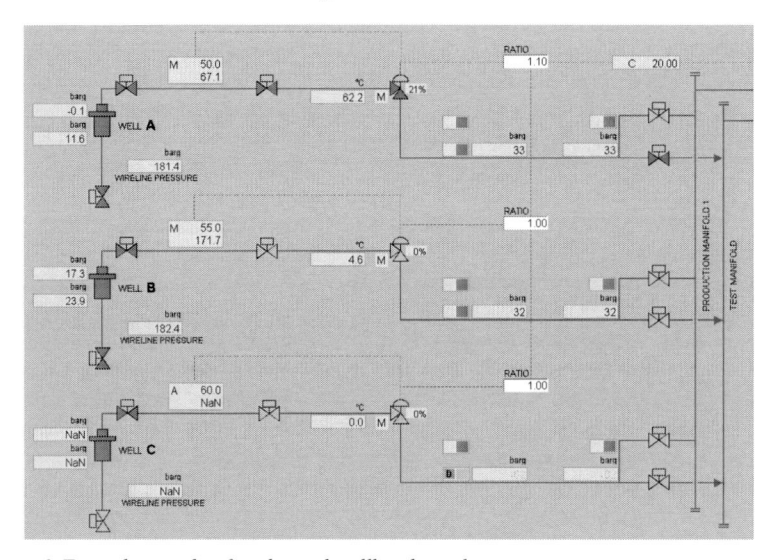

Figure 6. Typical example of traditional wellhead graphic

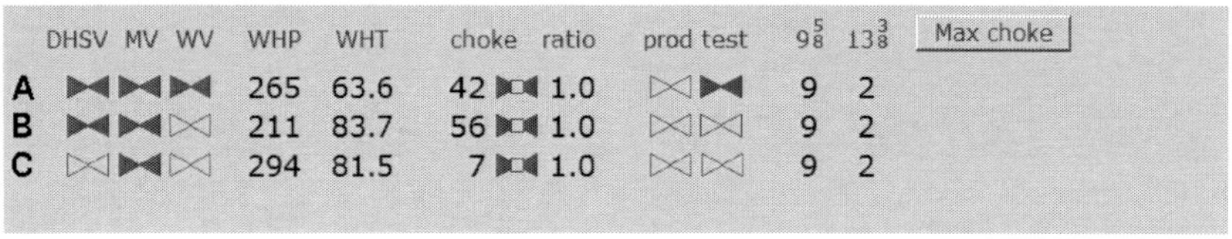

	DHSV	MV	WV	WHP	WHT	choke	ratio	prod test	$9\frac{5}{8}$	$13\frac{3}{8}$	Max choke
A	▶◀	▶◀	▶◀	265	63.6	42 ▶◀	1.0	▷◁ ▶◀	9	2	
B	▶◀	▶◀	▷◁	211	83.7	56 ▶◀	1.0	▷◁ ▷◁	9	2	
C	▷◁	▶◀	▷◁	294	81.5	7 ▶◀	1.0	▷◁ ▷◁	9	2	

Figure 7. Redesigned wellhead graphic

Conclusion - lessons learned

The aim of this paper is to review HF issues related to the move to shore of operator tasks. The authors learned several lessons, which will be indicated here in no particular order.

1. Control room design, i.e. layout and workplaces, is not much different from any other control room project. In case of combining two or more 24/7 off-shore control rooms, all traditional advantages are there, such as work load optimization (staff reduction) and easier communication between operators.
2. It may be difficult to find staffing for off-shore work. Operators may develop health problems (just by aging) that would not allow them going off-shore by helicopter. After many years off-shore, some operators just want to work closer to home (on-shore). And finally, well trained technicians are becoming scarce in industry.
3. At sea, a lot of maintenance will be going on. Question is whether local operators need local control to do an adequate job, and/or what role the on-shore control room should play. Communication is limited to telephone lines. Traditional radio communication between remote control room and field is difficult compared to, for example, a refinery. At the latter, field operators have easy radio contact and they walk into the central control room every now and then. Can this be achieved at a large distance using modern communication technology? Is there someone in the local control room? If so, why not take over control completely from on-shore? Answers will differ from case to case, and can only be given by looking deeper into the operator tasks and developing a process operations philosophy, describing when/how to organize tasks allocation best.
4. Regarding process control and supervision tasks, three things changed over the years:
    - Data transmission changed from CCTV-camera observation of analogue control panels to 100% on-shore availability of controlled variables. The major problem is a lack of context information, in particular related to safety issues. Is the Man Over Board alarm real, or part of an obligatory safety test? Is the gas alarm real, or just because there is a specific wind fall on the sensors at one side of the platform?
    - Contract management has become a new task. Nowadays, transport is separated from buying, and there are many contracts to be handled, requiring specific operator knowledge. At the North Sea area, there is a large variance in production volumes over the day. This introduced a new task: production volume management, i.e. how to optimize gas production wells.

- Production volume management includes production well optimization. For example, well pressure decreases over time. At high selling prices, it may be worthwhile to start up a compressor unit and produce from partially depleted wells. This is a matter of cost benefit calculations, which can be considered a new control room operator task. Should this task be combined or integrated with process control and supervision?
- In case 2A the gas dispatching and commercial activities are concentrated in a separate section of the office and daily/hourly volumes are dictated to the CCR crew, using special developed integrated hydro carbon calculation programs.

The system ergonomics approach focuses amongst others on the analysis of operator tasks. The move to shore definitely involves a change or reallocation of tasks. The systematic approach uncovers these tasks aspects very effectively, as has been illustrated by the case studies. A difficulty can be found in the distance between project teams, consisting of HF professionals and on-shore engineering staff, and the operators at the platform control room. It is not easy to visit the operators on-site and it proved to be difficult to keep the operators informed on project progress and project outcomes.

Acknowledgement

The authors thank all operators and project participants for their valuable contributions to the project presented in this case study presentation. They all contribute to the development of insight in Human Factors of moving off-shore operator tasks to on-shore control rooms.

References

Bullemer, P., Vernon Reising, D., Burns, C., Hajdukiewicz, J. and Andrzejewski, J. (2008), *ASM Consortium Guidelines – Effective Operator Display Design*. Houston, Honeywell International Inc./ASM Consortium.
De Looze, M., & Pikaar R.N. (2006). Meeting diversity in ergonomics. *Applied Ergonomics, 37*, 389-390.
EEMUA (1999). *Alarm systems: A guide to Design, Management en Procurement. Publication nr. 191*. London: The Engineering Equipment and Materials Users' Association.
EEMUA (2002). *Process plant control desks utilising human-computer interface, a guide to design, operational an d human interface issues. Publication nr. 201*. London: The Engineering Equipment and Materials Users' Association.
Henderson, J., Wright, K., & Brazier, A. (2002). *Human factors aspects of remote operation in process plants. Contract Research Report 432/2002*. HSE Books, Norwich. UK: Health & Safety Executive.
ISO 11064 (1998-2007), *Ergonomic Design of Control Centres – multi part standard*. Geneva, Switserland: International Organization for Standardization.
Kragt, H. (1992). *Enhancing Industrial Performance*. London: Taylor & Francis.
Pikaar, R.N., Rijnsdorp, J.E., Remijn, S.L.M., Lenior, T.M.J., Mulder, H.W.B., Van Bruggen, T.A.M., & Kragt, H. (1997). *Ergonomics in Process Control rooms,*

Part 3: The Analyses. Guidance Report M2657 X 97. The Hague, WIB International Users' Association.

Pikaar, R.N., Rijnsdorp, J.E., Remijn, S.L.M., Lenior, T.M.J., Mulder, H.W.B., Van Bruggen, T.A.M., & Kragt, H. (1998). *Ergonomics in Process Control rooms, Part 2: Design Guideline. Report M2656 X 98.* The Hague, WIB International Users' Association.

Pikaar, R.N. (2007). New challenges: Ergonomics in Engineering Projects. In R.N. Pikaar, E.A.P. Koningsveld, and P.J.M. Settels (eds.), *Meeting Diversity in Ergonomics* (pp. 29-64). Amsterdam, Elsevier.

Pikaar, R.N. (2012). HMI Conventions for Process Control Graphics. Accepted for publication in the Proceedings of the 18th IEA World Congress on Ergonomics. Recife, Brazil.

Rijnsdorp, J.E. (1991). *Integrated Process Control and Automation.* Amsterdam: Elsevier.

Van Erp, J. (1999). Situation awareness: theory, metrics and support. In: R.N. Pikaar. *Ergonomie in Uitvoering, de digitale mens* (pp 95-105). Utrecht, the Netherlands: Nederlandse Vereniging voor Ergonomie.

Cognitive Task Analysis – a relevant method for the development of simulation training in surgery

Norman Geissler, Anke Hoffmeier, Susanne Kotzsch, Stephanie Trapp,
Nadine Riemenschneider, & Werner Korb
HTWK Leipzig - University of Applied Sciences
Leipzig, Germany

Abstract

The insufficient amount of adequate practical experience for German surgeons in training has been criticised (Schröder et al., 2009). Realistic haptic simulations are needed to enhance the training and to improve the patients' safety. Initial research of the literature and several observations in operating rooms indicated the relevance of discectomy simulation training for neurosurgeons. A Cognitive Task Analysis (CTA) was performed to define a realistic and helpful scenario-based simulation. Important technical skills, cognitive aspects (e.g. strategies of decision-making), and implicit behavioural knowledge essential for performing high-quality surgery were analysed. The CTA was done in iterative cycles based on the results of the interviews with surgical experts and the progress of the developmental process. The results of the CTA are used as the basis for the development of a simulation system for surgical training. The relevance of analysing cognitive processes (especially evaluation of the surgical steps) has been proven and the important elements of a realistic surgical simulator (i.e. bleeding) were analysed. Due to the close interdisciplinary cooperation of engineers and psychologists, together with surgeons and the CTA-based user-centred design, the first validation showed that the concept of the simulator is highly relevant for surgical training.

Introduction

The German Medical Association points out that there is a shortage of young professionals in the surgical disciplines (Bundesärztekammer, 2009). Today, surgical training is mainly based on learning by doing, i.e. trainees are supervised by experienced surgeons during their first interventions based on the principle „see one, do one ..." (Lossing et al., 1992). In this process the responsibility and autonomy of the resident slowly progresses over years. Many interventions are needed (Jost & Klar, 2008) before the learning curve is flattened and complications are minimized (Kim et al., 2005). The training in the OR is not risk-free for the patients. Another problem is that, based on surveys of the Professional Association of German Surgeons (BDC), 63% of the surgeons in training do less than three interventions a week with supervision and 20% perform less than one surgery a week (Schröder et al., 2009). Training on human cadavers is cost-intensive, the properties of tissue may change post-mortem and their availability is not always sufficient – in addition to the ethical aspects. Further, the simulation of bleeding is not possible. Moreover, the complexity of surgical procedures is increasing based on technological innovations (navigation and mechatronics). Further, in many hospitals, the personnel numbers have been drastically reduced in the past few years. There is a need for additional supervised surgical training for practical and cognitive skills. Surgical simulators have been developed corresponding to this need.

Simulators

The current development history of surgical simulation starts at the end of the 1980`s (Satava, 2008). Surgical simulators are used to train residents (a) in the performance of critical steps in a surgical procedure, (b) the proper use of surgical equipment and (c) the management of exceptional cases and complications. The vast majority of surgical simulators are based on virtual reality and computer graphics (Satava, 2008). These simulators are technically and algorithmically complex due to the difficulty in modelling the physics of patient anatomy, especially pathologic cases. Innovative Surgical Training Technologies (ISTT) has identified another approach to model the patient, based on synthetic and non-human organic materials. This allows a correct simulation of the patient's anatomy and the surgical process. A relevant advance of physical simulators to virtual simulation systems is the use of real surgical instruments in an identical approach as performed in the operating room (OR) so that the surgeon can see the identical manipulative impact on the simulated anatomical structures (Korb et al., 2011; Grunert et al., 2006). The goal of the ISTT research project is the development of a high fidelity patient phantom for surgical training of spinal surgery (discectomy procedures) including a training concept (Korb et al., 2011). This interdisciplinary approach (surgeons, psychologists, educational scientists, engineers, computer scientists) is realized in close cooperation with the University Hospital Leipzig and the Medical Faculty of the Universität Leipzig. The first step was to perform a Cognitive Task Analysis (CTA) to confirm or reject the above stated specifications for a simulator. In a second step, the simulator was designed and modelled by different materials as well as with synthetic blood. The final steps included a continuous validation of the prototype simulator with clinical experts in an iterative development process.

Cognitive Task Analysis

In a first step a CTA was performed to identify, describe and detail cognitive processes, decision-making and physical actions involved in successful performance (Johnson et al., 2006) of a surgical procedure. CTA is a "…set of methods for identifying cognitive skills, or mental demands, needed to perform a task proficiently" (Militello & Hutton, 1998). The aim of a CTA is to collect data that can be used as an educational resource for trainees and could provide the basis for a training curriculum (Johnson et al., 2006). One specific method is the Applied Cognitive Task Analysis (ACTA) developed by Militello et al. (1997; Militello & Hutton, 1998). This method allows to elicit expert knowledge from Subject Matter Experts (SMEs) working in a dynamic decision making environment. ACTA is divided into four different phases: (1) a Task Diagram, (2) a Knowledge Audit, (3) a Simulation Interview and (4) a Cognitive Demands Table. An additional method was found necessary because of the missing opportunity of a knowledge audit and a missing explicit iterative character of the method.

A further interesting approach for a CTA is from Zachary et al. (2000) who suggest:

1. Completing an a priori domain analysis, therefore to study relevant information about the domain before further planning the details of the CTA to have an acceptable expertise in the domain.
2. The definition of participants, settings, and examples / scenarios.
3. The recording of experts in a real or simulated problem solution, closely followed by verbal question-answering-protocols.
4. The analysis and the representation of the data (eventually repeat step 3).
5. Development of an applicable cognitive model (eventually repeat step 3 and 4 to analyse if all needed details and quality levels have been collected for the analysis).
6. Validation of the CTA results.

These approaches were combined to analyse the relevant aspects for the design of the training concept and the simulator for discectomy training in an iterative development cycle.

CTA for the development of the discectomy simulator

Based on Zachary et al. (2000), an extensive a priori domain analysis was completed in which literature research, 36 observations in the OR and discussions with surgeons were included. To analyse the observations, a standardised OR form was developed that had to be completed after each observation and to be made anonymous. Based on a domain analysis, the decision for the surgical intervention was discectomy. This surgical intervention should be trained in simulator training. One of the reasons why discectomy was chosen is that it is a type of intervertebral disc operation that is one of the twenty most common operations in Germany (Gesundheitsberichterstattung des Bundes, 2009). There were around 161,000 cases of excisions of sick intervertebral disc tissue in 1996 (Statistisches Bundesamt Deutschland, 2009). Most of the cases were on the lumbal spine. One result of the observations was a task diagram of the surgical workflow. Based on this decision for the surgical procedure and the workflow, a concept for a guideline for a partly standardised face-to-face interview was designed. The interview guideline included the topics:

- Surgical procedure (i.e. practical experience, preparation of an operation, challenges, dealing with complications)
- Training (i.e. which materials and books are used, own training, own goals for training, relevant aspects for a simulator and trainings concept)
- Task analysis (analysis and comments of the surgical workflow, challenges especially for novices, possible consequences)
- Scenario presentation (evaluation and advice of the scenario concept).

The interview guideline was validated by a medical expert before the study. In the first study, the SMEs were primarily selected according to the principles of intensity sampling (Patton, 2002). Based on the definition of intensity sampling of searching for excellent examples we made interviews with a) experienced surgeons and b) surgeons in training to receive in depth information. For the definition and identification of the sample and the questions the prior exploratory work (i.e. literature, visits to the OR) was needed. The group of the experienced surgeons was needed for their expertise of the needed qualification and the group of the surgeons in training was needed for the analysis of their goals and explicit needs since they will be one explicit target group for the developed training.

The tasks, which require cognitive skills, i.e. assessment and problem solving, were identified based on this study. Next the SMEs were asked to explain these in detail and relevance for the simulator to be developed. The interviews with 17 surgeons in a multicenter design were analysed with MaxQDA. Observations both in situ and through the use of video data and in-depth interviews with experts were combined (Johnson et al., 2006). Individual interviews were necessary to avoid hierarchy effects and to provide a safe and anonymous atmosphere (Wu, 2000). The structure in German hospitals is in many cases quite hierarchical therefore exists a high risk that a surgeon who is subordinate in the hierarchy will be less outspoken in his opinion in the presence of a supervisor. Based on the interdisciplinary structure of the project the other groups were informed about the results so that the first prototypes of the simulator could be built.

The other groups are the bioengineering group (i.e. engineers, product designer) which is working on the design of the simulator materials, and the sensory group (engineers) which is working on the blood pump and sensors for the simulator. The main challenge of the project was the design of a clinically and anatomically correct training simulator. The first study showed that the surgeons see the development of a realistic simulator for a discectomy as very relevant, especially the optical and haptic characteristics were emphasised. A further step involved attending courses as well as studying anatomic specimens together with the Institute of Anatomy at the Universität Leipzig. Based on these derived specifications, the anatomic structures were built with (or using) synthetic and non-human organic materials (polyurethane, epoxy, modified silicone, modified gelatine, modified latex) with different moulding procedures. The following structures were modelled: ligamenta flava; lumbar vertebrae; herniated spinal disc consisting of annulus fibrosus and prolapsed nucleus pulposus; epidural space filled with adipose tissue; dura containing spinal nerves and cerebrospinal fluid. Further, a pumping system with a flow rate controller was developed to infuse theatre blood into the phantom to increase the realism of the training session (Korb et al., 2011). As part of an iterative development process, the validation of the prototype was performed in two steps:

(1) The first step of the validation was performed with two independent neurosurgeons with a similar degree of expertise (practical experience) based on a structured interview. The two surgeons performed the discectomy including a laminotomy and removing of the prolaps for decompression. The SMEs were able to proceed with the simulated surgery at the same time as the researcher was able to elicit information from the SMEs without interrupting the simulation, based on a think aloud technique. The ISTT researchers were analysing the comments and

answers of the surgeons and completed the validation with the open questions of the previously designed structured interview guideline after the simulated surgery. The validation was recorded on video and the files were transliterated and further analysed with MaxQDA. This procedure was chosen to enable the SME to fully concentrate on the high-fidelity simulation which was necessary for a realistic working condition.

(2) In a second step, a validation study with an improved prototype was performed at the German Spine Congress (DWG 2010) in Bremen. At the DWG a convenience sampling (Patton, 2002) was taken, the congress visitors were able to perform several surgical steps of the discectomy, and then a short structured interview was performed. All in all, eight surgeons from six different hospitals were interviewed at the congress.

The clinical experts agreed that this model approach is useful for the training of surgeons. The lamina, ligamenta flava, epidural connective tissue and the intervertebral disc were evaluated (without exception) as realistic. The dura and its haptic feel were evaluated as realistic by 50%. In detail, they observed that the complex structure of the dura should be improved. Based on the CTA the simulator does not include the first cutting of the skin. The necessity of integrated bleeding was emphasised and the demonstrated integrated bleeding was evaluated (without exception) as realistic. The evaluation proved that, with the exception of the dura, the simulator is a high fidelity simulation. The dura was improved based on the received feedback. This is relevant because fidelity is a combination of realism and comprehensiveness (Jones et al., 1985). The simulator is realistic because it makes use of the real surgical instruments of the task. The comprehensiveness is given based on the definition that it means "…the degree of completeness and accuracy of representation of all functions, environmental characteristics, situational factors, and external events that are present in the target system or affect its function" (Jones et al., 1985).

CTA for the training concept

A modern concept based on educational psychology is needed to define objective feedback criteria for the results of the participants, this is not the standard in surgery and had to be defined based on the different surgical approaches. Therefore, a golden standard for the surgical results has to be defined with surgeons, this is not just a problem for surgical training but also for surgery in the reality based on differing surgical approaches. The goal of the ISTT project is to develop teaching and learning concepts for the simulation of surgical interventions which can be evaluated. Therefore, certified training concepts have to be developed, which will make the education and hopefully surgery in general more standardised and efficient.

To reach this goal the definition of relevant tasks in cooperation with medical experts is a mandatory precondition. A relevant taxonomy of tasks for the training was developed by Rasmussen (Wentink et al., 2003), that distinguishes three cognitive behavioural levels:

- behaviour based on sensomotoric skills, i.e. wound closure
- rule based behaviour: complex and repeating tasks, i.e. inspection of the intervertebral disc
- knowledge based behaviour: i.e. in an unclear emergency situation with no rules for every detail for example an unexpected intensive bleeding

The implementation of these three levels of the Rasmussen model (Wentink et al., 2003) is the foundation of the training concept, a program to direct training sessions and the interactions of the trainees with the simulator.

For the definition of the training sessions the explicit identification and categorisation of the cognitive preconditions for the working on complex tasks is relevant. That means that every surgery has to be closely observed and analysed (see above). This analysis was done using the aforementioned observations and further face to face interviews with the above mentioned 17 surgeons in a multicenter approach in Germany. A relevant difference between experts and novices is the higher relevance of feedback for the novices. The detailed qualitative analysis of the 17 interviews will be published in a further publication. There was a specification of the relevant aspects for the training, especially the instruction and the relevant briefing and debriefing for the participants. The scenarios for the simulator were also defined. A scenario can be defined as stories that are illustrating a sequence of actions carried out by intelligent agents (Alexander & Maiden, 2004). A relevant precondition was the validated surgical workflow. Within each scenario, each task was examined and a decision was made on whether the task would continue to exist in the future simulation. Another result is the relevance of a surgical and educational trainer during the training. The purpose of the interviews and the validation interviews was to elicit information on how the SMEs operate within the task being investigated, and what the relevant aspects for the design of the simulator and training concept are.

Conclusion

The development of realistic scenario based training with simulators has a high relevance, because of the existing training situation of surgeons in Germany and the existing challenges i.e. the lack of validated training possibilities. The relevance of improving the training for the surgical community was also discussed on several scientific conferences i.e. the first World Congress on Surgical Training in Gothenburg (Sweden) 2011. The Cognitive Task Analysis is a relevant method for the development of training in medicine (Velmahos et al., 2004). Based on a CTA, including observations in the OR, interviews and discussions with surgeons, a prototype for a *"scenario based simulation training"* was developed.

The CTA approach used, based on the CTA methods of Militello et al. (1998) and Zachary et al. (2000), was very useful for the analysis of the necessary aspects for scenario based simulation consisting of the simulator and the educational concept. The data collected for the development of the discectomy simulator was valid, because the SMEs mental operations and judgments, their knowledge about the familiar task and its sub-domains could be analysed (Hoffman, 1987).

The results of the validation studies show the necessity of realistic training simulators for the improvement in surgical training of spinal disc surgery. The prototype system has characteristics well suited for training. Nevertheless, the dura should be improved and sensors integrated onto the spinal nerves. The sensors would allow trainers and trainees to receive visual feedback during a simulation session. The necessity of an integrated bleeding was emphasized and the demonstrated integrated bleeding was evaluated (without exception) as realistic. The specifications for synthetic simulators should (1) include correct optical properties, as well as a realistic haptic feeling of the surface; (2) be robust and capable, in order to be used for repeated surgical training; (3) include target structures and structures at risk (4) enable the trainer to analyse and save the injury of structures at risk for the debriefing; (4) provide simulation of bleeding, as well as management of the bleeding.

By using the high-fidelity simulation it could be analysed that the SMEs interacted with the scenario to a high degree. In this respect, the simulation had a high enough fidelity for the respondents to be in a realistic working atmosphere with the familiar task and to interact with the simulation in a way similar to how they would have worked in real life. The developed simulator and the training in the current state are not artificial. This is relevant because if the task deviates too much from the usual problem-solving situation then "…the less it tells the knowledge engineer about the usual sequences of mental operation and judgments" (Hoffman 1987). In surgery, as described for other domains (Edwards et al., 2004), a high-fidelity simulation requires a high knowledge of the domain being studied. Besides the needed knowledge a very high level of resources, in our case a minimum of 2 years with further experiences in the field and the financial resources for a large interdisciplinary team of researchers, is needed for a successful project.

The possible benefits, i.e. the targeted improvement of patient safety based on the improved training possibilities (Schröder et al., 2009), justify these resources. A surgical skills lab course based on the results of a CTA and using inanimate models can, as shown in other studies (Velmahos et al., 2004), improve the knowledge and technical skills of new surgical interns. The CTA showed the relevant aspects for the training concept especially concerning the wanted feedback for the learning process, a clear structure and the wanted independence during the procedure. More data must be accumulated to convince administrators of the university and general hospitals that the regular training of their surgeons in a surgical skills lab on validated simulators with an educational training concept (including debriefing) is necessary and should become part of the curriculum. The next steps include advancing the prototype as part of an iterative development process and integrating a didactic scenario-based concept for the surgical training modules. The scenarios are discussed with medical experts and a validation of the scenario based training with the next generation of the high fidelity simulator is planned in a skills lab course with neurosurgeons in February 2012.

Acknowledgements

The authors of this paper would specifically like to thank all participating surgeons especially the University Hospital Leipzig Department of Neurosurgery and Prof.

Meixensberger, Dr. Adermann, Dr. Lindner and Dr. Dengl for their participation in this study and for letting us borrow time and space from their very valuable work.

References

Alexander, I.F. & Maiden, N. (2004). *Scenarios, stories, use cases*. Chichester: John Wiley & Sons.

Bundesärztekammer (2009). Die ärztliche Versorgung in der Bundesrepublik Deutschland, Ärztestatistik vom 31.12.2009, BÄK 12.04.2011.

Brodje, A., Prison, J., Jenvald J., & Dahlman, J. (2011). Applied Cognitive Task Analysis as a tool for analyzing work demands in a C4I environment: a case study using a mid-fidelity simulation. In D. de Waard, N. Gerard, L. Onnasch, R. Wiczorek, and D. Manzey (Eds.), *Human Centred Automation* (pp. 313-326). Maastricht, the Netherlands: Shaker Publishing.

Edwards, J.S., Alifantis, T., Hurrion, R.D., Ladbrook, J., Robinson, S. & Waller, A. (2004). Using a simulation model for knowledge elicitation and knowledge management. *Simulation Modelling Practice and Theory, 12*, 527-540.

Gesundheitsberichterstattung des Bundes (2009). Die 50 häufigsten Operationen der vollstationären Patientinnen und Patienten in Krankenhäusern – Bericht von 2009, *gbe-bund.de, 07.04.2011*.

Grunert, R., Strauß, G., Moeckel, H., Hofer, M, Pössneck, A., Fickweiler, U., Thalheim, M., Schmiedel, R., Jannin, P., Schulz, T., Öken, J., Dietz, A., & Korb, W. (2006). ElePhant - An anatomical Electronic Phantom as Rapid Prototyping Simulation-system for otorhino-laryngoscopic-surgery. *IEEE International Conference on Engineering in Medicine and Biology (EMBC)*, 312-313.

Hoffman, R.R. (1987). The Problem of Extracting the Knowledge of Experts from the Perspective of Experimental-Psychology. *AI Magazine, 8*, 53-67.

Johnson, S., Healey, A., Evans, J., Murphy, M., Crawshaw, M., & Gould, D. (2006). Physical and cognitive task analysis in interventional radiology. *Clinical Radiology, 61, 97–103*.

Jones, E.R., Hennessy, R.T., & Deutsch, S. (1985). *Human Factors Aspects of Simulation*. (Report NOOOI4-85-G-0093). Washington DC: National Research Council.

Jost, J.O. & Klar, E. (2008). Mindestmengen in der Chirurgie, Chirurgische *Gastroenterologie, 24*, 270-270.

Kim, S.S., Lau, S.T., Lee, S.L., & Waldhausen, J.H. (2005). The learning curve associated with laparoscopic pyloromyotomy. *Journal of Laparoendoscopic & Advanced Surgical Techniques, 15*, 474-477.

Korb, W., Sturm, M., Andrack, B., Bausch, G., Geissler, N., Handwerk, J., Müller, M., Seifert, A., Steinke, H., & Meixensberger, J. (2011). Development and Validation of a Prototype for Training of Discectomy. *International Journal of Computer Assisted Radiology and Surgery, Vol 6, Suppl 1*, 121-122

Lossing, A., Hatswell, E.M Gilas. T., Reznick, R.K., & Smith, L.C. (1992). A technical-skills course for 1st year residents in general surgery: a descriptive study, *Canadian Journal of Surgery, 35*, 536–540.

Militello, L.G., Hutton, R.J.B., Pliske, R.M., Knight, B.J., & Klein, G. (1997). *Applied Cognitive Task Analysis (ACTA) Methodology.* (Report NPRDC TN-98-4). San Diego, USA: Navy Personnel Research and Development Center.

Militello, L.G. (1998). Learning to think like a user: using cognitive task analysis to meet today's health care design challenges, *Biomedical Instrumentation & Technology, 32*, 535–540.

Militello, L.G. & Hutton, R.J.B. (1998). Applied cognitive task analysis (ACTA): a practitioner's toolkit for understanding cognitive task demands, *Ergonomics, 41*, 1618-1641.

Patton, M.Q. (2002). *Qualitative Research and Evaluation Methods.* Thousand Oaks: Sage Publications Inc.

Schröder, W., Krones, C., & Ansorg, J. (2009). Akquise von chirurgischem Nachwuchs was ist zu tun? *BDC Online – 01.03.2009,* Berufsverband der Chirurgen, http://www.bdc.de/index_level3.jsp?documentid=4BE4C5D385F061A5C1257 57F0030CFBC&form=Dokumente 12.04.2011.

Satava, R. M. (2008). Historical Review of Surgical Simulation – A Personal Perspective. *World Journal of Surgery, 32*, 141-148.

Statistisches Bundesamt (2009). Deutschland Die 20 häufigsten Operationen der vollstationär behandelten Patienten insgesamt, destatis.de, 07.04.2011

Velmahos, G.C., Toutouzas, K.G., Sillin, L.F., Chan, L., Clark, R.E., Theodorou, D., & Maupin, F. (2004). Cognitive task analysis for teaching technical skills in an inanimate surgical skills laboratory, *The American Journal of Surgery, 187*, 114–119.

Wentink, M., Stassen, L.P., Alwayn, I., Hosman, R.J., & Stassen, H.G. (2003). Rasmussen's model of human behavior in laparoscopy training. *Surgical endoscopy, 17*, 1241-1246.

Wu, A. (2000). Medical error: the second victim. The doctor who makes the mistake needs help too. *British Medical Journal, 320*, 726-727.

Zachary, W., Ryder, J.M., & Hicinbothom, J.H. (2000). Building Cognitive Task Analyses and Modelsof a Decision-making Team in a Complex Real-Time Environment. In J.M. Schraagen, S.F. Chipman, and V.L. Shalin (Eds.) *Cognitive Task Analysis* (pp. 365–383). Mahwah, NJ: Lawrence Erlbaum Associates, Inc.

Cognitive performance limitations in operating rooms

Nicki Marquardt[1], Kristian Gerstmeyer[2], Christian Treffenstädt[3], & Ricarda Gades-Büttrich[4]
[1]Rhine Waal University of Applied Sciences, Kamp-Lintfort
[2]Augenpraxis-Klinik Minden
[3]Georg-August University of Goettingen
[4]Leuphana University of Lüneburg
Germany

Abstract

Currently there is a lack of validated and applied models concerning cognitive performance limitations and human error in medical work environments. The dirty dozen model (Dupont, 1997), for instance, an established concept of human performance limitations and error causation in aviation maintenance, was applied to surgical context – especially to ophthalmology. The 12 categories presented in this concept are: lack of resources, complacency, lack of teamwork, stress, lack of communication, distraction, lack of knowledge, lack of awareness, lack of assertiveness, fatigue, social norms and pressure. Roughly the whole population of surgically practicing ophthalmologists in Germany (N = 1063) was surveyed in regard to the relevance of various performance limiting factors. The questionnaire included a quantitative as well as a qualitative section, where participants were able to state experienced examples for each category. So, this study concerned the general perceptions and judgments of surgeons on their own as well as team-based cognitive performance limitations during surgery. The response rate of this survey was about 20%. The results indicate that pressure, lack of communication and stress are the most considerable categories. A factor analysis based on these 12 categories was performed. The results of this analysis were the two factors organisational context and social interaction. Thus, the results indicate a strong negative impact of organisational and social factors on the cognitive performance of surgeons in operating rooms.

Introduction

Human error in medical care causes many lethal incidents (Calland et al., 2002), about half of which are estimated to be preventable (Mishra et al., 2008). Patient safety studies conducted in the United States show that more than 60.000 Americans die each year due to adverse events while being hospitalised (Brennan et al., 2004; Thomas et al., 2000; Awad et al., 2005). Moreover, other research even suggests that up to 98000 U.S. patients die each year from preventable medical errors (Sexton, Thomas & Helmreich, 2000). In addition, Thomas and colleagues' study (2000) shows that almost fifty per cent of all adverse events (46.1 %) occur in operating

In D. de Waard, N. Merat, A.H. Jamson, Y. Barnard, and O.M.J. Carsten (Eds.) (2012). *Human Factors of Systems and Technology* (pp. 317 - 326). Maastricht, the Netherlands: Shaker Publishing.

rooms during surgery. The complexity and technical requirements of surgical interventions are assumed to be comparable to high demanding work environments in other fields of work (Carthey et al., 2001; Sexton et al., 2000; Reader et al., 2006). Consistent with recent argumentations (Mishra et al., 2008; Reader et al., 2006) the current study therefore considers the operating room to be a high-risk work place. The key factor to successful surgery is the surgeon's actual performance, which is dependent not only on technical skills but also on various other abilities (Carter, 2003). McDonald et al. (1995) conducted a study comparing three factors of readiness for performing a surgical operation. Their results show that, compared to technical and physical preparations, surgeons believe that the state of mental readiness is the most influential factor for an excellent performance in the operating room. A full focus on the operating procedures, the anticipation of upcoming work steps and potential complications as well as the ability to manage distractions quickly and efficiently were further factors which surgeons stated to be important prerequisites of excellent performance. In general, these aspects relate to cognitive performance, which is defined as the aggregated performance of several cognitive functions and processes including aspects of perception, attention and deliberate thinking (Budde & Barkowsky, 2008). Introducing the theoretical framework of *team cognition*, recent approaches to teamwork and team performance additionally emphasise a major influence of cognitive processes and hence cognitive performance on effective communication and coordination within teams (Salas et al., 2008). Altogether, past research has supported our view, regarding cognitive performance as an important success factor in the operating room.

Those aspects of performance in surgical interventions, that do not reflect the technical skills of the surgeon or the operating room personnel in general, are commonly subsumed under the domain of *non-technical performance* (Flin & Maran, 2004), which among other things includes teamwork, leadership, situation awareness, decision making, task management and communication (Yule et al., 2006; Yule et al., 2008). Although cognitive and non-technical performance are by definition similar to each other, the former of both concepts is broader and reflects a more general set of abilities. Hence, it is our belief that cognitive performance not only affects non-technical performance but also the actual technical performance of an operating surgeon, e.g. via aspects of attention and concentration. Supporting this line of reasoning, Mishra and colleagues (2008) were able to show that cognitive performance and technical performance in a set of surgical interventions were related to each other. Taken as a whole it can be reasonably argued that an insufficient cognitive performance, by diminishing non-technical as well as technical performance, increases the risk of human error induced incidents and accidents. Accordingly, surgeons report that cognitive performance limitations are a cause of error in more than half of all surgical incidents (Gawande et al., 2003). An analysis of medical malpractice claims reveals human errors arising from deficient cognitive performance to be involved in over sixty per cent of all cases (Rogers et al., 2006). Examples of performance limitations leading to human error in these studies were misjudgments, insufficient vigilance, memory failure, communication breakdown or distractions. An overview of the different theoretical concepts and their previously described relationships is given in Figure 1.

cognitive performance limitations 319

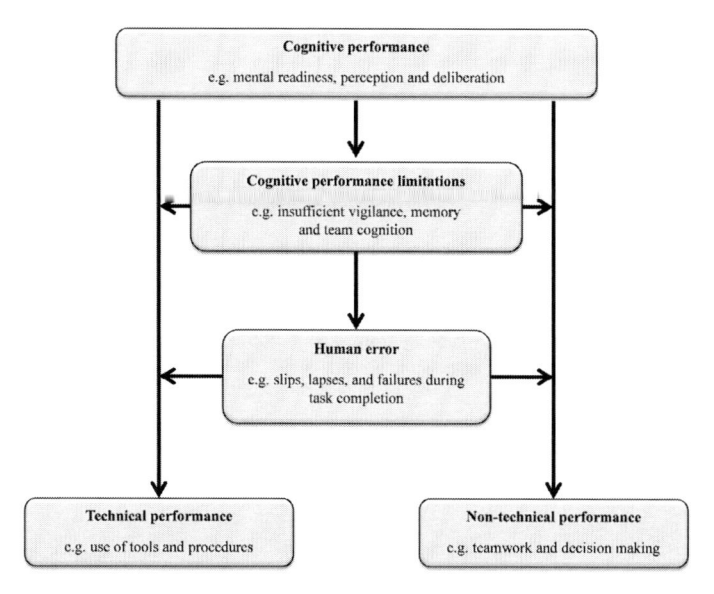

Figure 1. Assumed relations between relevant theoretical concepts

Research on human error in medical care is strongly influenced by research in other fields of work with equally highly demanding work environments. Especially insights derived from the aviation industry are often successfully transferred to the medical field (Calland et al., 2002; Sexton et al., 2000). In this context, models of error causation and safety related non-technical skills originating from aviation have been adapted to fit the requirements of specific medical contexts, such as for operating room staff (using HFACS; ElBardissi et al., 2007), surgeons alone (Non-technical Skills for Surgeons NOTSS; Yule et al., 2008) or anesthetists (Anesthetists Non-technical Skills, ANTS; Flin & Maran, 2004). Although these types of results are generally encouraging, there is currently no predominant model of human error in medical care.

In line with the approaches that are mentioned above to integrate theories of human factors coming from aviation into the medical field the current study introduces a comprehensive categorisation of human error causation – the *dirty dozen* – to error research in medical work environments. The dirty dozen was first introduced by Dupont (1997). It is based on Dupont's practical experiences with his *Human Performance in Maintenance* (HPIM) workshops which are a form of *crew resource management* (CRM) training program. Assigned by Transport Canada (the Canadian ministry of transportation) he developed the HPIM workshops to sensitise aircraft maintenance crews for human error and its' sources (Dupont, 1998). His experiences resulted in the systematisation of human error causes into twelve distinct categories (see table 1).

Table 1. Descriptions of the dirty dozen categories of human error

Lack of communication	Missing exchange of information regarding the state of work
Lack of teamwork	Missing coordination of work flow and low cooperation
Complacency	Not enough diligence and low accuracy while working
Social norms	Disregard of safety interests and safety instructions
Pressure	Time pressure, high pressure to perform
Lack of awareness	Low situation awareness, overlooking of security threats
Stress	Impediments in performance through strain, fears or worries
Fatigue	Weariness or exhaustion because of shift-work or medical condition
Distraction	Work interruptions, external distractors
Lack of resources	Missing or damaged equipment, lack of personnel, lack of information
Lack of knowledge	Lack of experience or work-related knowledge, out-dated knowledge
Lack of assertiveness	Repressing of concerns to avoid conflict, scared of confrontations

Many causes for human error presented by the dirty dozen are related to cognitive performance by resembling individual cognitive limitations (e.g. lack of awareness, distraction, complacency) or team cognition limitations (e.g. lack of communication, lack of teamwork), both of which we assume to negatively impact technical and non-technical performance in the operating room by decreasing the general cognitive performance of operating surgeons. The dirty dozen categorisation was directly derived from practical experiences and is easy to adapt for different fields of application. It was tested and successfully transferred to the defense industry (Marquardt & Höger, 2007) as well as the automotive industry (Marquardt, Robelski, & Höger, 2010) and several other industries (Marquardt, Gades, & Robelski, in press). Although not all categories of the dirty dozen are mutually exclusive, it covers a wide range of possible causations for human error and hence qualifies as global human error systematisation in the medical field. The current study presents a first trial of the dirty dozen within human error research applied to the medical working environment.

Method

Participants

Tasks and cognitive requirements within the field of medicine are very heterogeneous. On one hand, there are routine jobs but also complex surgical operations under normal conditions. On the other hand, there is anaesthesia and operations in emergency situations. In order to adapt the dirty dozen model from aviation to medicine and to gain comparable results it was important find a homogeneous sector within the broad field of medicine. Therefore, the field of surgically practicing ophthalmology was selected as an adequate working environment. Ophthalmologists are a suitable population because the need for precise work – due to working with microscopes and laser – usually requires high cognitive performance and motor skills. Hence, roughly the whole population of surgically practicing ophthalmologists in Germany was surveyed regarding the relevance of different performance limiting factors. Specifically, 1063 surgically practicing ophthalmologists received an anonymous questionnaire. The response rate was about 20 % (N= 215 operators). The demographic data were: sex: 75,3% male, 18,6% female (6,0% no particulars) and age: 61,9% above 46 years, 34% below 46 years (4,2% no particulars). Compared to response rates in marketing research surveys and official statistics concerning demographic aspects of the population of ophthalmologists this sample can be seen as relatively representative.

Materials

The questionnaire included a quantitative as well as a qualitative section, where participants were able to state self-experienced examples for each category. The categories were based on the 12 dirty dozen causes of human error. Within the quantitative section prototypical descriptions, behavioural markers and indicators for each category were presented. The category *lack of teamwork*, for example, was described by elements such as

- insufficient understanding of cooperation
- bad cooperation among team members
- deficient task coordination

A five-point Likert-Scale was provided to assess the negative impact of each category on cognitive performance of the operating room team. Within the qualitative section of the questionnaire the surveyed operators had the opportunity to state specific examples of adverse medical events.

Results

Quantitative results

The mean values of the 12 categories are presented in figure 1. Higher values mean a stronger negative impact of each category on cognitive performance within the operating room team. As can be seen in figure 1, the surveyed surgeons felt that categories are very similar in their impact on cognitive performance. The range of mean values of all 12 categories is between 3.4 and 2.4. As a consequence there is no overall dominant performance limiting factor.

In order to find a latent structure behind the dirty dozen, a factor analysis (PCA) of all twelve categories was performed. The factor analysis yielded two factors with eigenvalues greater than 1 for each and explained 59 % of the variance. The rotated factor solutions – only factor loadings of ≥ .5 were accepted – led to the following factors: *Social Interaction* and *Organisational Context*. As can be seen in table 2, the factor referring to *Social Interaction* reflected categories such as *lack of teamwork* (.80), *lack of communication* (.78) and *social norms* (.83). The second factor was interpreted as comprising demanding but insufficient work conditions such as *fatigue* (.83), *stress* (.86) and *pressure* (.76). Nevertheless, some dirty dozen categories as e.g. lack of resources or distraction revealed weak factor loading. Those categories could not be associated distinctively with one of these two performance limiting factors.

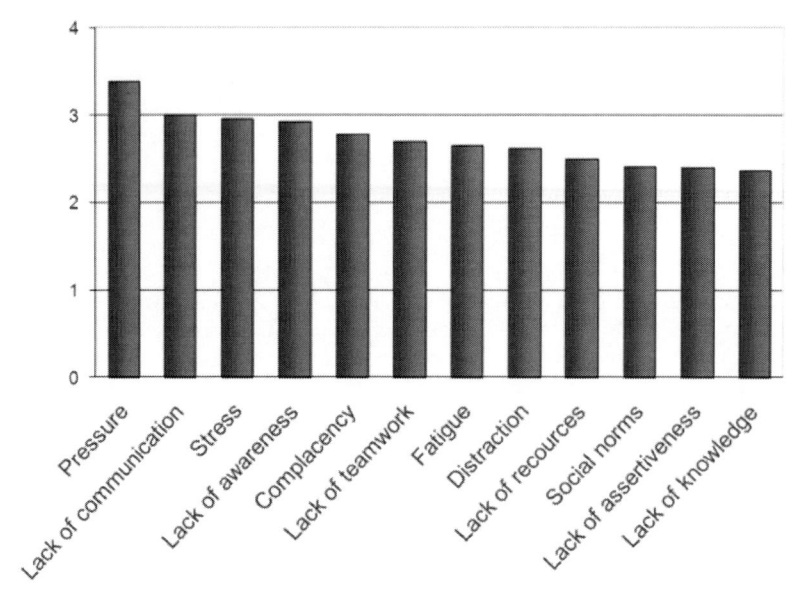

Figure 2. Categories for cognitive limitations and adverse medical events within the operating room

Table 2. Rotated (VARIMAX) factor loadings of specific categories

	Performance limiting factors	
	Social Interaction	Organisational Context
Lack of Communication	0.776	0.099
Lack of Teamwork	0.803	0.124
Complacency	0.727	0.256
Social Norms	0.839	0.183
Pressure	0.186	0.759
Lack of Awareness	0.656	0.296
Stress	0.163	0.856
Fatigue	0.251	0.826
Distraction	0.482	0.355
Lack of Resources	0.482	0.490
Lack of Knowledge	0.673	0.281
Lack of Assertiveness	0.616	0.358

Qualitative results

As noted above the second part of the questionnaire was a qualitative section. The surveyed operators were asked to describe examples of adverse medical events as precisely as possible. There were huge differences in the relative frequency of each category. Roughly 13% of all stated examples, for instance, referred to *pressure* whereas only 3% referred to *lack of assertiveness*. In all, the relative distribution of these categories was similar to those of the quantitative section. Although, it was not always easy to categorize examples distinctively, categories such as *pressure, lack of communication* and *lack of awareness* were stated most frequently. Table 3 summarises the most prototypical examples of qualitative statements concerning causes of medical incidents.

Table 3. Examples of causes for adverse medical events stated by the surveyed operators

Lack of communication	"…a doctor explains to the nurse, which instruments he will need in the upcoming surgery. The nurses have a changeover but no information is transferred. Outcome: the doctor has the wrong instruments at hand…"
Lack of teamwork	"…social conflicts between the nurses have such negative impact on the atmosphere in the operating room, that the participating nurses are agitated and absent-minded…"
Complacency	"…the operating doctor ignores the patients' upcoming uneasiness and proceeds with an 'it will be ok' attitude…"
Social norms	"… sometimes the implementation of new hygienic quality standards is deferred because existing processes have been affiliated over long-term periods…"
Pressure	"…threads are being used much longer than appropriate, as a result surgery is compounded and safety risks for the patient might occur…"
Lack of awareness	"…the already laying out lens is to be replaced by another kind, the routinised and experienced nurse confounds the dioptre value…"
Stress	"…cost-cutting leads to shifts up to 16 hours, which leads to perceived stress…"
Fatigue	"…oftentimes there is no possibility to benefit the breaks, guaranteed according to the tariff; the speed of work cannot be self-controlled…"
Distraction	"…the anaesthetist lectures his young colleague loudly; the operating doctor is distracted and makes a mistake…"
Lack of resources	"…missing or time-worn instruments; multiple usage of surgical knives, made for one time use; diamond knives are often in bad condition…"
Lack of knowledge	"…still practicing, older doctors sometimes show a delay in diagnostic decisions…"
Lack of assertiveness	"…young attendings refuse to object against experienced nurses and deal with the presented instruments, even if they are not most suitable for the procedure…"

It is obvious, as shown in table 3, that small events (e.g. the indirect distraction of the operator by the anaesthetist) can reduce cognitive performance and finally cause human error and medical incidents. Therefore, on one hand the roots of cognitive performance limitations can be found within the *organisational context* of the operating room team (e.g. stress, pressure, lack of resources) and on the other hand within the social interaction of the team itself. As stated in table 3, a changeover of the nurses paired with insufficient communication can result in handing over wrong instruments within the operating room.

Discussion

The results of the present study support the position, that human error in the operating room as in other fields of work can be caused by many contributing factors (Gawande et al., 2003; Reason, 1990; Rogers et al., 2006), therefore justifying a systems approach to patient safety in the medical field (Calland et al., 2002). Ratings of possible error causations do not only indicate that there is no single predominant cause for human error but also that none of the possible error causations can be interpreted as irrelevant in the medical field. This fact emphasises how important a wide ranged error concept such as the dirty dozen is. In addition, the factor analysis confirms the assumption, that human error causation can be related to different underlying aspects of cognitive performance limitations. These aspects can be separated into the categories of individual and team cognition. While the factor *social interaction* relates to limitations regarding team cognition, the factor *organisational context* includes many aspects of individual cognitive performance limitations. This distinction should be analyzed more thoroughly in future studies. This view is supported by Catchpole and colleagues (2006), who found latent failures and small problems to impair patient safety in otherwise successful operations. Thus, the present findings encourage the dirty dozen approach to the medical field. Like other error categorisations the dirty dozen can be applied well to the high demanding work environment of operating surgeons. This success can be attributed to the approach's broadness and high adaptability which has been demonstrated in many fields of work.

Despite the findings of the current study, there are some theoretical and methodological weaknesses that must be addressed. On the one hand the above mentioned broadness of the dirty dozen may be helpful in terms of adaptability across diverse industries and work environments. On the other hand, the categories of this model are not very specific and make it hard to separate different error causes accurately. In addition, there are no detailed assumptions according to underlying individual and team-based cognitive processes. Future research should focus on improving the dirty dozen approach to obtain a categorisation of human error causes with high discriminability and a sound theoretical foundation. Apart from limitations in the theoretical approach there are also methodological difficulties. The current study should be seen as a first approach in investigating cognitive performance limitations in operating rooms in the field of ophthalmology. Due to the explorative design of the quantitative and qualitative section of the used questionnaire criteria like internal consistency could not be calculated. Therefore, future studies should construct instruments which address this concern and also validate the distinctive limiting factors of cognitive performance underlying this model. Furthermore, those studies could prove the applicability of the dirty dozen on other medical areas than the operating rooms of ophthalmology and other medical care personnel, e.g. operating room and general hospital nurses. An investigation of human error causes from a surgery nurse point of view – based on this dirty dozen concept – is planned in the near future. Another possible future application of the dirty dozen in the medical field could be the construction of a *critical incidents reporting system* (CIRS) based on the twelve categories of human error. Error reporting systems provide an important data source and hence can lead to further insights into the causation of human error in medical work environments. On a long-term perspective these insights can help to improve current safety trainings and allow us to develop new error focused concepts of work redesign to enhance patient safety.

References

Awad, S.S., Fagan, S.P., Bellows, C., Albo, D., Green-Rashad, B., De la Garza, M., & Berger, D.H. (2005). Bridging the communication gap in the operating room with medical team training. *The American Journal of Surgery, 190*, 770-774.

Brennan, T.A., Leape, L.L., Laird, N.M., Hebert, L., Localio, A.R., Lawthers, A.G., Newhouse, J.P., Weiler, P. C., & Hiatt, H.H. (2004) Incidence of adverse events and negligence in hospitalized patients: results of the Harvard Medical Practice Study I. *Quality and Safety in Health Care, 13*, 145-151.

Budde, S.H. & Barkowsky, T. (2008). *A Framework for Individual Cognitive Performance Assessment in Real-time for Elderly Users* (Association for the Advancement of Artificial Intelligence, Eds.). (Technical Report FS-08-02).Menlo Park, California: The AAAI Press.

Calland, J., Guerlain, S., Adams, R., Tribble, C., Foley, E., & Chekan, E. (2002). A systems approach to surgical safety. *Surgical Endoscopy, 16*, 1005-1014.

Carter, D. (2003). The surgeon as a risk factor. *British Medical Journal, 326*, 832-833.

Carthey, J., de Leval, M.R., & Reason, J.T. (2001). The human factor in cardiac surgery: errors and near misses in a high technology medical domain. *The Annals of Thoracic Surgery, 72*, 300-305.

Catchpole, K.R., Giddings, A.E.B., Wilkinson, M., Hirst, G., Dale, T., & de Leval, M.R. (2007). Improving patient safety by identifying latent failures in successful operations. *Surgery, 142*, 102-110.

Dupont, G. (1997). The Dirty Dozen Errors in Maintenance, paper presented at the 11th FAA-AAM meeting in Human Factors in Aviation Maintenance and Inspection, March, San Diego, CA.

Dupont, G. (1998). Human Factors Training in the Training Schools, *The 12th Symposium on Human Factors in Aviation Maintenance* (pp. 77–83). Civil Aviation Authority

ElBardissi, A.W., Wiegmann, D.A., Dearani, J.A., Daly, R.C. & Sundt III, T.M. (2007). Application of the Human Factors Analysis and Classification System Methodology to the Cardiovascular Surgery Operating Room. *The Annals of Thoracic Surgery, 83*, 1412-1419.

Flin, R. & Maran, N. (2004). Identifying and training non-technical skills for teams in acute medicine. *Quality and Safety in Health Care, 13* (supply 1), 80-84.

Gawande, A.A., Zinner, M.J., Studdert, D.M., & Brennan, T.A. (2003). Analysis of errors reported by surgeons at three teaching hospitals. *Surgery, 133*, 614-621.

Marquardt, N., Gades, R., & Robelski, S. (in press). Implicit social cognition and safety culture. *Human Factors and Ergonomics in Manufacturing and Service Industries.*

Marquardt, N. & Höger, R. (2007). The structure of contributing factors of human error in safety-critical industries. In D. de Waard, G.R.J. Hockey, P. Nickel, and K. Brookhuis (Eds.), *Human Factors Issues in Complex System Performance* (pp. 67–71). Maastricht: Shaker Publishing.

Marquardt, N., Robelski, S., & Höger, R. (2010). Crew Resource Management Training Within the Automotive Industry: Does It Work? *Human Factors, 52*, 308-315.

McDonald, J., Orlick, T., & Letts, M. (1995). Mental Readiness in Surgeons and Its Links to Performance Excellence in Surgery. *Journal of Pediatric Orthopaedics, 15*, 691-697.

Mishra, A., Catchpole, K., Dale, T., & McCulloch, P. (2008). The influence of non-technical performance on technical outcome in laparoscopic cholecystectomy. *Surgical Endoscopy, 22*, 68-73.

Reader, T., Flin, R., Lauche, K., & Cuthbertson, B.H. (2006). Non-technical skills in the intensive care unit. *British Journal of Anaesthesia, 96*, 551-559.

Reason, J.T. (1990). *Human error*. Cambridge: Cambridge University Press.

Rogers Jr, S.O., Gawande, A.A., Kwaan, M., Puopolo, A.L., Yoon, C., Brennan, T.A., & Studdert, D.M. (2006). Analysis of surgical errors in closed malpractice claims at 4 liability insurers. *Surgery, 140*, 25-33.

Salas, E., Cooke, N J. & Rosen, M A. (2008). On Teams, Teamwork, and Team Performance: Discoveries and Developments. *Human Factors, 50*, 540-547.

Sexton, J.B., Thomas, E.J., & Helmreich, R.L. (2000). Error, stress, and teamwork in medicine and aviation: cross sectional surveys. *British Medical Journal, 320*, 745-749.

Thomas, E.J., Studdert, D.M., Burstin, H.R., Orav, E.J., Zeena, T., Williams, E.J., Howard, K.M., Weiler, P.C., & Brennan, T. A. (2000). Incidence and Types of Adverse Events and Negligent Care in Utah and Colorado. *Medical Care, 38*, 261-271.

Yule, S., Flin, R., Paterson-Brown, S., & Maran, N. (2006). Non-technical skills for surgeons in the operating room: A review of the literature. *Surgery, 139*, 140-149.

Yule, S., Flin, R., Maran, N., Rowley, D., Youngson, G., & Paterson-Brown, S. (2008). Surgeons' Non-technical Skills in the Operating Room: Reliability Testing of the NOTSS Behavior Rating System. *World Journal of Surgery, 32*, 548-556.

Collecting battlefield information using a multimodal personal digital assistant

Stas Simon Krupenia[1], Mathilde Cuizinaud[1], Tijmen Muller[2], & Anja H. van der Hulst[2]
[1] Thales Research and Technology Netherlands
[2] Netherlands Organization for Applied Scientific Research (TNO)
The Netherlands

Abstract

In a Network Centric battlefield, the information available for analysis and distribution is limited by the information collected from the environment. We examined how soldiers collect information from the battlefield using a multimodal Personal Digital Assistant (PDA) containing five options (modalities) for data collection: photo, video, icon, text, and audio. Twenty male Polish soldiers completed a simulated reconnaissance mission in a virtual environment during which they were commanded to collect information using the PDA. Results indicated that soldiers were more likely to use a single modality than a combination of modalities and were more likely to use photos and videos than the other modalities. The audio-visual properties of the events had a small influence on modality choice, but only to the extent that transient were captured more quickly. In general, participants appeared to adopt a minimalistic interaction style, one possible explanation is that soldiers chose instead to invest greater effort in observation and threat detection than in data collection. Within a networked environment, the pattern of data collection has system-wide implications regarding situational awareness and decision making. The pattern of data collection observed in the current research applies to other similar domains such as search and rescue, and disaster management.

Introduction

The project 'Smart Information for Mission Success' (SIMS) aims to improve military Force Protection planning, execution, assessment and training by delivering a proof of concept system to support the dissemination of battlefield information (Duistermaat, et al., 2011). A necessary pre-curser to information dissemination is information collection. Consistent with the Every Soldier is a Sensory (ES2) concept (U.S. Army, 2008) the process of information collection within SIMS is completed in-part by regular soldiers within a platoon or company. A defining feature of the ES2 concept is that *all* soldiers are expected to observe critical battlefield details and to report their experience, perception and judgements (U.S. Army, 2008).

Although the situation is changing, currently the most prevalent tools used to support data collection (excluding within specialist units) are a pen and paper (Dalziel, 1998). When using a pen and paper, soldiers write down their observations after an event has occurred (as opposed to during the event) and during debriefing, refer to these notes to describe the battlefield. The information conveyed during debriefing enters the intelligence cycle and influences future mission planning. The SIMS project examines how to support the collection, analysis, and dissemination of information. In the current study, we examined how soldiers would use a multimodal data collection Personal Digital Assistant (PDA) to record battlefield information when told that the collected information would be used for mission debriefing and for real time communication with Command and Control (C2). The generic PDA used in the current study could record and store five modalities: Photos, Videos, Audio, Text, and Icons. Similar to Ashish et al., (2008) 'multimodal' in the current manuscript is used to refer to different 'media types' as opposed to the more common understanding as auditory, visual, tactile, etc.

In constructing hypotheses about how soldiers would use the PDA to collect information, only limited literature was available for review. It is surprising that although much research exists defining the information needed to support situational awareness at tactical levels (for multiple domains) little research can be found that describes how this data can be collected from the environment. Additionally, attempts to define the interaction preferences at the 'sharp end' (soldier, fire-fighter…) focus more on the information access preferences than on inputting information. For example, van de Ven et al. (2008) discuss how to support Crisis Management via a network centric approach and focus on maximising system effectiveness given unreliable data. A complimentary approach is to identify methods to support the more refined, reliable, entry of information into the system.

One of the few attempts to define a data collection device for reconnaissance was by Dalziel (1998) who delivered an associated list of technical requirements. It is unclear, however, how Dalziel (1998) developed the technical solutions for the functional requirements. For example, the functions 'provide information that is responsive and timely, and 'provide information that is accurate, clear and concise' were addressed through a limited set of technical solutions (a digital still camera and wireless communication). Furthermore, there was no assessment of how soldiers would use such technologies if available.

Although some general literature exists investigating how people collect multimodal data, this research focuses on data collection to support personal memory recall (Czerwinski, et al., 2006), and generally do not consider human-computer interaction. One meta-analysis that examined differences between Hand Held Computer devices and pen-and-paper methods for data collection was Lane et al. (2006). These authors conducted a review of medical studies where patients were required to collect personal medical information that was then submitted back to the research team. Of interest to Lane et al. was how user preferences, data accuracy, adherence to the data collection schedule, and time to complete data collection, would differ between the two 'technology' conditions. Of an initial 201 potentially relevant studies, the authors found only nine studies that included a randomized controlled trial of the two technologies. From their meta-analysis, Lane et al. concluded that: data accuracy was approximately equal for the two technologies, but that patients prefer to use, are quicker, and adhere better to the data collection schedule using the hand held device. An additional difficulty in using prior work to construct hypotheses for the current research is that existing literature considers only simple contexts (e.g. conversations), and includes a limited set of data collection options (Kalnikaitė & Whittaker, 2007; Kalnikaitė, et al., 2010). The literature on 'life-logging' (Kalnikaitė, et al., 2010) focuses on the collection of data for personal long term memory archival but does not involve critical situations, post event review and assessment, or the real time review of the captured data (C2).

Research questions

We investigated how soldiers will use a multimodal PDA to collect battlefield information. The four questions were: (1) will soldiers use one modality type more frequently than another modality, (2) is there a correlation between the modality type used and the audio visual properties of the events, (3) will soldiers use one modality or a combination of modalities to capture information, and (4) is there a correlation between the number of modalities used and the audio-visual properties of the event? Given that only a limited amount of prior research has investigated data collection with multimodal handheld devices during operations in critical environments, the current research can be considered exploratory. Four null hypotheses were constructed for the research questions: (1) There will be no differences in the frequency of the modalities used, (2) there will be no correlation between the audio visual properties of the events and the modality used, (3) there will be no stable pattern in the number of modalities used to capture information, and (4) there will be no correlation between the audio-visual properties of an event and the number of modalities selected.

Method

Design

Participants completed two simulated reconnaissance missions in a virtual military environment called Virtual Battle Space 2© (VBS2; Bohemia Interactive). In Mission 1, the participant was the commander of a two-person reconnaissance team. The research goal of Mission 1 was to identify *when* (for what kinds of events) soldiers would collect information. The results of Mission 1 are discussed elsewhere (Krupenia et al., in press) and are thus not reported here. Unless otherwise specified all references to 'the mission' are to Mission 2. In Mission 2 the participant was the subordinate soldier in a two-person reconnaissance team with an experimenter playing the role of commander. The VBS2 setting was the same for both missions and the route travelled by the participants in the two missions was similar. Some of the events presented in Mission 2 were only meaningful when considering the events in Mission 1. It is important to note that in Mission 2 it was not the participant's responsibility to decide whether or not to collect information (the commander's role) but only to collect data when told to do so by the commander.

Experiment description and experimenter roles

The experiment was conducted at the National Defence Academy of Poland (Warsaw). Participants were seated in front of two laptop computers: one for VBS2 (18.3" Toshiba Qosmio) and one for the PDA (17.3" Dell XPS; see Figure 1). The experimenter playing the role of commander (Experiment Confederate) sat to the right and behind the participant. A second experimenter sat to the participant's left and led the participant through the experiment. A third experimenter sat opposite the participant at a separate desk and coordinated the VBS2 events. When required, a Polish Major assisted with English-Polish translations.

Figure 1. Picture showing experimental set-up; view of participant VBS2 computer and participant PDA computer

Participants

Twenty male Polish soldiers (eighteen Captains, two Majors) between the ages of 31 and 50 (average = 38.24; SD = 4.33) voluntarily participated having an average of 18.57 years of military experience (SD = 4.24). All participants had completed at least one university degree (18 Master's degrees, one Bachelor's degree). Participants were reasonably comfortable with technology; 47.37% of the participants reported that they feel "somewhat comfortable" (second most comfortable option on a five-point scale) using a computer and 47.37% of participants either owned or had used a smart phone.

VBS2 scenario and events

The mission was set in a fictitious African village. Participants were told that in two days, the local government building and church would be used to host a meeting of local leaders and UN representatives. The participant and the Experimenter

Confederate's role were to conduct a reconnaissance mission to support the generation of an area protection plan. The Experiment Confederate led the participant to fourteen events and ordered the participant to collect information about each VBS2 event (e.g. see Figure 2). To create realistic events a list of potential events were extracted from initial interviews conducted with soldiers. These potential events were reviewed by persons with military experience, and with persons experienced at designing virtual scenarios for military participants. As a result, three different Event Types were identified (Hostile Indicators, Population, and Geography). The Event Types influenced the Icons available on the PDA. Each VBS2 event was also categorised as having an audio quality (No Sound, General Sound, Direct speech) and a visual quality (Visually Static, Visually Dynamic). The list of the events, and their audio-visual properties, is presented in Table 1.

Figure 2. Screenshot of VBS2 Event 1 (See Table 1)

Table 1. List of the fourteen VBS2 events, including their Audio properties (No Sound, General Sound, Direct Speech), and Visual Properties (Visual Static, Visual Dynamic), chronologically ordered

Event	Description	Audio	Visual
1	A lot of dirt next to the shed	No Sound	Static
2	River not flowing as fast today as last time [Misison1]	General	Dynamic
3	Some stuff under the bridge	No Sound	Static
4	There is a car on the other side of the bridge	No Sound	Static
5	Markets have totally gone	General	Dynamic
6	View that there are a lot of people around	General	Dynamic
7	One person talking loudly from within crowd	Speech	Dynamic
8	Multi-ethnic meeting near the church	General	Dynamic
9	Police are guarding only front of building	No Sound	Dynamic
10	Public speaker telling people that they will have to move out of the area for the visit	General	Dynamic
11	Inspect church entrances/exits	No Sound	Static
12	A hole in the ground	No Sound	Static
13	Political posters in church	No Sound	Static
14	Direct speech with church leader	Speech	Dynamic

PDA interface

The digitally presented PDA included five data collection options: photo, video, icon, text, and audio (Figures 3 and 4). When given the order to collect information, participants had to select the modality of their choice (including multiple modalities) to capture the event. There were five PDA sections: (1) two tabs for switching between map view (Figure 3) and icon view (Figure 4), (2) a space for entering text, (3) a 'ribbon' with four iconic buttons to use for capturing data, (4) a map, and (5) an icon page. The data collection buttons were, a camera (for still images); a video camera (moving images including VBS2 audio and participant commentary); a microphone (VBS2 audio and participant commentary); and a person with a speech/text bubble (text). When participants selected the "Markers" tab, nine icons were presented: three were generic (did not convey any pre-defined meaning) and the rest were grouped into Event Type (Geography, Population, and Hostile Indicators). For each Event Type there were two icon colour options, green (low urgency) and red/purple (high urgency).

Figure 3. Screenshot of the PDA

Questionnaires

Three questionnaires were used. A demographics questionnaire was used to obtain background information about the participant sample. Because changes in mental state can influence performance (Hockey, 1997), a Mental State questionnaire was used to probe Stress, Fatigue, Concentration, Motivation, Vigilance and Cognitive Overload on a scale from 1 (do not feel at all) to 7 (feel this a lot) prior, during, and after the mission.

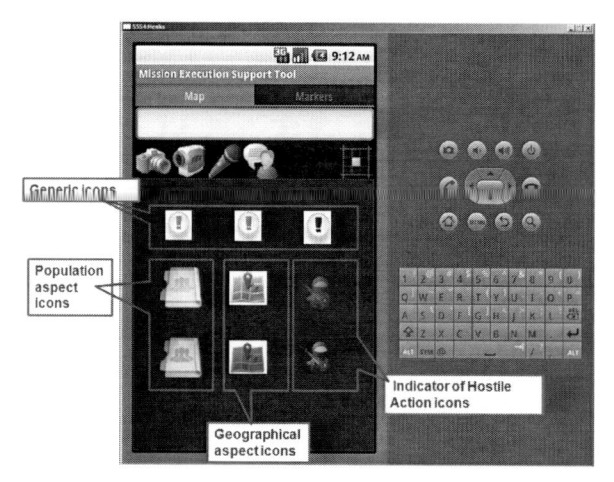

Collecting battlefield information using a PDA 333

Figure 4. Screenshot of the PDA screen showing the nine icons available after clicking on the Markers button

The usability questionnaire was used to probe PDA usability and the rationale for modality selection. During the mission, the modality selected by the participant for each event was recorded. This record was given to the participant as part of the Usability Questionnaire. For each event and corresponding mode of data capture, participants were asked "Why did you choose this modality to record this event?" Five answer options answers were available: (A) Because it was the best modality for collecting the data, (B) By default, because the other modalities were less good, (C) I have no specific rationale, I could have chosen another modality [please specify], (D) I would have preferred a combination of modalities [please specify]. (E) Other [please explain]. Participants also rated how comfortable, and frustrated, they were when using the PDA (on a seven point scale where 1 = "not at all" and 7 = "very much"). Participants were also asked to provide any other comments.

Procedure

In advance of the experiment each participant had received and was asked to (1) read and sign the information and consent document, and (2) complete the demographics questionnaire. Participants that had not completed this in advance were given time to do so at the experiment. Participants were then briefed on, and completed Mission 1. Before starting Mission 2, participants completed a PDA training session. The five modalities were explained together with how to interact with the PDA. Next, participants completed the mental state questionnaire. Participants then received the mission briefing and task instructions. To stimulate realistic PDA interactions, participants were told that the data they collect would be made immediately available to C2 and will be used for mission debriefing. At the end of the mission, participants completed the mental state questionnaire (this time probing how they felt during the mission and 'now'—at the end of the mission). Participants then completed the usability questionnaire. Finally, participants were thanked, given an experiment debriefing document, and a small gift.

Results

Data were analyzed using PASW© Statistics 18 statistical package (IBM). Analysis of the results was conducted on nineteen participants (the data from one participant was removed due to a technical failure).

Mental State Questionnaires

Before Mission 1, participants were relaxed, motivated, and attentive. The averages (and standard deviations) for the pre-Mission 1 Mental State probes were; stressed, 2.11 (1.15); tired, 2.68 (1.63); motivated, 4.21 (1.75); concentrating, 4.42 (1.71). During the mission, concentration, motivation, overload, and vigilance increased by one point and returned to the pre-mission values after the mission. Stress increased by about half a point during the mission and returned to the pre-mission values at the end.

Usability Questionnaire

Participants reported being comfortable using the PDA (average = 4.74, SD = 1.10) but felt some frustration (average = 2.11, SD=1.10), mostly because they wanted additional familiarization. Some participants commented that they would have liked additional functionalities (e.g. thermal imaging). From the usability questionnaire it was also clear that Photo and Video were most preferred, followed by Audio, Icons and lastly, Text. This subjective data was replicated in the performance data (Figure 5). Responses to the question "why did you choose this modality to capture this event" were, A = 73.5%, B = 8.62%, C = 16.4%, D = 5.15%, and E = 5.68%.

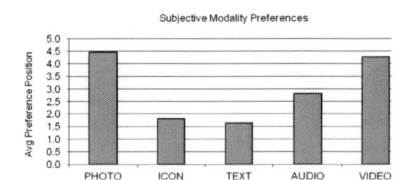

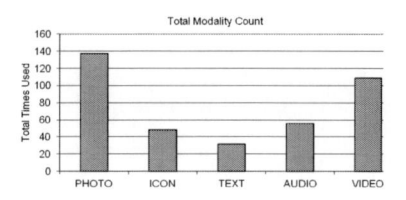

Figure 5. Average subjective preference position order for each modality (left). Total number of times each modality was used (right)

Performance data

As mentioned in the introduction, there were four research questions. The first question was whether participants were more likely to use one modality versus another to capture information. A summary of these results is presented in Figure 5. To test the null hypothesis a repeated measures analysis of variance ($_{rm}$ANOVA; repeated covariance type: AR1) was conducted to test for differences in modality use with participants and Modality (5 levels) as fixed effects and Participants as a random effect. The $_{rm}$ANOVA indicated a significant main effect of Modality, $F(4) = 18.627$, $p < 0.001$. Follow-up pairwise comparisons (with a Bonferroni correction for alpha) revealed that participants were equally likely to use Photo and Video ($p =$

0.502) and were more likely to use these modalities than the remaining options. There was no difference between the use of the other modalities.

The second question was if there was a correlation between the modality selected and the audio visual properties of the event. To test the null hypothesis, ten loglinear analyses were conducted to identify any correlation between the audio properties of an event and the use of a modality, and between visual properties and the use of a modality. Of the ten analyses, five were rejected because the loglinear model produced a non-significant likelihood ratio. The retained analysis (for all the likelihood ratio of the loglinear model was, $\chi^2 (0) = 0$, $p = 1$) and follow-up comparisons (Chi Square) are presented in Table 2.

Table 2. Results of five retained loglinear analyses showing overall effect of Audio and Visual Properties on the use of Photo, Video, Audio including follow-up pairwise comparisons showing effect of No Sound, (General) Sound, Speech, Visual Static, and Visual Dynamic events on increasing, decreasing, (or no effect) on the use of each Modality Type

	Photo	Video	Audio
Audio	$\chi^2 (2) = 50.02, p < .001$	$\chi^2 (2) = 20.60, p < .001$	
- No sound	Increases use, $p < .001$	Decreases use, $p < .001$	
- Sound	Decreases use, $p < .001$	No influence, $p = 0.08$	
- Speech	No influence, $p = 0.748$	Decreases use, $p < .025$	
Visual	$\chi^2 (1) = 10.036, p < .005$	$\chi^2 (1) = 12.331, p < .001$	$\chi^2 (1) = 10.036, p < .005$
- Static	Increases use, $p < .001$	Decreases use, $p < .001$	Decreases use, $p < .001$
- Dynamic	Decreases use, $p < .001$	No influence, $p = 0.794$	Decreases use, $p < .001$

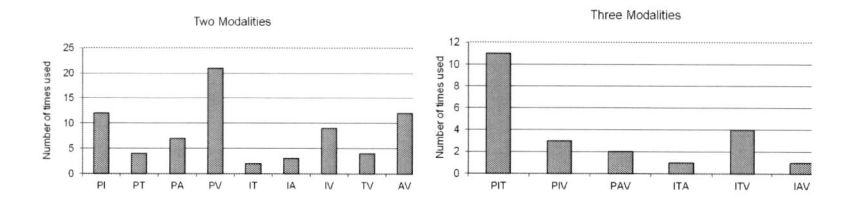

Figure 6. The distribution of use of modalities, when: two modalities were used (left), and three modalities were used (right); P denotes photo, I denotes Icon, T denotes Text, A denotes Audio, V denotes Video

The third question was whether participants would use one, or a combination of modalities, to capture information. A summary of these results is presented in Figure 6. To test the null hypothesis an $_{rm}$ANOVA was conducted to test for differences in Number of Modalities Used use with participants and Number of Modalities Used (4 levels) as fixed effects and Participants as a random effect. Results of the $_{rm}$ANOVA indicated a significant main effect of Number of Modalities Used, $F(3) = 73.327$, $p < .001$. Follow up pairwise comparisons (using a Bonferroni correction for alpha) indicated that participants were more likely to use a single modality than two, three, or four modalities (for all comparisons, $p < .001$). Additionally, participants were more likely to use two modalities than three ($p < .005$), or four modalities ($p < .001$). There was no difference between using three or four modalities ($p = 1.000$).

The fourth question was whether there would be a correlation between the number of modalities used and the audio-visual properties of the event. To test the null hypothesis, six individual loglinear analyses were conducted to identify any possible correlation between the audio properties of an event and the use one, two, or three modalities, and between visual properties and the use of one, two, or three modalities. The results of these analyses, however, were unreliable because either the chi square assumptions were not met or because the final model did not sufficiently fit the data. Consequently, the results of the loglinear analysis are not reported. To understand the likelihood of selecting a number of modalities on the basis of the audio-visual properties of events, the percentage use of a specific number of modalities for the three audio and two visual properties is presented in Figures 7 and 8. A visual inspection of the two figures suggests that Direct Speech increases the likelihood of using One modality at the cost of using Two modalities and that compared to visually static events, when events were dynamic, the likelihood of using One modality increases at the cost of using two and three modalities.

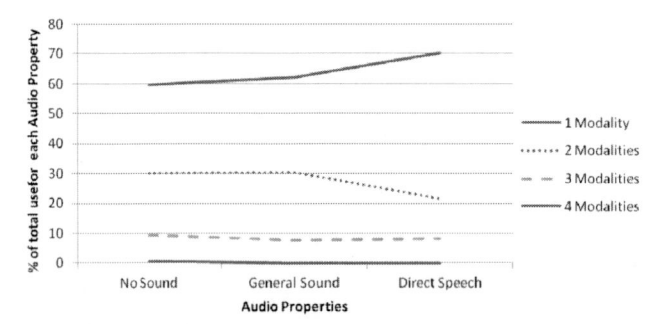

Figure 7. Percentage use of One, Two, Three, and Four modalities for No Sound, General Sound, and Direct Speech events

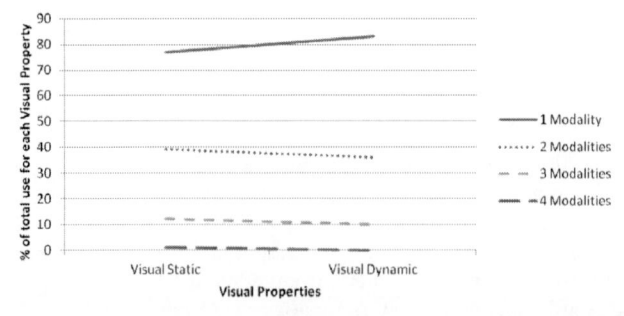

Figure 8. Percentage use of One, Two, Three, and Four modalities for Visual Static and Visual Dynamic events

Discussion

The four questions presented in the introduction will be discussed. The first question was whether or not participants were more likely to use one modality versus another to capture information. For this question, the null hypothesis was rejected. The results indicate a clear subjective and objective preference for using photos and videos to capture information. Collecting photos was the easiest data collection method (requiring a single button press) followed by video (two button presses; the same as for audio). Indeed, it appears likely that the rank ordered use of Photos, Videos, and then Audio matches closely the amount of interaction required with the PDA. Additionally, it can be argued that the use of video rather than audio is because video was equally good at capturing audio as was audio. It is perhaps not surprising that soldiers would prefer to interact in such a minimalist way, given that in any potentially hostile environment a soldiers' primary task is to observe the environment for potential (kinetic) threats. Interestingly, Mitchell et al. (2006) also noted that regardless of any interface technology, for soldiers in the field attention must be "keenly directed and maintain on the immediate threat environment" (p7) because incorrect attention management can be deadly. The results reported in the current study are also consistent with Mitchell, et al. who reported that the ability to capture and send photos was one of seven features requested by soldiers for novel display prototypes. Two alternative explanations for the observed results are that the soldiers were 'lazy' or that performance was influenced by the PDA design characteristics. Although no definitive evidence can be provided to argue against the 'laziness' hypotheses, anecdotally, it was observed that participants found the task engaging and enjoyable. It is also acknowledged that it is not (always) possible to disentangle interaction patterns from product design, however, given that for about 74% of responses, participants said that they selected their modality because "it was the best modality for collecting the data" we argue that the interaction patterns observed were not strongly determined by the PDA design.

The second question was if there would be a correlation between the modality selected and the audio visual properties of the event. The null hypothesis was rejected. At least some of the audio-visual properties of the events influenced the modality of data collection and the pattern of results observed was mostly 'logical'. The modalities that could capture specific audio-visual properties where used appropriately (e.g. silent events were captured with photos and not videos). Surprisingly, video was not used for direct speech. One explanation is that participants' may have assessed the speech as being irrelevant for the mission goals. Alternatively, participants may have felt that the use of video camera in this specific setting was culturally inappropriate.

The third question was whether soldiers would use one modality or a combination of modalities to capture information. The null hypothesis was rejected. The results clearly indicate a preference for the use of one modality versus two, and for two versus three. This pattern of results is consistent with the argument presented earlier that soldiers have adopted a minimalist interaction style—preferring to use the device in a way that either minimally detracts from their primary tasks.

The fourth question was whether there would be a correlation between the number of modalities used to capture an event and the audio-visual properties of the event. The null hypothesis cannot be confidently rejected. The results, though somewhat inconclusive, suggest that audio visual properties have a minor influence on the number of modalities used. One explanation is that the audio visual properties determine an event's permanency/transiency which in turn influences the number and type of modalities selected. Visually dynamic and audio events are transitory and capturing these temporary events requires swift action. Data can be collected more quickly using a single modality than combinations.

Four limitations regarding the current research have been identified. First, the simulated PDA did not fully capture the interaction constraints and possibilities associated with a real PDA. For example, participants left the mouse curser hovering over the Photo button which enabled data capture without looking towards the PDA. Second, the accessibility of the five modalities was not equal—to use icons, participants had to press two buttons. Third, although an attempt was made to create a fully factorized list of events involving differing audio-and visual properties this was abandoned because some events were too artificial for the scenario and it was important to promote realistic PDA interaction. Finally, although not a limitation of the current study, in regards to the practical implications of the data reported here, it is important to note that patterns of human-machine interaction can change over time as users become more familiar with, or 'adapted' to, technology (Woods, 1993; Sarter, 2006).

Conclusion

We investigated soldier preferences for collecting battlefield information using a multimodal data collection device. The clear preference for collecting a single modality (either photo or video) suggests that participants adopted an interaction pattern that minimally detracted from their primary tasks of observation and threat detection. Despite being told that the data collected would be used (1) by C2 to monitor the mission, and (2) to support mission debriefing, it appears that by adopting the 'self-preservation' style of interaction soldiers may not have considered the impact of their data collection strategies for other persons within the networked environment. Alternatively, soldiers considered this impact, but gave short term self-preservation a higher priority. These findings suggests that practical implementation of ES2 will be difficult without the assistance of 'smarter' automated data capture technologies. It is also possible that participants felt that the information collected was sufficient for C2 and mission debriefing. It is also encouraging to know that modality use was 'logical' in that the options for recording audio and visual dynamic events were used appropriately. It also appears that the permanence of an event influences the choice of modalities used to capture the event—transient events must be captured quickly and thus only a single modality is used.

Finally, it is likely that in other similar safety-critical environments, the data collection strategies of participants at the 'sharp end' of the system will be such that they minimally distract from the primary responsibilities. If the data collected using the SIMS PDA is to be used to support planning, execution, assessment and training, then the impact of the data collection patterns reported here must be considered.

Specifically, the impact of the data collection patterns observed in the current study on C2 and mission debriefing must be examined.

Acknowledgements

This research was supported by European Defence Agency Research Grant A-0934-RT-GC and by a VRC Corporation Grant (VRC.23) awarded via the Human Factors and Ergonomics Association Europe Chapter to M. Cuizinaud (EC/2011.12).The research would not have been possible without contributions from the TNO for PDA Development (Henk Henderson), and general support (Lesley Jacobs, Lotte van Lier, and Niek Schmitz). Additional gratitude is extended to the National Defence Academy of Poland (Major Tomasz Kacała and Professor Jarek Wolejszo) for supporting participant recruitment and reviewing the experiment for relevance to military personnel.

References

Ashish, N., Eguchi, R., Hegde, R., Huyck, C., Kalashnikov, D., Mehrotra, S., Smyth, P., & Venkatasubramanian, N. (2008). Situation Awareness Technologies for Disaster Response. In H. Chen, E. Reid, J. Sinai, A. Silke, and B. Ganor. *Terrorism Informatics: Knowledge Management and Data Mining for Homeland Security* (pp. 517-544). New York, USA: Springer.

Czerwinski, M., Gage, D., Gemmell, J., Marshall, C., Perez-Quinonesis, M., Skeels, M., & Catarci, T. (2006). Digital memories in an era of ubiquitous computing and abundant storage. *Communication of the ACM, 49*(1), 44–50.

Dalziel, R.P. (1998). *Improving the engineer reconnaissance reporting process through the use of digital imagery and handheld computers.* Unpublished Master of Science thesis in Information Technology Management, Naval Postgraduate School, Monterey, CA.

Duistermaat, M., Schmitz, N., Jacobs, L., Faye, J-P., Bertucat, E., Krupenia, S., Clot, V., Dymowski, W., Zak, M., Museux, N., & Raynal, P. (2011). SIMS - Smart Information for Mission Success. Enhancing joint mission planning and training. Presented at ITEC conference, 10-12 May 2011, Cologne.

Hockey, G.R.J. (1997). Compensatory control in the regulation of human performance under stress and high workload: A cognitive-energetical framework. *Biological Psychology, 45*, 73-93.

Kalnikaitė, V., Sellen, A., Whittaker, S., & Kirk, D. (2010). Now let me see where I was: Understanding how Lifelogs mediate memory. In *Proceedings of ACM Conference on Human Factors in Computing Systems* (CHI 2010), (pp. 2045-2054). New York, USA: ACM Press.

Kalnikaitė, V., & Whittaker, S. (2007). Does taking notes help you remember better? Exploring how note taking relates to memory. *Workshop on Supporting Human Memory with Interactive Systems. British HCI Conference* (pp. 33-36). London: British Computer Society

Krupenia, S., Cuizinaud, M., Muller, T., & Van der Hulst, A. (in press). Identifying battlefield information collection strategies to support 'every soldier is a sensor' training. Submitted to the *56th Annual Meeting of the Human Factors and Ergonomics Society*, Boston, USA.

Lane, S.J., Heddle, N.M., Arnold, E., & Walker, I. (2006). A review of randomized controlled trials comparing the effectiveness of hand held computers with paper methods for data collection. *BMC Medical Informatics and Decision Making, 6,* 23-33.

Mitchell, K. B., Sampson, J.B., Short, M., & Wilson, W. (2006). Display option for dismounted infantry: Flexible display center human factors preliminary user survey. Report for the U.S. Army Research, Development and Engineering Command, Natick Soldier Systems Center. Report NATICK/TR-07/007.

Sarter, N.B. (2006). Multimodal information presentation: Design guidance and research challenges. *International Journal of Industrial Ergonomics, 36,* 439-445.

U.S. Army. (2008). The Warrior Ethos and Soldier Combat Skills. Field Manual 3-21.75, Washington, D.C: Headquarters, Dept. of the Army.

Van de Ven, J., van Rijk, R., Essens, P., & Frinking, E. (2008). Network Centric Operations in crisis management. In *Proceedings of the 5th International ISCRAM Conference* (pp. 764-773). Brussels, Belgium: Academic & Scientific Publishers

Woods, D.D. (1993). The price of flexibility. In *Proceedings of the First International Conference on Intelligent User* Interfaces (pp. 19-25). New York, USA: ACM Press.

Modelling

Developing a unified model of driving behaviour for cars and trains

Björn Peters, Anna Vadeby, Åsa Forsman, & Andreas Tapani
VTI Swedish National Road and Transport Research Institute
Sweden

Abstract

A unified model of driver behaviour and driver interaction with innovative technologies was developed in the European project ITERATE. The model aims to be applicable for all surface transport modes. As a basis of the model development it was assumed that underlying factors influencing human behaviour such as age, gender, culture etc. are constant between transport modes. The model can be of great use when designing innovative technologies since it will allow for assessment and tuning of the systems in a safe and controllable environment without use in real traffic. This paper presents the results of a set of driving simulator experiments carried out to support the model development process. The experiments are unique in the sense that common scenarios were run on two identical portable driving simulator platforms circulated among project partners across five countries as well as full scale train and car driving simulators. This allowed a large number of subjects to take part in the experiment. An important finding from the experiments was that country/culture was found to be a significant factor for almost all performance indicators in both car and train experiments. Furthermore, it seems like small scale simulators provide comparable results as more advanced simulators.

Introduction

A unified model of driver behaviour (UMD) was developed within the ITERATE project based on a literature review of driver behaviour models (Oppenheim, et al., 2010a, Oppenheim, et al., 2010b), see Figure 1. The UMD basically shows that there are a number of factors (*culture, personality, state, experience, and workload*) that have an impact on driver behaviour and interaction with support systems. A review of innovative technologies, i.e. driver assistance systems such as Intelligent Speed Adaptation (ISA) for the road traffic domain, the European Rail Traffic Management System (ERTMS) for the rail domain, and also for the maritime domain was also conducted in the project (Lai, et al., 2010, Barnard, et al., 2010a). Based on these reviews and considering the UMD, a large set of hypotheses were formulated on how car drivers and train operators will behave and interact with support systems depending on the underlying factors influencing human behaviour (Barnard, et al., 2010b). To test these hypotheses and thereby aiming to verify the theoretically developed UMD, a set of road and rail driving simulator experiments were conducted. Results of the driving simulator experiments were also used to estimate parameters for a numerical simulation model (Hjälmdahl & Amantini, 2012).

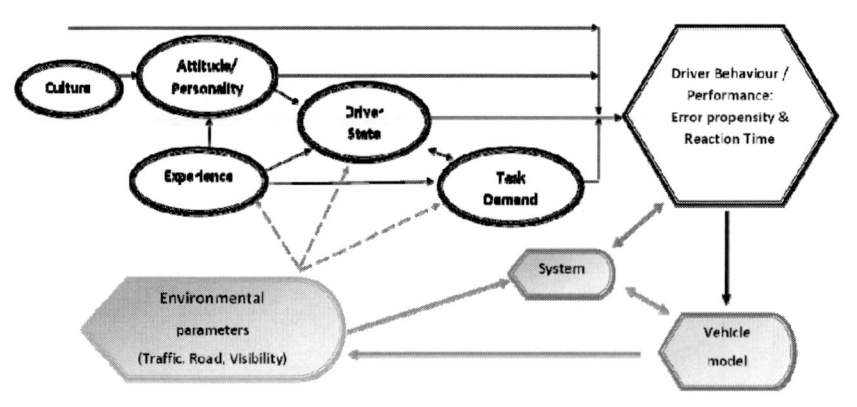

Figure 1. The UMD model

The ITERATE driving simulator experiments are unique in the sense that common scenarios were run on a common portable car/train driving platform as well as on full scale train and car driving simulators. Two identical portable driving simulator platforms were circulated among the project partners across five countries allowing a large number of participants to take part in the experiments. This paper presents the main results from these experiments.

Experimental design

A mixed design with four between factors (culture, experience, driver state, and personality) and one within factor (workload) was applied for all experiments. Culture was included as a factor with 5 levels represented by the five countries (Sweden, England, France, Italy and Israel) as cultural differences (even within Europe) has shown to be of importance (Özkan, et al., 2006, Özkan & Lajunen, 2011). Experience was a factor with two levels (novice and experienced) determined by the number of years active as a train driver or the number of years holding a car driving licence. Fatigue was also a factor with two levels (alert, fatigued) with the "post lunch dip" as the fatigue condition. Personality in terms of sensation seeking (measured by the Brief Sensation Seeking Scale (BSSS) (Hoyle et al. 2002)) was not actually controlled for by screening subjects but rather by forming groups of drivers based on BSSS scores. Workload was manipulated by a secondary counting backwards task in three difficulty levels (low, medium and high). Workload and fatigue manipulations had been tried out in pilot experiments.

Method

Car drivers drove with support systems that would warn if speeding and or driving too close to a lead vehicle, train drivers had a system that showed current maximum allowed speed and warned if driving too fast. A selection was made among the hypotheses formulated in Barnard et al., (2010b) to be tested in the experiments e.g.:

- Experienced drivers will drive faster but receive fewer warnings from the support system (speed and distance)
- High sensation seekers will drive faster than low sensation seekers and get more warnings
- Fatigued drivers will drive slower than alert drivers but rely on the system to warn and get more warnings
- Low workload would result in higher speed, while low and high workload in curves would provide more warnings

As a complement to the hypothesis testing, exploratory data analysis in the form of a cluster analysis was conducted to study alternative underlying factors controlling driver behaviour, see (Forsman et al., 2011).

Apparatus

The three different types of simulators used in the experiments are depicted in Figure 2. Two identical simple mobile simulators were circulated among partners to minimize the difference between test sites. Two advanced simulators (a large scale motion-base car simulator in Leeds, Great Britain, and a train simulator with a mock-up of a train driver's cab at VTI, Sweden) were also used as reference to determine if simple simulators yield the same results as more complex ones. Simulator software and driving tasks were the same independent of simulator. Participants were though not the same (between subject design).

Figure 2. Simulators used (top left - small scale car simulator, top right - small scale train simulator, bottom left – full scale car simulator, bottom right - full scale train simulator)

Participants

In total 183 car divers and 110 train operators participated in the simulator experiments (Table 1 & 2). The distribution was not ideal, i.e. few female train drivers, few novices, few French train drivers. The target was 16 in each cell. However, in total 76% of the target was reached.

Table 1. Distribution of car drivers with respect to country, gender, experience, and driver state

Country	Number	Gender		Experience		State	
		Female	Male	Experienced	Novice	Alert	Fatigue
France	32	7	25	16	16	16	16
Great Britain	30	15	15	16	14	15	15
Israel	31	16	15	16	15	16	15
Italy	27	11	16	20	7	16	11
Sweden	34	16	18	26	8	14	20
Total mobile simulator	154	65	89	94	60	77	77
Advanced simulator GB	29	17	12	15	14	14	15
Grand total	183	82	101	109	74	91	92

Table 2. Distribution of train drivers with respect to country, gender, experience, and driver state

Country	Number	Gender		Experience		State	
		Female	Male	Experienced	Novice	Alert	Fatigue
France	6	0	6	6		4	2
Great Britain	19	3	16	11	8	10	9
Israel	14	0	14	11	3	7	7
Italy	18	1	17	16	2	8	10
Sweden	21	1	20	16	5	10	11
Total mobile simulator	78	5	73	60	18	39	39
Advanced simulator SE	32	4	28	21	11	16	16
Grand total	110	9	101	81	29	55	55

Driving tasks

The car driving task was divided in two parts, driving on a two-lane rural road with an Intelligent Speed Adaptation System (ISA) and motorway driving with a Forward Collision Warning System (FCW). The ISA part of the car experiment included negotiating speed limit changes and sharp curves, as well as driving through villages and a school zone. In the FCW condition, the driver encountered events with a lane changing truck, a road work with a lane drop, the sudden braking of a car in front and breakdown of a downstream vehicle. However, it turned out that in most cases the FCW situations were not critical enough for the warning to be triggered. Thus, the results from this part are excluded in this paper but can be found in in the project deliverable (Forsman et al., 2011).

During the train driving experiment, the driver interacted with a simplified version of the European Rail Traffic Management System (ERTMS) called ETCS (European Train Control System) which provided information to the driver concerning speed management. Participants drove approx. 80 minutes according to a given timetable. The scenario included several changes in speed and nine stations with stops. Speed changes and stops included different speed alignments, i.e. slopes, and levels of permitted speeds.

Performance indicators

During normal car driving and driving through curves, school zone and villages, the performance indicators were mean speed and number of warnings. When approaching curves, villages and a school, the performance indicators were spot speeds at certain locations before the curve/village/school, mean speed passing a village or school zone and number of warnings. Performance indicators for the train drivers were speed deviations from permitted speed at station stops and speed reduction events. Furthermore, warning duration (due to over speed) was also used as a dependent variable.

Statistical analysis

Two basic ANOVA models were used (with/without workload). The model included the following main factors: *Country, Experience, Fatigue, Sensation Seeker, Gender, Age, Order, Subject* (*Country x Experience x Fatigue x Sensation, Seeker x Gender x Age x Order*) and *workload*. A priority list for interesting two-way interactions to be included in the model was made. No higher order interactions were included. When comparisons between different levels of the factors were made, these were based on the estimated marginal means which compensate for an unbalanced design if that was the case.

In the validation part, small scale (portable) simulators versus full scale simulators, only drivers from Great Britain were considered for the car simulator and only Swedish train drivers for the train simulator. The model described above was used with a slight modification: the factor *country* was removed and a factor that describes *simulator type* was added. The main effect of simulator type answers whether or not there are differences in level between simulator types for different performance indicators. In general a significance level of 5% was used in the analysis.

Results

The results presented here is a subset of what can be found the project deliverable (Forsman et al., 2011).

Car drivers

For most measures during normal driving and driving through curves, four of the main factors were significant: *Country, Gender, Workload, and Subject*. The significant effect of subject was expected since it is known from previous experience that there are large individual differences of different subjects in driver performance. The only significant interaction effects were *Country × gender* during normal driving and *Sensation Seeker × Workload* at curve sign, entry and apex.

The mean speed levels for different countries are shown in Table 3. The signed speed limit was 80 km/h both at normal driving and through curves. It can be seen that during normal driving, the drivers from Israel had the relatively highest mean speed and drivers from Italy the lowest. When driving through curves, drivers from Israel had the highest speed at curve sign, but at curve entry they tend to slow down and through the curve (apex) they have among the lowest speeds. The Swedish drivers tended to have among the lowest speeds at all spots in the curves.

Table 3. Analysis of variance – adjusted mean values, mean speed at normal driving and speed at various locations in relations to curves (km/h)

	Normal driving	Speed through curves			
Country	Mean speed	Mean speed	At sign	Entry	Apex
GB	73.47	57.30	72.42	55.09	57.97
SE	75.50	57.16	72.07	53.46	57.88
FR	75.93	62.14	74.79	60.76	62.55
IT	71.58	61.82	72.09	61.42	63.68
IL	76.27	57.12	76.55	58.36	57.68

Table 4 shows that during normal driving, male drivers drove more than 2 km/h faster than female drivers. At curves the difference is even larger, male drivers drove about 4 km/h faster than female drivers. There were minor differences between men and women in Great Britain, Sweden and France, while in Italy and Israel male drivers drove considerably faster than women, see Figure 3.

Table 4. Analysis of variance – adjusted mean values for gender (km/h)

	Normal driving	Speed through curves			
Gender	Mean speed	Mean speed	At sign	Entry	Apex
Female	73.41	56.27	71.24	55.08	57.14
Male	75.59	61.14	75.36	59.64	61.98

Table 5 shows that during normal driving and high workload the speeds were about 2 km/h lower than at low and medium workload. Through curves (mean speed, entry and apex) there was a tendency that the speeds were higher at high workload.

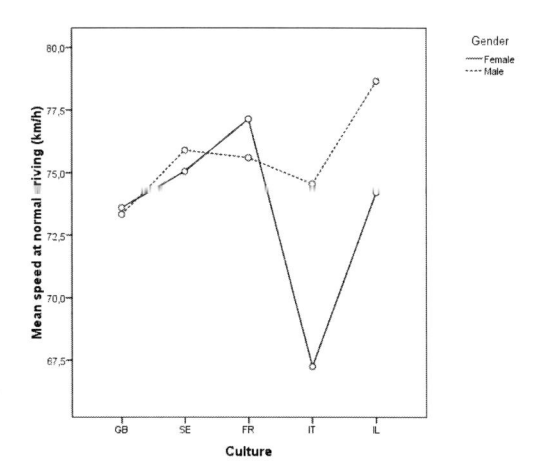

Figure 3. Adjusted mean speed for Country and Gender at normal driving (km/h)

Table 5. Analysis of variance – adjusted mean values for workload (km/h)

	Normal driving	*Speed through curves*			
Workload	*Mean speed*	*Mean speed*	*At sign*	*Entry*	*Apex*
Low	75.79	58.33	75.10	56.16	58.85
Medium	75.11	58.96	72.04	57.62	59.74
High	73.08	59.88	73.65	59.30	61.15

In figure 4, adjusted mean speeds for workload and sensation seekers at curve entry are shown. The difference between high and low sensation seekers was smaller during high workload. There were similar tendencies for speeds at the other locations were spot speeds were measured. When the number of warnings was analysed similar results as those for speeds were obtained.

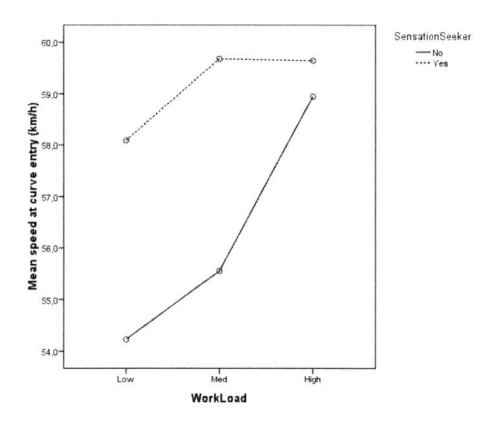

Figure 4. Adjusted mean speeds for Workload and Sensation Seekers at curve entry (km/h)

Speed behaviour through villages and school zones were also studied. Only results from the villages are included here, but the results were similar at school zones. When driving through the village, *Country, Order* and *Subject* are the main factor effects that are significant at sign visible and entry of village. When studying the mean speed through village, *Country* was the only significant main effect. At the spot where the village-sign was visible, the interaction effects *Country × Experience* and *Sensation Seeker × Experience* were significant. The mean speed levels for different countries and at various locations in relation to the village are shown in table 6. The drivers from Israel have the highest speeds and drivers from Sweden, Italy and Great Britain tend to drive slower, French drivers are in between. Looking at the first column in table 6, showing the mean speed through the village (speed limit 50 km/h) for drivers from different countries, drivers from Great Britain and Italy drove at about 46 km/h while drivers from Israel had a mean speed close to 53 km/h. Table 7 shows that, similar to normal driving and driving through curves, male drivers tend to drive faster than female drivers.

Table 6. At villages: analysis of variance – adjusted mean values for different countries (km/h)

Country	\multicolumn{4}{c}{*Speed at villages*}
	Mean speed	**Sign visible**	**Entry**	**Apex**
GB	45.72	70.15	51.66	53.03
SE	47.29	70.55	51.35	56.81
FR	48.95	73.18	55.01	60.35
IT	45.70	68.54	53.20	54.79
IL	52.73	77.17	61.07	60.02

Table 7. At villages, analysis of variance – adjusted mean values for gender (km/h)

Gender	*Speed at villages*			
	Mean speed	**Sign visible**	**Entry**	**Apex**
Female	47.15	70.85	53.13	55.44
Male	48.86	72.82	55.39	58.35

Figure 5 shows the adjusted mean speeds through the villages. There were minor differences in speed choice between novice and experienced drivers in all countries except for Israel. In Israel, the mean speeds through villages were about 10 km/h higher for experienced than novice drivers.

The adjusted mean speeds where village sign was visible for *Sensation Seeker* and *Experience* are illustrated in Figure 6. If classified as a sensation seeker, experienced drivers had a higher speed than novice, but if classified as a non-sensation seeker, novice drivers had higher speeds than experienced.

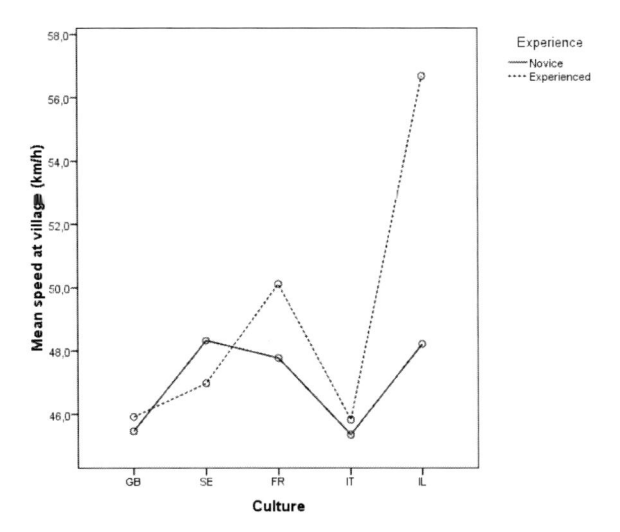

Figure 5. Adjusted mean speed for Culture and Experience. mean speed through village

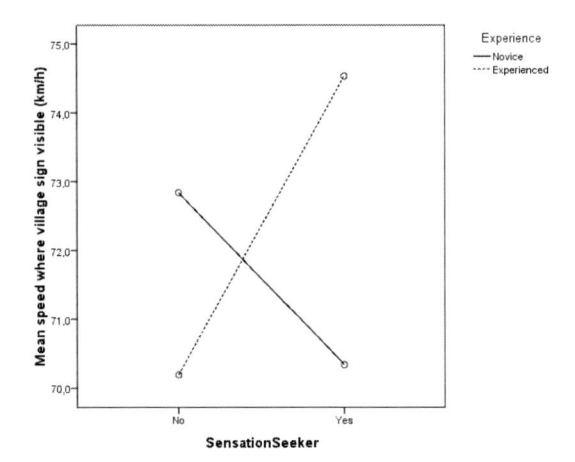

Figure 6. Adjusted mean speed for Sensation Seeker and Experience where village sign visible

Car drivers – small scale portable simulator versus full scale simulators

Data from both types of car simulators were incorporated in the same model (only drivers from Great Britain). In general, two of the main factors turned out significant: *simulator* and *workload*. None of the interaction effects with simulator were significant. This means that it does not seem that for example becoming more experienced, fatigued, changing workload etc. affects the behaviour in the portable simulator differently than in the full scale simulator. In table 8, adjusted mean speeds for the two simulator types are shown. The mean speed during normal driving and driving through curves is about 3-4 km/h higher in the full scale simulator compared to the portable one. The speed difference between the portable and full motion simulator is almost 9 km/h at curve entry and about 3 km/h at curve apex and exit.

Similar to the results in the portable simulator when all five countries were included, a higher workload during normal driving seem to lower the speeds. Furthermore, for normal driving there is no difference in speed choice between low and medium workload (Table 9).

Table 8. Analysis of variance – adjusted mean values for simulator type (km/h)

	Normal driving	Speed through curves			
Type of simulator	Mean speed	Mean speed	At sign	Entry	Apex
Portable	72.77	57.13	72.27	54.93	57.76
Full motion	75.96	61.00	74.81	63.30	60.60

Table 9. Analysis of variance – adjusted mean values for workload at normal driving and in curves (km/h)

	Normal driving	Speed through curves			
Workload	Mean speed	Mean speed	At sign	Entry	Apex
Low	75.84	58.91	74.77	57.72	58.87
Medium	75.51	59.42	72.95	60.11	59.78
High	71.50	58.84	72.69	58.83	58.76

When analysing the speeds driving through the village and school zone, the pattern was the same as for normal driving and driving through curves; the speed is about 3-4 km/h higher in the full simulator trial than in the portable one.

Train drivers

Speed reduction events occurred during two in principle different situations, at station stops and during driving (no stop). Five speed metrics were used; spot speed when information on future new speed limit was given (PI), mean speed during approach to new speed limit (SI), spot speed at initiation of speed reduction (PII), mean speed during speed reduction (SII), and speed at event end (PIII). Furthermore, warning duration during speed reductions and station stops was also used in the analysis. Gender was not included in the analysis as there were so few female train drivers.

It was found that during station stops three of the main factors were significant: *Sensation Seeker* (PI, SI, PII), *country* (All except PIII) and *workload* (All metrics). For some of the metrics there were significant interactions: *Country* x *Workload* (PI), *Experience* x *Country* (PI), and *Experience* x *Sensation Seeker* (PIII). Furthermore, for speed reduction while driving two main factors were significant: *country* (All except PI), *Workload* (PII). Interactions were found for: *Experience* x *Country* (PII, SII, PIII), *Country* x *Workload* (SI), and *Experience* x *Workload* (SI).

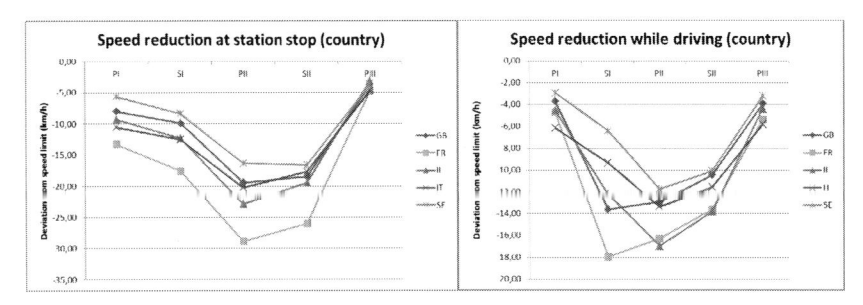

Figure 7. Adjusted mean deviation from speed limit per country (low value closer to the limit) during station stop (left panel) and while driving (km/h)

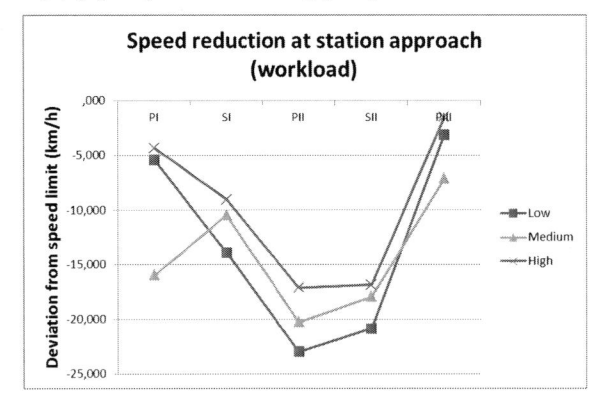

Figure 8. Adjusted mean deviation from speed limit for Workload (low value closer to the limit) during station stop (km/h)

Warnings were given if the driver was over speed limit. Thus, it was considered interesting to include warning duration (seconds/km) in the analysis. Four metrics were used: total warning duration for all conditions, warning duration during speed reduction for station stops, warning duration during speed reduction while driving and warning duration for station stops. There were main effects of *experience (speed reduction while driving)*, *Country* and *Workload (not speed reduction for station stops)*. Furthermore, there were significant interactions between *Experience x Country (three metrics)* and *Experience x Sensation Seeker (one metric)*.

Figure 9 depictures some differences between countries, e.g. Israeli train drivers experienced in total the longest duration of warnings and also during station stops. Train drivers from Great Britain and Italy encountered the longest warning duration during speed reduction for station stops. It can also be seen that French and Swedish train drivers evoked the least warnings.

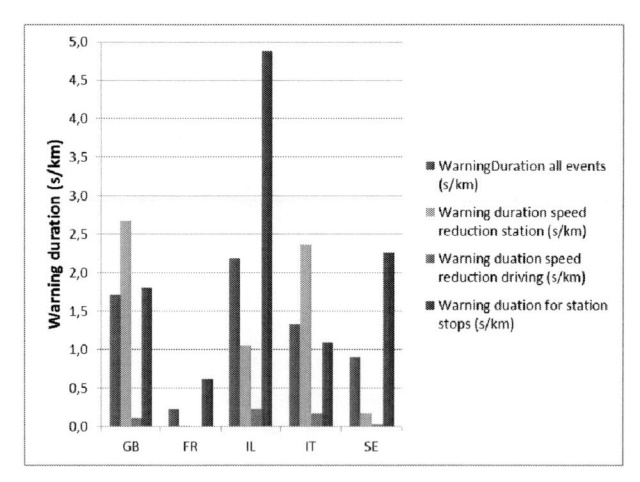

Figure 9. Adjusted mean warning duration (seconds/km) per country

Figure 10 shows warning durations under the three workload conditions (low, medium and high) for the four situations. The warning duration was under high workload compared to low and medium for three of the four situations. However, the difference in warning duration during speed reduction while driving seems much less.

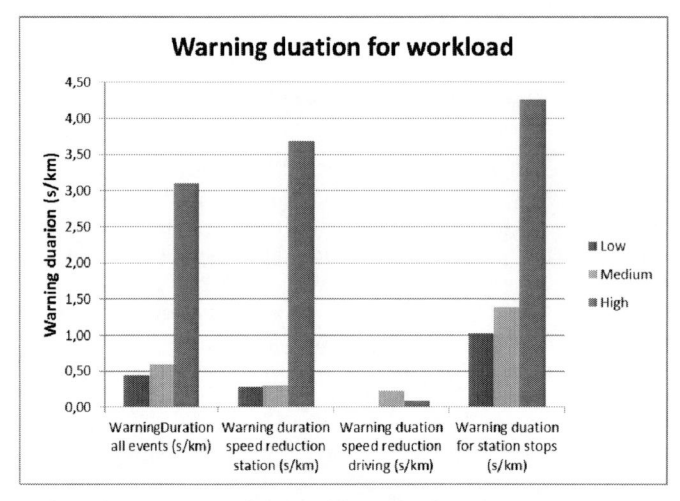

Figure 10. Adjusted mean warning duration for workload

Finally, the comparison between the small scale train simulator and the full scale simulator revealed no significant differences between the simulators with respect to speed performance. Only Swedish train drivers were included in this analysis. In total 53 professional drivers were included. Concerning warning durations it was found that drivers in the full scale simulator had warnings with significantly longer durations when considering the total drive but not for the specific events (i.e. speed reductions). The rationale behind this does not seem self-evident and call for further analysis. In general, there were no interaction effects found between type of simulator and other factors. Thus, no major differences between the two simulator types were observed within the context they were used here.

Discussion and conclusions

The results from both the car and train experiments show that the factor *Country* is statistically significant for almost all performance indicators. This was an interesting finding which requires further investigation. Traffic culture was captured by a questionnaire developed by Özkan et al. (2006), might give some further insight but it has not yet been analysed.

However, participants in the study may not be representative of drivers in general in the different countries. Furthermore, it might well be that it is not nationality per se that cause the observed difference but rather some underlying factors. It was for instance found that Israeli drivers were in general younger and that male Israeli drivers seem to score highest on the BSSS. Furthermore, it should be noted that train drivers were not just from different countries but also from different professional work cultures which could explain some of the differences found. The impact of different work cultures on driving behaviour has been found to influence driving behaviour (Li & Itoh, 2011).

Three levels of *Workload* were evaluated: low, medium and, high. The results showed that driving under different workloads affects car drivers' speed choice in both normal driving and driving through curves. In normal driving, the mean speed was about the same for low and medium workload, and lower for high workload. This confirms to what was expected. Workload had also an impact on train drivers speed choice so that they drove close to the speed limit (i.e. higher speed) under high workload compared to low and medium. Warning duration was longer during high workload in general and during speed reduction for station stops. Thus, it can be concluded that workload is a factor that has an important impact on driving performance and thus traffic safety which per se is not new (de Waard, 1996) and it is a factor that should be included in a model of driver behaviour.

Gender was another important factor for car divers which affected some components of the drive. The results showed that males drove faster during normal driving and when driving through curves, but not when driving through villages or when passing school. In normal driving and when driving through curves, males drivers generally drive faster than female drivers. The results indicate that gender is a factor that needs to be considered in a further analysis. However, it can be questioned if it is gender per se that can explain the differences found in driving behaviour or if it is a more complex phenomena e.g. masculinity and femininity (i.e. both males and females can have masculinity and femininity aspects in their characters) that can turn out as an explaining factor (Bem, 1974; Stets & Burke, 2000).

When validating the portable car simulator against the full motion car, the comparison showed different levels of speed for every scenario studied. The speeds were about 3-4 km/h higher in the advanced simulator. In general, there were no interaction effects between type of simulator and other factors, meaning that we cannot see that for example becoming more experienced, fatigued or changing workload affects the behaviour in the portable simulator in another way than the full motion simulator. The comparison between the portable train simulator and the full scale simulator revealed in general no main effects that could be attributed to differences in the simulators. However, the single difference with respect warning duration should be further investigated.

Neither the hypotheses that high sensation seekers choose higher speed than low sensation seekers could not be confirmed, nor that experienced drivers chose higher speed than inexperienced or that fatigued drivers chose higher speeds. However, the differences in BSSS between subjects classified as high and low sensation seekers were rather small. This was considered in a follow-up study in the same project by screening car drivers for high and low sensation seekers. Concerning fatigue, no difference in sleepiness as measured by the Karolinska Sleepiness Scale (Reyner & Horne, 1997) could be found before the simulator drive, only after. This indicates that the post lunch dip design to induce fatigue might not have been successful. Drivers seem to choose lower speed during high workload compared to low and medium workload at normal driving, but the opposite when driving through curves. Furthermore, it seems like small scale simulators can provide comparable results to full motion simulators.

References

Barnard, Y., Lai, F., Carsten, O., Merat, N., Hjälmdahl, M., Dukic, T., Vanderhaegen, F., Polet, P., Enjalbert, S., Hasewinkel, H., Lützhöft, M. Kircher, A., Kecklund, L. & Dimgård, M. (2010a). *Selection of operator support systems across modes* (Project deliverable 2.2): ITERATE (IT for Error Remediation And Trapping Emergencies) Consortium.

Barnard, Y., Lai, F., Carsten, O., Merat, N., Hjälmdahl, M., Dukic, T., Wallén Warner, H., Enjalbert, S., Pichon, M. & Vanderhaegen, F. (2010b). *Specification of test procedures for the simulator experiments* (Project deliverable 3.1): ITERATE (IT for Error Remediation And Trapping Emergencies) Consortium.

Bem, S. (1974). The measurement of psychological androgyny. *Journal of Consulting and Clinical Psychology, 42*, 155-162.

De Waard, D. (1996). *The Measurement of Drivers' Mental Workload*. Doctoral Thesis, University of Groningen, Groningen. Retrieved from http://www.home.zonnet.nl/waard2/mwl.htm

Hjälmdahl. M. & Amantini, A. (2012). Implementation of a Unified Model of Driver into numerical algorithms for a predictive simulation of behaviour in different transportation contexts. In D. de Waard, N. Merat, A.H. Jamson, Y. Barnard, and O.M.J. Carsten (Eds.) *Human Factors of Systems and Technology* (pp. 359-372). Maastricht, the Netherlands: Shaker Publishing

Lai. F., Barnard., Y., Merat, N., Carsten, O., Lutzhoft, M., Kecklund, L., Enjalbert, S., Vanderhaegen, F. & Pichon, M. (2010). *Review of existing technologies and systems supporting the operator* (Project deliverable 2.1): ITERATE (IT for Error Remediation And Trapping Emergencies) Consortium.

Li, Y. & Itho, K. (2012) Contributing Factors to Driving Errors in Trucking Industry: Drivers' Individual, Task and Organizational Attributes. In D. de Waard, N. Merat, A.H. Jamson, Y. Barnard, and O.M.J. Carsten (Eds.) (2012). *Human Factors of Systems and Technology* (pp. 213-221). Maastricht, the Netherlands: Shaker Publishing.

Oppenheim, I., Shinar, D., Carsten, O., Barnard, Y., Lai, F., Vanderhaegen, F., Polet, P., Enjalbert, S., Pichon, M., Hasewinkel, H., Lützhöft, M., Kircher, A. & Kecklund, L. (2010a). *Critical review of models and parameters for Driver models in different surface transport systems and in different safety critical situations* (Project deliverable 1.1): ITERATE (IT for Error Remediation And Trapping Emergencies) Consortium.

Oppenheim, I., Shinar, D., Enjalbert, S., Dahyot, R., Pichon, M., Ouedraogo, A., Lützhöft, M., Carsten, O., Hjälmdahl, M., & Cacciabue, C. (2010b). *Description of Unified Model of Driver behaviour (UMD) and definition of key parameters for specific application to different surface transport domains of application.* (Project deliverable 1.2): ITERATE (IT for Error Remediation And Trapping Emergencies) Consortium.

Reyner, L.A. & Horne, J.A., 1997. Suppression of sleepiness in drivers: combination of caffeine with a short nap. *Psychophysiology, 34*, 721–725.

Stets, J. E. & Burke, P. J. (2000) *Femininity/Masculinity*. In E.F. Borgatta and J.V. Rhonda (Eds.) *Encyclopedia of Sociology*, Revised Edition (pp. 997-1005). New York: Macmillan

Forsman. Å., Vadeby. A., Yahya, M-R., Tapani. A., Enjalbert, S., Cassani, M., Armantini. A., Lai, L., Kecklund, L. & Arvidsson, M. (2011) *Results from the analysis and input to development and validation of the statistical models.* (Project deliverable 5.1). ITERATE (IT for Error Remediation And Trapping Emergencies) Consortium.

Özkan, T., Lajunen, T., Wallén Warner, H., & Tzamalouka, G. (2006). *Traffic climates and driver behaviour in four countries: Finland, Greece, Sweden and Turkey*. Paper presented at the 26th International Congress of Applied Psychology, Athens, Greece, 16-21 July.

Özkan, T., & Lajunen, T. (2011). Person and Environment: Traffic Culture. In B. Porter (Ed.), *Handbook of Traffic Psychology* (pp. 179 - 192). London: Academic Press.

Implementation of a Unified Model of Driver into numerical algorithms for a predictive simulation of behaviour in different transportation contexts

Magnus Hjälmdahl[1], Aladino Amantini[2], & Pietro C. Cacciabue[2]
[1]VTI, Linköping, Sweden
[2]KITE Solutions, Laveno Mombello (VA), Italy

Abstract

The research presented in this paper stems from the hypothesis that "driver" behaviour is based on certain fundamental characteristics that can be shared amongst different working contexts. Consequently, it is possible to define a "Unified Model of Driver" (UMD) that captures the basic aspects of behaviour of a human being in control of a vehicle. The variation associated to diverse contexts is obtained simply by modifying the parameters that affect the fundamental modelling correlations. Following this hypothesis, a research initiative has been performed within an EU funded Project that has studied automotive, rail and ship domains through a theoretical development, associated to substantial experimental activity in the three domains. The experiments leading to the implementation of the computerised simulation approach consisted of close to 300 subjects from five different countries, probably making it the largest controlled experiment in the transport domain. In this paper, the simulation tool resulting from the research work and experimental activity is presented. A number of simulation runs are presented demonstrating the feasibility of the approach.

Introduction

Modelling human behaviour has become a necessary endeavour since the eighties when the consideration of human-machine interaction became integral part of the design process of "modern" technologies and their control processes.

Following the initial models based on the representation uniquely of human performances, due to the reason that machines were governed essentially through manual control, the cognitive aspects and the functions of perception, interpretation, planning decision making etc. have become much more important as automation has gradually replaced manual control to become the predominant actor in plant and system management.

Modelling implies a theoretical representation of the phenomena under consideration. In the case of human behaviour, modelling this means considering the mechanism that govern the above mentioned cognitive and behavioural functions as well as the associated mental processes and knowledge base, typical of a human

In D. de Waard, N. Merat, A.H. Jamson, Y. Barnard, and O.M.J. Carsten (Eds.) (2012). *Human Factors of Systems and Technology* (pp. 359 - 372). Maastricht, the Netherlands: Shaker Publishing.

being. Furthermore, in order to implement human-machine interaction in design and safety analysis, the model of human behaviour must be transformed into a "simulation" capable of "expressing in a numerical or logical representation" the theoretical "equations" typical of models and paradigms.

This step of implementation of a simulation approach can be effectively carried out in association with field observations which enable to assess the dependence between the critical variables that govern the model and certain characteristics or environmental conditions that affect behaviour (both cognitive and operational). Therefore, the development of a simulation approach requires the consideration of a sound theoretical approach as well as a consolidated experimental design of field and experimental observations.

A simulation approach can be developed in different ways, primarily consisting in two large families of approaches: micro-simulation and mathematical correlations. The micro-simulations imply the development of a set of equations and numerical solutions that describe the cognitive functions and processes. These are coupled to parameters that depend on the field data collected during the experiments. On the other side, the mathematical correlations consist of a set of logical and mathematical expressions whose coefficients are assessed by regression analysis from the observed behaviours.

Both types of simulations enable to predict behaviour simply by applying the numerical solutions or by extrapolating the mathematical correlations outside the observed experimental conditions, to dynamic changes of the environment. The model and simulation described in this paper belongs to the second family of development and stems from the hypothesis that "driver" behaviour is based on certain fundamental characteristics that can be shared amongst different working contexts. Consequently, it is possible to define a "Unified Model of Driver" (UMD) that captures the basic aspects of behaviour of a human being in control of a vehicle. The variation associated to diverse contexts is obtained simply by modifying the parameters that affect the fundamental modelling correlations.

Following this hypothesis, a research initiative has been performed within the EU funded project ITERATE that has studied automotive, rail and ship domains through a theoretical development, associated to substantial experimental activity in the three domains. The experiments leading to the implementation of the computerised simulation approach consist of close to 300 subjects from five different countries, probably making it the largest controlled experiment in the transport domain.

In this paper, the simulation tool resulting from the research work and experimental activity is presented. The basic theoretical framework is briefly introduced and the simulation approach is discussed in detail, showing the numerical implementation of the basic human behaviour correlations and the mechanism that enables to differentiate amongst the three different domains, while maintaining the same theoretical architecture. This paper focuses on the work related to car drivers and train operators since those two modes are the most developed with regard to the numerical model. The work related to ships is presented in another paper at this conference. A number of simulation runs are presented here demonstrating the feasibility of the approach.

Finally, it is important to note that the feedback of the experimental activity on the model is of paramount importance in such a research initiative. The simulation instrument has fully accounted for the outcome of experimental results. However, the discussion on such activities falls outside the scope of this paper and it is only referenced in this paper, while it is discussed extensively in one of the projects deliverables (Hjälmdahl, 2012).

Unified Model of Driver

The UMD that has been developed within ITERATE has derived from a comprehensive literature search and several workshops on the topic within the project consortium. The UMD is described in figure 1, for a more detailed description of how the model has been developed and the considerations taken refer to (Oppenheim, 2010, Shinar, 2010).

In order to be a useful tool, the selected model should include as inputs, factors that have been shown to influence risk, risk-taking and errors. The selected driver variables chosen as input for the model and for the experiments are:

ATT - Attitudes/personality (Sensation Seeking) - especially relevant for the road vehicles, For other transport modes this is of less relevance because they employ professional drivers who are recruited under restrict conditions and therefore the presence of sensation seekers among drivers might be rather low. Personality traits may have negative influence on driving performance. Many articles show correlation between sensation seeking and some aspects of risky driving.
EXP - Experience (Hazard Perception Skills) – relevant to all modes of transport. Hazard perception skills have been found to correlate with crash risk.
DS - Driver State (Fatigue) – relevant to all modes of transport. Monotony of road environment has an adverse effect on driver performance and fatigue caused by driving in complex road had the greatest impact on driving behaviour.
TD - Task Demand (Subjective workload) – also important within all transport modes, Task demand arises out of a combination of environmental features (complexity of traffic, weather, and light conditions), other road users' behaviour, and characteristics of the vehicle; not necessarily in the same level of importance for the different transport modes.
CULT - Culture (country) - Common to all transport modes, Most Studies dealing with cross cultural differences in driving found significant effect on behaviour and performance variables.

There are, of course, some differences between the different transport modes concerning these parameters, but they were deemed to be sufficient to give a reasonable cover of most of the important and relevant factors.

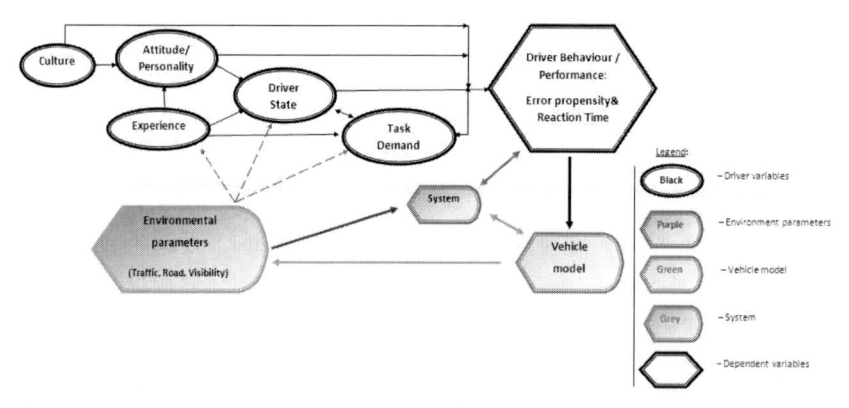

Figure 1. The Unified Model of Driver behaviour (UMD) developed within ITERATE

Method

In this paper there are two separate lines of work that are performed and then merged to provide the results. One is the experimental work providing the data on driver behaviour needed to tune the model and the other is the development and tuning of the numerical model.

Simulator studies

To be able to take the model from the theoretical model described above to an executable numerical model a large scale experiment, encompassing both cars and trains, was carried out. A number of considerations were established for the experiments such as comparability which means that the train and car experiments should have as much commonality as possible and the scenarios should be viable for the different countries involved. Further the coverage of model parameters should be as complete as possible and there should be the possibility of studying parameters in combination in order to study interactions. There were also a lot of practicalities considered to make sure that the experiment could be carried out identically in all five countries.

For this means two identical portable simulators, designed for both cars and trains, where developed. The simulators consisted of a PC, a 40" 1920x1080 monitor with a stand, a 15" 1366x768 monitor, a gaming steering wheel, a "gaming" train controller and a GameRacer seat adjusted to a less "racy" position including a stand for the steering wheel/train controller. The equipment was custom fitted to the two stands and assembly instructions including heights and inter-distances were specified. Two freight boxes were also constructed to fit each simulator. One of the simulators was shipped between Sweden and Italy while the other was shipped from the UK to France and then further to Israel and back.

Figure 2. One of the portable simulators on display at a conference in Italy. In the background the wooden freight box can be seen

One of the conditions described above was that the experiments should cover as many of the model parameters as possible and also that it should be possible to study parameters in combinations in order to study interactions. In order to achieve this, an experimental design allowing for all parameters to be tested in one experiment was developed illustrated by figure 3, where the within driver/train operator factor is shown inside the figures and between driver/train operator factors are shown outside. The drivers' sensation seeking was established by using the Brief Sensation Seeking Scale (Hoyle et al., 2002), Culture was given by the subjects' nationality while experience was based on years of driving experience and controlled by the recruitment. Fatigue was more difficult to include in the experimental design without breaking the consideration that it should be easy to carry out the experiment. The attempt in ITERATE is based on the post lunch dip where half the subjects drove in the afternoon after having lunch. In addition their experiment started with them watching a 20 minute long "boring drive". The idea was that circadian rhythm (Åkerstedt & Folkard, 1997) in combination with digesting lunch should lead to some level of fatigue. Fatigue was measured by the Karolinska Sleepiness Scale, KSS (Gillberg, Kecklund, & Åkerstedt, 1994) before and after the drive. Workload was varied within each subject in three levels by a counting backwards task. For the high workload the drivers/train operators had to count backwards in steps by seven, and for the medium workload the counted in steps by three. There was also the low workload condition which was driving without any secondary task.

To be able to draw any conclusions on effects of the various model parameters and even interaction effects a large number of subjects were needed, and that was indeed the case in the ITERATE experiment. According to the plan each of the five countries should run 32 car drivers and 32 train operator which would give a total of 160 car drivers and 160 train operators. In addition to this another 32 car drivers and 32 train operators were tested in full scale driving simulators; this to be able to test the validity of the portable simulators. Due to various circumstances the ambitious goal 2x32 driver in each country was not always fulfilled, especially for train drivers which turned out to be hard to recruit. In all the experiments resulted in 154 car drivers in the portable simulators and 29 in the full scale simulator while the corresponding figures for trains were 78 and 32. All in all 293 subjects participated in the experiment

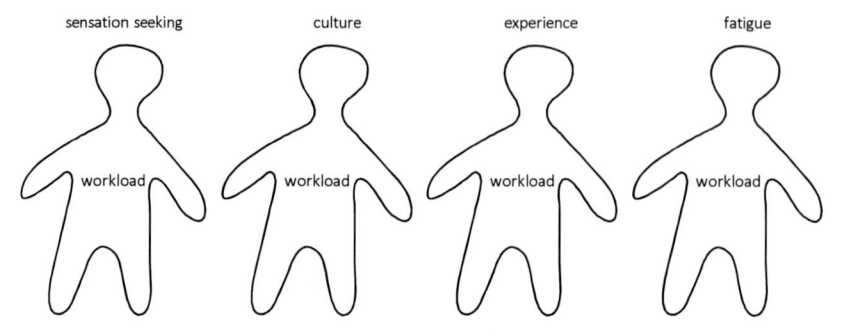

Figure 3. Overall experimental design

The scenarios in the experiment were designed as to be able to test driver behaviour and especially driver behaviour in interaction with driver support systems. An effort was made to find support systems and scenarios that would be comparable between the three modes of transport. The decision in ITERATE was to go for systems related to speed adaptation and Forward collision warning and the reason thereof is that those where systems that had commonalities across modes and it was possible to come up with somewhat comparable scenarios. In Lai (2010) and Barnard (2010) the process leading up to this decision is explained.

Development and tuning of the numerical model

In order to implement the UMD in a simulation of Driver Vehicle Environment (DVE) interaction it is necessary that the cognitive functions and processes that generate the actions of the driver are considered. These are implemented in the simulation in a quite simple architecture that attempts to capture the complex and realistic descriptive driver behaviour, i.e., a driver is affected by personal attitudes, motivational aspects, workload, etc. Moreover, the overall requirements of the simulation are of being predictive and fast running, accounting eventually also for dynamic interactions, human errors, and adaptive behaviour.

The most suitable and simple approach for representing driver behaviour during the performance of normative and descriptive activities is to apply a simple "Task Analysis" (Kirwan & Ainsworth, 1992). The accuracy of the analysis and description of the tasks that are performed by the driver defines also the granularity of the simulation (Michon, 1985). By carrying out a driving Task Analysis, the performance of a driver can be formalised and structured in a sequence of goals and actions that are carried out during the interaction with the vehicle and environment.

In addition, in order to implement the decision making process and the execution of actions, a very simple formulation has been chosen, based on the driver assessment of the minimum time to reach the objective: going from the starting point of the journey to its destination in the minimum allowable time. In other words, the vehicle is "driven" at the highest "intended" speed which is, in general, a function of the "maximum" allowed speed v_{max}. This depends on the rules and regulations and contextual conditions, i.e., for the automotive environment, speed limits and traffic/road conditions. The meaning of maximum desired speed varies according to the application domain and can depend on:

- Driver parameters
- Speed limits
- Vehicle characteristics
- Time Scheduling

In this simulation approach, rather than focusing on the detailed decision making process, the basic concept of similarity between different transportation modes of the UMD has been developed.

In general, various driving modes can be simulated according to the choice and execution of different tasks. These vary from vehicle to vehicle, i.e., car, train or ship, and can be summarized, in a first instance, in the following 4 main tasks:

1. Cruise
2. Overtake / Lane change
3. Follow
4. Start / Stop

The execution of these tasks is represented by means of several variables and functions that account for driver personal aspects, for the vehicle characteristics and for the environmental conditions. In order to capture the effects of the 5 basic parameters that affect behaviour (CULT, EXP, DS, TD and ATT), the Simulator for Model of Unified Driver (SiMUD) implements the UMD model according to the following conceptual elements:

1. The values of each parameter are implemented by means of discrete quantities.
2. These have been selected in such a way to enable the association of the essential variables affected by driver behaviour, i.e., speed, distance, acceleration and braking mode, with reasonably expected values.
3. These formulations are then tuned according to the results of the experimental observations. In the current project, the experiments have been carried out in the ITERATE portable simulator as described in Peters et al. (2011).
4. This process and the use of the ITERATE portable simulators enabled to differentiate between types of drivers as well as between the 5 national cultures involved in the project.

The following discrete values of the parameters have been associated to different types of driver characteristics and contextual and personal conditions (Table 1).

Table 1. Basic constants affecting driver behaviour

ATT	0 = Prudent		1/8 = Sensation seeker
EXP	-1/8 = Novice		1/8 = Experienced
DS	1/8 = Alert		-1/8 = Fatigued
TD	1/8 = Low	0 = Medium	-1/8 = High

CULT is considered separately and is not intended to have a numeric value, but is instead used to identify the nationality of the driver. The basic assumption of the simulation approach is that a generic expression (F_i) can be considered and utilized throughout the simulation to characterise all essential quantities associated to driver behaviour, e.g., speed, gas and brake activities, distances from obstacles and leading vehicles, stop and start performances etc. The functions F_i take the following form:

$$F_i = K_i(CULT) + \alpha_i(CULT)*C_1(ATT, EXP) + \gamma_i(CULT)*C_2(DS) + \beta_i(CULT)*C_3(TD) \quad \text{EQ 1}$$

where:
The quantities C_1, C_2, and C_3 are constants that depend on the characteristics of the driver:

$$\text{if} \rightarrow ATT = 1/8 \Rightarrow C_1(ATT, EXP) = 1/4$$
$$\text{if} \rightarrow ATT = 0 \Rightarrow C_1(ATT, EXP) = EXP$$
$$C_2(DS) = DS$$
$$C_3(TD) = TD$$

The constant term K_i and the coefficients α_i, β_i, and γ_i enable to differentiate between drivers according to their culture and, within cultures, to their personal characteristics.

The functions F_i take different values according to the specific activity of the driver. Their values have been determined by interpolating the experimental results with the formulations that have been assumed for each simulated activity. Some formulations utilized in the SiMUD are shown in table 2. The left side of the equations contains the quantity calculated, while the right side shows the expression containing the function F_i, generated from the experimental tests. The value 1.6 for the evaluation of the intended distance has been assigned from literature studies (Treiber, Hennecke, & Helbing, 2000). More complex formulations for the accelerator and brake pedals have been developed, but are not reported in table 2 for brevity.

Table 2. Some formulation utilized in the SiMUD simulation

Automotive domain formulations	Rail domain formulations
$V_{intended-car} = F_{car-int-speed} \cdot V_{max-allowed-car}$ $\text{reaction\_distance}_{car} = \dfrac{\text{visibility}}{F_{car-react-dist}}$	$V_{intended-train} = F_{train-int-speed} \cdot V_{max-allowed-train}$ $\text{reaction\_distance}_{train} = \dfrac{\text{visibility}}{F_{train-react-dist}}$

$$\text{intended\_distance}_{car} = \frac{1.6 * \text{speed}_{car}}{F_{car-int-dist}} \qquad acc = F_{train-cruise-acc} \cdot \left(1 - \frac{\text{speed}_{train}}{\text{inten\_speed}}\right)$$

Table 3 contains the coefficients K, α, γ and β that are evaluated from the regression analysis on the data collected during the experiments for some of the F_i functions. The values between parentheses represent the standard deviations.

It is to be noted that only linear regression analysis has been performed. This has generated certain values of standard deviations that are relevant and seem to indicate that a non-linear type of correlation could have been considered in those cases. However, it has been assumed that for this first stage of development of the SiMUD, the results obtained with linear correlations are sufficient to perform simulations and evaluations of different behaviours.

Table 3. Results of the regression analyses on the data collected during the experiments

F	CULT	K	α	γ	β
$F_{car-int-speed}$	FR	0.957 (0.000)	0.036 (0.001)	-0.077 (0.002)	0.002 (0.002)
	IL	0.960 (0.001)	0.228 (0.003)	-0.065 (0.003)	0.165 (0.004)
	IT	0.892 (0.001)	0.104 (0.003)	-0.187 (0.002)	-0.121 (0.003)
	SE	0.967 (0.000)	-0.105 (0.001)	0.047 (0.001)	-0.003 (0.002)
	UK	0.918 (0.000)	0.011 (0.001)	0.071 (0.001)	-0.019 (0.002)
$F_{car-intended-dist}$	FR	0.831 (0.002)	-0.115 (0.011)	0.187 (0.012)	-0.238 (0.018)
	IL	0.621 (0.001)	0.447 (0.004)	0.558 (0.006)	0.451 (0.008)
	IT	0.387 (0.018)	0.899 (0.081)	-0.107 (0.055)	0.602 (0.092)
	SE	0.695 (0.001)	-0.389 (0.005)	0.300 (0.004)	0.509 (0.006)
	UK	0.671 (0.001)	-0.371 (0.004)	-0.101 (0.005)	1.120 (0.010)
$F_{car-reaction-dist}$	FR	1.148 (0.043)	-0.082 (0.226)	-0.459 (0.264)	0.889 (0.414)
	IL	1.256 (0.089)	0.177 (0.421)	-0.315 (0.519)	0.784 (0.798)
	IT	1.208 (0.175)	0.516 (0.777)	-0.975 (0.547)	2.105 (0.873)
	SE	1.158 (0.041)	-0.243 (0.224)	0.230 (0.190)	0.576 (0.295)
	UK	1.240 (0.062)	-0.496 (0.352)	-0.820 (0.439)	1.484 (0.697)

Results

Two sets of results are presented hereafter: two demonstration scenarios, a car cruise and a train cruise and stop configurations; a cruise and follow car scenario utilising the correlations developed with reference to the experimental results.

Case study 1: Demonstration scenarios

In order to show the simulation capability of the SiMUD tool, two demonstration scenarios are presented in figure 4: a car cruise scenario and a train cruise and stop scenario. The driver characteristics are: prudent, experienced, fatigued, with low task demand. The coefficients K, α, γ and β and the F_i functions have been set to standard values not correlated to the experiments.

The car driving task considers a dangerous curve at about 3000 m of the scenario, with a speed limit of 50 Km/h, and a second section of the road, at about 5000 m, with a speed limit of 30 Km/h. The driver, when approaching a dangerous curve, reacts initially by decreasing the intended speed (v_{int}) simply releasing the pressure on the accelerator pedal. When the turning is completed, the cruising speed is re-established. At about 5000 m, the speed limit of 30 Km/h is reached by decelerating and braking according to need.

The train driving task shows the behaviour of train driver when performing Cruise and Stop Tasks. Three speed limit sections are present and a stop at a station is foreseen. Figure 4 shows how the driver operates on the controllers (accelerator and brake) in order to reduce speed, eventually reaching the stop at the station.

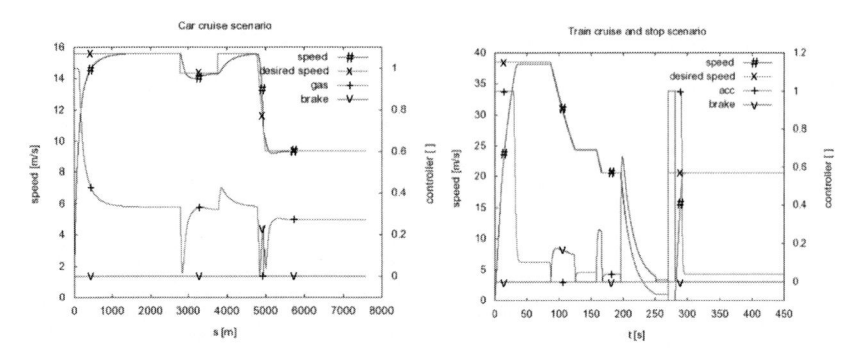

Figure 4. Speed and controllers behaviour while performing cruise task (car scenario - left) and cruise and stop task (train scenario - right)

Case study 2: Car cruise and follow scenario with correlations to experiments

The case study discussed hereafter implements the coefficients K, α, γ and β that are evaluated from the regression analysis on the data collected during the experiments (Table 3). The road configuration is the one used for the experiments. The traffic scenario includes a cruise and a follow tasks: in the initial section of the road, with a speed limit of 110 Km/h (30,55 m/s), no other vehicle is present and the ego-vehicle can travel at the intended speed; then a leading vehicle is encountered, which travels at a constant speed of 72 Km/h, much lower than the one of ego-vehicle. Therefore, a switch from "cruise" to "follow" mode occurs. No "overtaking" mode is considered in the present version of SiMUD.

The driver is assumed to be experienced, sensation seeker, and alert with low task demand conditions. Therefore the three constants C_1, C_2, and C_3 are set to the value 1/8 (EQ.1).

On the basis of speed limits and driver parameters it was expected that the simulation would show speed well above the limits, with an aggressive driving by the ego-vehicle, especially as the driver was alert with low task demand. Different cultural behaviours were to be compared.

The results are shown in figure 5. For each national culture, namely Israeli, French, Italian, Swedish and British, the following variables are reported: desired speed (marked x) and current speed (marked #), controllers activity (gas and brake pedal positions). Differences between cultures are noticeable. However, in general, the expected sensation seeker behaviours are less relevant than expected, especially in terms of speed, i.e., all drivers respect quite closely the speed limit or travel at a lower speed. All drivers modulate their speed while approaching the leading vehicle.

Israeli and French drivers are the fastest drivers and reach the leading vehicle earlier than the others. The most controlled drivers are the Italians, as they reach the leading vehicle later than others and have a relatively low intended speed. The behaviour of the British and Swedish drivers is very similar, to the Italian drivers and is located in between the others.

The activities on the gas and brake pedals show more prominent cultural differences and, in general, are more in line with the expected sensation seeker behaviour. All drivers "play" quite substantially with the brake and the accelerator. In general, all sensation seekers act on the gas and brake pedal in an aggressive way, by accelerating or braking prominently and several times. Contrary to all other drivers, the Israeli drivers brake only once, in a much more prominent than all other drivers. Italian drivers modulate their speed by acting on the gas and brake pedals five times before stabilising the car "follow" mode, reached much later than all other drivers.

This set of results may be considered slightly unrealistic, as experience shows that many drivers tend to keep speeds higher than the speed limit, and frequently greater than 5% above the limit.

However, it is important to note here that the overall simulation approach is based on the attempt to adapt the formulations and numerical expressions to the results of the experiments and not to capture what may be a fully realistic driving behaviour in actual traffic conditions. In other words, the outcome of the simulations depend heavily on the coefficients (K, α, β and γ), i.e., the F_i functions, evaluated from the regression analysis. These results are in agreement with the outcome of the analyses and evaluations of the data collected during the experimental sessions (Peters et al., 2011).

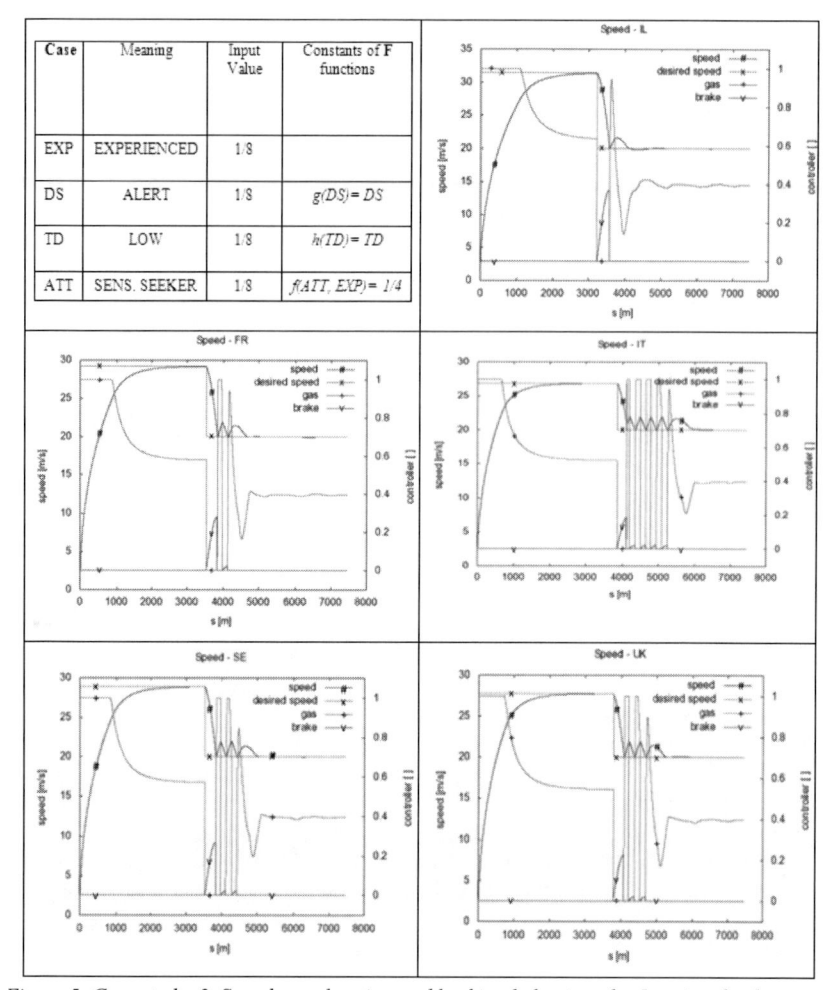

Case	Meaning	Input Value	Constants of F functions
EXP	EXPERIENCED	1/8	
DS	ALERT	1/8	$g(DS) = DS$
TD	LOW	1/8	$h(TD) = TD$
ATT	SENS. SEEKER	1/8	$f(ATT, EXP) = 1/4$

Figure 5. Case study: 2. Speed, acceleration and braking behaviour for 5 national cultures

Discussion and conclusions

This paper has presented how a simulation tool SiMUD can be developed and completed according to the objectives of being able to incorporate the Unified Model of Driver and to reproduce the results of the experimental findings. The results obtained from the analyses of the test cases show that it is possible to unify the behavioural processes of drivers of surface vehicles, with a set of formulations that can be adapted to the specific mode of transport, while maintaining the same modelling architecture. Several simulation runs have been performed, assigning generic values to the coefficients that affect the essential variables characterising driver behaviour. The results of these runs showed reasonable responses of the drivers in terms of speed, braking and/or accelerating performances in reaction to specific demands of the traffic and the environment, and as a function of several driver characteristics, i.e., attitudes, experience, and driver state. As far as the capability of the simulation platform to adapt to the results of the experiments, only the automotive environment has been analysed in this paper. The corresponding linear regression correlations have been performed, focusing on specific sections of the experimental settings, avoiding specific situations where a linear regression might be inadequate or totally inappropriate.

Results were obtained from two Case Studies where different national cultures were simulated and drivers with a variety of personal characteristics were considered. Two specific tasks were analysed, namely a "cruise" and "car following" condition, which enabled the evaluation of the maximum intended speed and the braking or speed control behaviour of different drivers. The findings support the main hypothesis that that behavioural performance in different surface transport domains can be assimilated and therefore the same formulations developed for the automotive domain can be utilised for the rail environment.

The simulation was then further adapted by means of the linear regression to the results of the experiments and the results showed that SiMUD is able to reproduce the results of the experiments. Therefore the goal of developing a simulation adapted to the results of experiments has been fulfilled.

At the same time the results of the case studies show that there is room for improvement, especially in terms of maximum intended speed but also in terms of regulatory improvements of the use of the brake pedal. One reason for this may be that the SiMUD is adapted to driver behaviour when driving fairly simple simulators which mean that speed choice and braking patterns can differ somewhat from actual driving behaviour. This set of considerations demonstrates that the simulation has been developed according to plans. However, two issues remain open: the capability of the simulation to reproduce the behaviour of a train operator and a helmsman of a ship and also the ability to adapt the formulations to more realistic behaviour.

The first issue will be tackled and possibly resolved already within the ITERATE Project, as the activity of validation aims precisely to this objective. Therefore, possible imprecisions or inadequate formulation will be corrected during this process. As far as the adequacy of the SiMUD to reproduce more realistic behaviour, as well as for developing the performance of other additional tasks, such as overtaking or roundabout behaviour etc., there is a need for additional development. This is presently outside the scope of the current project. However, the adaptation of the formulation and the performance of new linear regression expressions require simply the performance of further experiments and simulation development with minimal innovative work.

This would imply a new well defined experimental design and associated experimental observations and consequently the derivation of similar regression analyses as discussed in the course of the current research project. At the same time, the overall architecture of the UMD model and the implementation of the simulation platform and approach would not need further development, as they have both been developed with a sufficient perspective view that increases the value and power of the R&D work performed within this project.

Acknowledgements

The ITERATE project has received funding from the European Commission Seventh Framework Programme (FP7/2007-2013) under grant agreement n°218496

References

Barnard, Y. (2010). *Selection of operator support systems across modes* (Project deliverable 2.2). ITERATE (IT for Error Remediation And Trapping Emergencies) Consortium, Linköping, Sweden, VTI.

Gillberg, M., Kecklund, G., & Åkerstedt, T. (1994). Relations between performance and subjective ratings of sleepiness during a night awake. *Sleep, 17*, 236-241.

Hjälmdahl M. (2012). *Validation of the UMD model for cars and trains* (Project deliverable 7.2). ITERATE (IT for Error Remediation And Trapping Emergencies) Consortium, Linköping, Sweden, VTI.

Hoyle, R.H., Stephenson, M.T., Palmgreen, P., Lorch, E.P., & Donohew, R.L. (2002). Reliability and validity of a brief measure of sensation seeking. *Personality and Individual Differences, 32*, 401 – 414.

Kirwan, B. & Ainsworth, L.K. (1992). Guide to task analysis. London: Taylor and Francis.

Lai. F. (2010). *Review of existing technologies and systems supporting the operator* (Project deliverable 2.1). ITERATE (IT for Error Remediation And Trapping Emergencies) Consortium, Linköping, Sweden, VTI.

Michon, J.A. (1985). A critical review of driver behaviour models: What do we know? what should we do? In L.A Evans and R.C. Schwing (Eds.) *Human Behaviour and Traffic Safety* (pp. 487-525). New York: Plenum Press.

Oppenheim. I. (2010). *Critical review of models and parameters for Driver models in different surface transport systems and in different safety critical situations* (Project deliverable 1.1). ITERATE (IT for Error Remediation And Trapping Emergencies) Consortium, Linköping, Sweden, VTI.

Peters, B., Forsman, L., Vadeby, A., & Tapani, A. (2011). Developing a unified model of driving behaviour for cars and trains. This Conference Proceedings.

Shinar, D. (2010). *Description of Unified Model of Driver behaviour (UMD) and definition of key parameters for specific application to different surface transport domains of application* (Project deliverable 1.2). ITERATE (IT for Error Remediation And Trapping Emergencies) Consortium, Linköping, Sweden, VTI.

Treiber, M., Hennecke, A., & Helbing, D., (2000). Congested traffic states in empirical observations and microscopic simulations. *Physical Review, 62*, 1805-1824

Åkerstedt, T. & Folkard, S. (1997). The three-process model of alertness and its extension to performance, sleep latency, and sleep length. *Chronobiology International, 14*, 115-123.

How is surrounding traffic complexity related to driver workload?

Evona Teh, Samantha Jamson, & Oliver Carsten
University of Leeds, UK

Abstract

Driving a car can involve extreme fluctuations in mental workload and some vehicle manufacturers are attempting to develop systems that manage workload. Such systems are needed to manage the attentional processing demands placed on the driver both from outside the vehicle and within the vehicle. As the modern driving task is a driver-vehicle-environment interaction, this research explores the possible methods in quantifying the workload imposed by the dynamic environment. In this study, different methods of assessing the demands placed on drivers by traffic variables, particularly traffic density and lane change effects were examined. The changes in the driving demand resulting from the surrounding traffic were measured using subjective ratings and tactile detection task. The analysis presented here was based on subjective ratings and detection task response times, compared with data obtained from the simulator. The results showed that both main effects of traffic density and lane change were found to be significantly affecting the driving task difficulty, while subjective ratings was found to be the most sensitive measure of driver workload.

Introduction

Today, there is a wide range of assistance systems available in the market, built to support the driver in the car. The increase of interaction of the driver with in-vehicle systems has increased the awareness of the risk of extra driver workload and distraction, simply because the driver has to divide his or her attention between the outer world and the system inside the vehicle. While all researchers would agree that driver workload and distraction are a major safety issue, the question still remains how do we measure workload, particularly during presence of any sudden increases in workload?

Most studies often looked at how adding secondary tasks interferes with driving, with some of these studies also manipulated the complexity of the drive insofar as they compare rural and urban driving (Cantin, Lavalliere, Simoneau, & Teasdale, 2009), low and high density traffic (Cnossen, Meijman, & Rothengatter, 2004, Verwey, 2000), etc. The focus is often on the dual-task manipulation and the effect magnitudes vary. This study used a more naturalistic approach exploiting the fact that driving task is a visual cognitive task and the driving difficulty is affected by environmental factors outside the vehicle, namely the roads and traffic. The goal of the study was to investigate the impact of challenges in different types of traffic, and also the social context of driving, i.e. the behaviour of other traffic users on road. The 100-car study, performed by Virginia Tech (Neale, Dingus, Klauer, Sudweeks, & Goodman, 2005), was the first large scale study that aimed at collecting pre-crash data in a naturalistic setting without any experimental conditions or experimenters present in the car. Inattention to the forward roadway was the primary factor in crashes and near-crashes and various authors emphasized that the traffic condition in the vicinity of a vehicle is one of the main external factors for driver performance. The demand of the surrounding traffic environment on the perceptual capacities of drivers is constantly changing and high traffic density can be expected to increase the attentional processing requirements of driving, even in professional drivers (Hanowksi, Hickman, Olson, & Bocanegra, 2009). Most research concluded that workload increases with traffic density or, as Zeitlin (1998) puts it, with increasing unpredictability of the traffic. Although some studies suggested there is an effect of traffic density on driver workload (Miura, 1986, De Waard, 1996, Verwey, 1993b; 2000) it is not a straightforward relationship and has not been explored in detail.

In an attempt to investigate the effect of traffic density on driver's performance, Verwey (2000) conducted a field study based on time of day but found no significant effect. Therefore, due to the random nature of traffic in real driving situations, the current study was conducted in a driving simulator with the objective of generating three levels of traffic conditions for the scenario of a two-lane motorway. This represents a different approach to the study of mental workload as it does not require secondary tasks that are not a natural part of driving. The local traffic conditions investigated in this study is based on Vogel's (2002) study who found that two vehicles are linearly dependent on time headway (THW) for headways up to 6 seconds. Therefore, the number of vehicles within close vicinity is defined as the total number of vehicles on both lanes within 6 seconds THW.

The purpose of this study was to explore the attentional demands resulting from the surrounding traffic within close vicinity of a driver via a number of methods of measurements, including continuous and instantaneous measures. The study had the following aims: a) To investigate the effect of traffic density on drivers mental workload b) To evaluate and compare the sensitivity of continuous measures of workload (e.g. subjective ratings (SR)), and tactile detection task (TDT).

Method

Participants
Thirty six participants (18 men and 15 women; M_{age}= 37; $Range_{age}$ = 25-50 years old), holders of a valid driving licence for over five years, driving on average at least 10000 miles per year and normal working hours workers, were recruited.

Design and driving simulator
Three roads, each with a different traffic flow rate (low/medium/high) were designed and modelled in the University of Leeds Driving Simulator, see Figure 1. Each road was 38km long. The driving scenes simulate a two-lane divided motorway consisting of a varying number of cars located in the driver's field of view, all moving in the same direction as the driver's vehicle. There was no

oncoming traffic. The behaviour of the surrounding traffic was dynamically scripted to change lanes, overtake and stay in front of or behind a participant's vehicle. Since traffic density is positively correlated with traffic flow rate, the traffic density factor was manipulated to assess the influence of load in the drive comparing performance when there was an average traffic flow of 416 vehicles/lane/hour (low density), 810 vehicles/lane/hour (medium density) to 1654 vehicles/lane/hour (high density). Traffic density is defined as the number of vehicles per unit area of the roadway with positive correlation with flow rate. All three roads were counterbalanced so that both male and female groups experienced every combination of conditions.

Figure 1. The Leeds Driving Simulator

Participants were instructed to drive with an element of urgency to complete the drive whilst adhering to the traffic regulations (i.e. speed limit). There were two blocks of tasks in each road, in which participants were required to first conduct the SR task in the first block of the driver and then participants were required to conduct the tactile detection task (TDT) in the second block of the drive (Figure 2). For the SR task, participants were asked to provide a continuous rating of their driving demand using the 10-point rating scale (Figure 3). To measure the workload continuously, subjects provided the rating verbally at every 252 m road section when prompted by an audible tone. (Note: 252 m is the default road dimension in the driving simulator)

Upon completion of the first task, participants were then required to provide response to the tactile stimulus. The detection task was presented via a small vibrating mechanism of the size 5.8 cm x 5.8 cm x 2.5 cm (as shown in Figure 4), strapped on the driver's seat, placed directly below participants' left thigh outside their clothing. Subjects received a short vibration pulse (i.e. 1 s) from this equipment at certain intervals and responses were given by pressing the button nearest to the left index finger on the steering wheel. Detection performances were measured in terms of response time and hit rate.

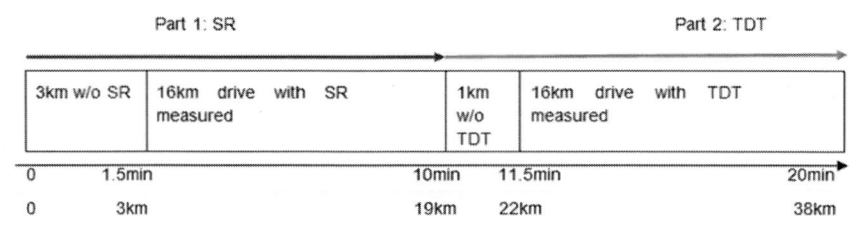

	Part 1: SR			Part 2: TDT
3km w/o SR	16km drive with SR measured	1km w/o TDT	16km drive with TDT measured

```
0       1.5min                      10min   11.5min                  20min
0       3km                         19km    22km                     38km
```

Figure 2. Order of tasks administration within each 20 minutes drive

EASY				MODERATE				DIFFICULT
1	2	3	4	5	6	7	8	9	10

Figure 3. 10-point rating scale of the continuous rating method

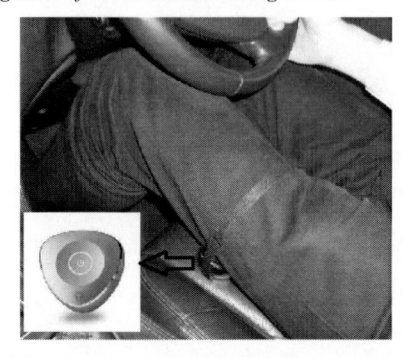

Figure 4. The position of the vibrating mechanism during study

Driver's visual behaviour was tracked using a Seeing Machines faceLAB v4 eye-trackers housed within the vehicle cab throughout the 38 km drive. However, only SR and TDT results will be discussed here.

Mental workload measures

Two types of dependent variables were measured; subjective measures and performance measures.

Subjective measures

Subject self-reported ratings were recorded during and after each drive. Within the drive, subject's self-reported ratings (SR) based on the 10-point rating scale were recorded continuously to assess driver's workload transition and sudden increases in workload during the drive. After the completion of each route, the two most commonly used techniques of subjective mental workload are administered; NASA-Task Load Index (NASA-TLX; Hart & Steveland, 1988) and RSME (Zijlstra, 1993). The NASA-TLX includes six subscales exploring the Mental Demand, Physical Demand, Temporal Demand, Own Performance, Effort, and Frustration Level. A weighting procedure is usually included in the complete NASA-TLX. However, "raw" NASA-TLX on the six subscales self-report measure was used in this study as they still yield similar information as the complete NASA-TLX (Byers, Bittner, &

Hill, 1989). Each subscale is a response scale of 10-cm long depicting the scale of 0 to 100, with the endpoints of the response scale labelled 'low' and 'high'. RTLX was used in the present study, as shown for instance in a study by Jahn, Oehme, Krems, & Gelau (2005) that revealed a higher sensitivity of RTLX ratings to small workload changes, specifically in mental demands and temporal demands.

The RSME is a uni-dimensional rating scale consisting of a 15 cm long vertical line marked at 1-cm intervals, with nine anchor points ranging from 'absolutely no effort' (close to the 0 point on the 0-150 point scale), to 'rather much effort' (approximately 57 on the scale) to 'extreme effort' (approximately 112 on the scale).

Performance measures

Within this group of methods, the participant's mental workload was inferred from their overt behaviour or performance, in particular response accuracy and response latency. Subject's performances were measured in terms of hit rate and response times (RT). A hit is defined as a response within 200-2000 ms from stimulus onset with any responses less than 200 ms are excluded due to unrealistically fast responses. Hit rate is defined as the number of hits divided by the total number of stimuli during a task. Response time is used as the main performance metric and hit rate must be above 70% for a data segment to qualify for analysis (Merat, Johansson, Engstrom, Chin, Nathan, & Victor, 2006). Measures of driving performance obtained from the driving simulator include driving speed and standard deviation of lane position.

Traffic behaviour measures

Each subject completed all three different traffic conditions with the same road layout. The average number of traffic flow increased from T1 (low traffic density; 416 vehicles/lane/hour) to T2 (medium traffic density; 810 vehicles/lane/hour) and T3 (high traffic density; 1654 vehicles/lane/hour). Estimates of current local traffic conditions were continuously measured over the distance of 252m, which included flow rate (number of vehicles/lane/hour) and lane changes (presence or absence within each 252m).

Figure 5. Varying traffic density conditions presented to driver. From left to right: low/medium/high traffic density

Results

Our primary hypothesis was that if the driver is affected by the surrounding traffic behaviour in terms of information processing and handling complexity, changes in attentional requirements would be reflected both on the subjective assessment of the workload and driving performance. Comparing the subjective ratings scores for each of the three traffics, the results showed that the overall experienced workload (based upon the RSME and mental demand component from NASA-RTLX and mean ratings) appeared to be higher for higher traffic density. One way repeated MANOVA using SPSS 17.0 was conducted on standardised RSME data, overall NASA RTLX data and mean ten-point ratings. Effect of type of conditions on RSME, overall NASA RTLX and mean ten-point ratings were significant, $F(6,30)=110.44$, $p<0.001$, partial eta squared=0.96, observed power=1.0. Statistics test results (Hotelling's Trace and Roy's Largest Root have the same value = 22.69) indicated that these measures are highly correlated. The bivariate pearson correlations between mean ten-point ratings with RSME (correlation = 0.72, $p<0.0001$) and overall NASA-RTLX (correlation= 0.739, $p<0.0001$) were significant. No difference were reported on comparison for the gender, thus suggesting both male and female drivers reported significantly more effort during T2 than during the T1 condition, and also more effort during T3 than during the T2 condition.

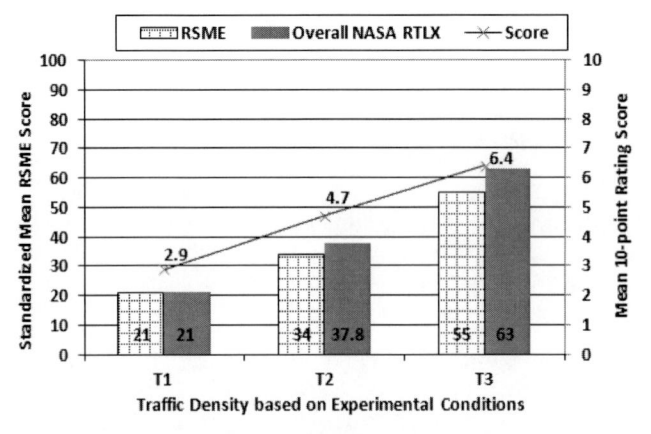

Figure 6. Graph for the comparison between mean subjective mental workload scores on the RSME, overall NASA RTLX and the mean ten-point ratings scores for the three conditions (means were standardized to 100 point scale for graphing purposes)

A bivariate correlation (Spearman) analysis between mean ratings and traffic flow (number of vehicles/lane/hour) shows a positive correlation of 0.633, significant at $p<0.01$ thus indicating that drivers generally experience more difficulty in higher traffic flow conditions (i.e. higher number of movements within driver's driving environment). For a better analysis of traffic density effect, data from each participant from each of the three roads were grouped in a database and mapped to three different levels of traffic flow rate. Data were subjected to SPSS ANOVA repeated measurement analyses, with factors involving density level (low, medium and high) and lane change (presence and absence). In case of sphericity violation, the Greenhouse-Geiser modification was used. Statistical result showed that the main effect of traffic density was significant, $F(1.59, 50.89)=114.30$, $p<0.001$ with partial eta squared= 0.781. Figure 8 showed an increase of ratings with presence of lane changes in each traffic density ($F(1,32)=29.75$, $p<0.001$ with partial eta squared=29.75). Hence, this indicates that driver perceived a higher driving task difficulty with the presence of lane change. The increases of rating across the scenarios for conditions with and without presence of lane changes are similar in trend. The interaction between traffic density and lane changes are statistically not significant thus indicating that the increase in ratings due to the presence of lane changes was not affected by whether the surrounding traffic density was high or low.

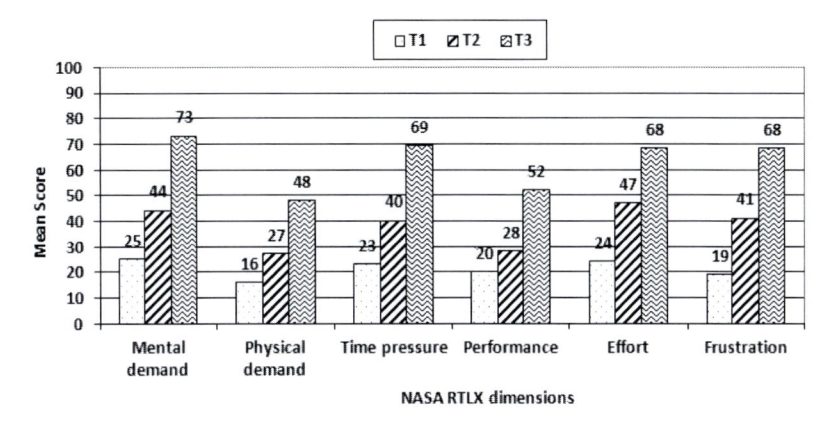

Figure 7. Mean results for the NASA-RTLX effort dimensions by NASA-RTLX dimension and three driving environments. T1 =low traffic density, T2 = medium traffic density, T3 = high traffic density

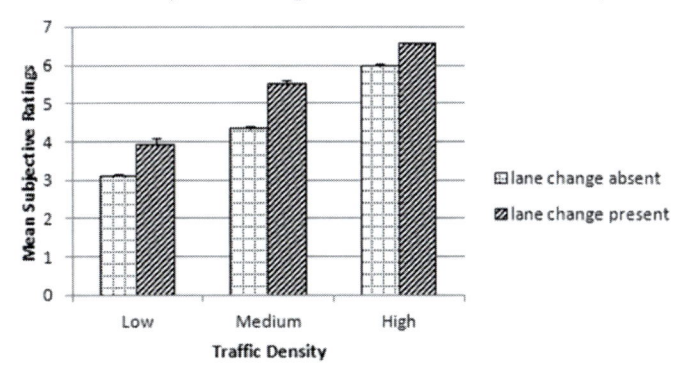

Figure 8. Mean subjective ratings as function of traffic density (low: less than 600 vehicle/lane/hour, medium: between 600-1200 vehicles/lane/hour, high: beyond 1200 vehicles/lane/hour) and presence of lane changes in subject's forward view

Driver's driving behaviour also differed across the three traffic density whereby mean driving speed was significantly reduced in high density traffic (mean speed difference between high and low traffic density = 13.9 mph). There was significant effect of other traffic user's lane change movements on participant's speed management as the speed variability of participant's vehicle increases with the higher complexity road environment (refer to Figure 10). This has been shown statistically for which mean speed was significantly lower with the presence of lane changes ($F(1,32)=97.95$, $p<0.001$ with partial eta squared= 0.69), while speed variability increases non-linearly with the presence of lane changes, thus indicating higher control of speed in higher density and with presence of lane changes ($F(1,31)= 49.90$, with partial eta squared= 0.62).

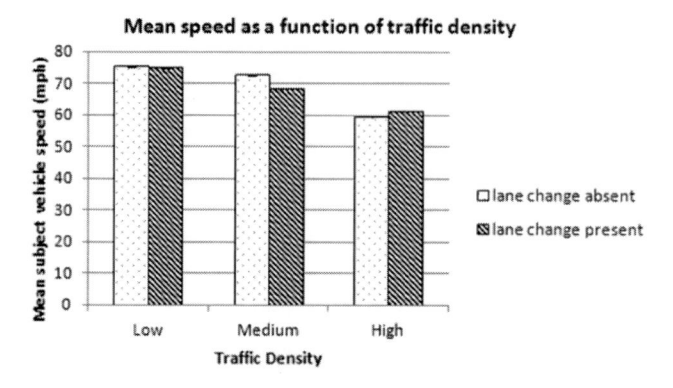

Figure 9. Mean subject vehicle speed as function of traffic density (low: less than 600 vehicle/lane/hour, medium: between 600-1200 vehicles/lane/hour, high: beyond 1200 vehicles/lane/hour) and presence of lane changes in subject's forward view

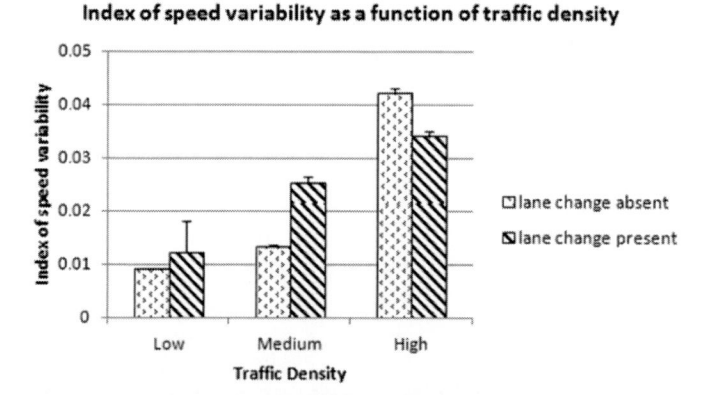

Figure 10. Index of speed variability (which is defined as ratio of standard deviation of speed/mean speed) as function of traffic density (low: less than 600 vehicle/lane/hour, medium: between 600-1200 vehicles/lane/hour, high: beyond 1200 vehicles/lane/hour) and presence of lane changes in subject's forward view

The standard deviation of lane position was assessed on straight segments of the road and was found to not significantly increase with increasing traffic density. Although Figure 11 shows a general increase in standard deviation of lane position from low to medium traffic density, however both the main effects of traffic density and lane changes effect on driver's lateral driving performance were statistically not significant.

Figure 11. Mean standard deviation of lane position as a function of traffic density (low: less than 600 vehicle/lane/hour, medium: between 600-1200 vehicles/lane/hour, high: beyond 1200 vehicles/lane/hour) and presence of lane changes in subject's forward view

Performances on the TDT between the three traffics density conditions were measured in terms of response time and hit rate. Repeated measures ANOVA showed that there were no significant differences in response times as a result of increasing traffic density from low to high (Figure 12) and also with the presence of lane changes. As illustrated in Figure 13, both main effects of traffic density and lane changes on TDT response times were non-significant thus indicating that TDT is not a sensitive measure of driver workload in this study.

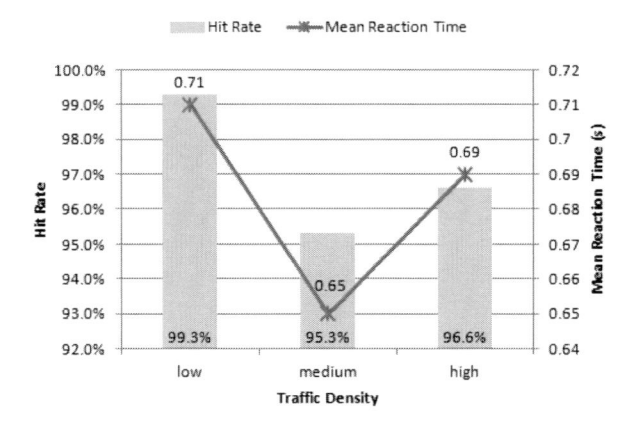

Figure 12. TDT mean response time and hit rate across the three traffic density conditions (low/medium/high)

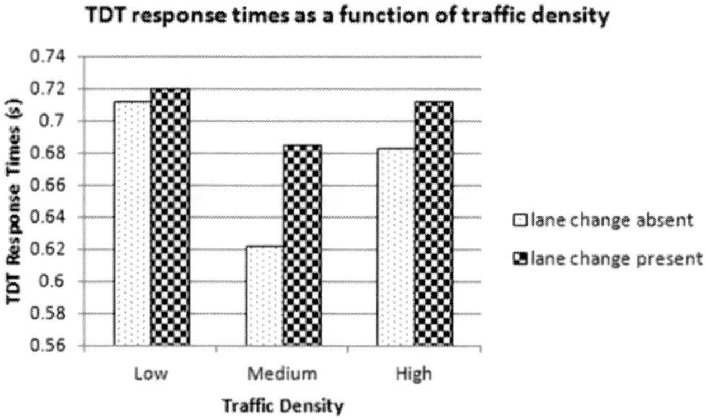

TDT response times as a function of traffic density

Figure 13. TDT mean response time as a function of traffic density (low: less than 600 vehicle/lane/hour, medium: between 600-1200 vehicles/lane/hour, high: beyond 1200 vehicles/lane/hour) and presence of lane changes in subject's forward view

Conclusion

This study explored how traffic flow might affect driver performance in a naturalistic manner with the following preliminary results; comparing RSME, NASA-RTLX and mean rating which showed that mean rating is a reliable measure of driver workload in various traffic density conditions, while TDT showed a non-linear V-shaped effect across low, medium and high traffic density. The study had shown that drivers report a general increase in driving task demand with increasing traffic flow, thus verifying the sensitivity of subjective measures. In addition, all six NASA-RTLX dimensions showed similar trend and variation across all three roads, and also the overall NASA-RTLX score were highly correlated with RSME. Therefore, using the uni-dimensional rating (i.e. RSME) as an estimate of overall driver workload is sufficient in this study. The effect of traffic density using TDT measure however, were found to be statistically insignificant thus indicating that TDT is not a sensitive measure of workload.

Effect of lane changes was shown with participants providing higher mean ratings in the presence of lane changes. On top of that, mean speed had reduced and speed variability increased non-linearly with the presence of lane changes, thus indicating higher level of control in speed among drivers was observed in higher density and with presence of lane changes. Therefore, this simulator study has shown that both factors of traffic density and presence of lane changes affect driving difficulty and the ten-point subjective rating method used is a sensitive measure of estimating driver's continuous workload in the simulated dynamic traffic environment.

References

Byers, J.C., Bittner, A.C. & Hill, S.G. (1989). Traditional and raw task load index (TLX) correlations: Are paired comparisons necessary? In A. Mittal (Ed.), *Advances in industrial ergonomics and safety* (pp. 481–485). London: Taylor & Francis.

Cantin, V., Lavalliere, M., Simoneau, M., & Teasdale, N. (2009). Mental workload when driving in the simulator: effects of driving complexity. *Accident Analysis and Prevention 41*, 763-771.

Cnossen, F., Meijman, T., & Rothengatter, T. (2004). Adaptive strategy changes as a function of task demands: a study of car drivers. *Ergonomics 47*, 218-236.

De Waard, D. (1996). The measurement of drivers' mental workload, Ph.D. thesis, University of Groningen.

Hanowksi, R., Hickman, J., Olson, R. & Bocanegra, J. (2009). Evaluating the 2003 revised hours-of-service regulations for truck dirvers: the impact of time-on-task on critical incident risk. *Accident Analysis and Prevention 41*, 268-275.

Hart, S. G. & Steveland, L. E. (1988). Development of NASA-TLX (Task Load Index): results of empirical and theoretical research. In P. A. Hancock, & N. Meshkati, *Human Mental Workload* (pp. 139-183). Amsterdam: Elsevier Science.

Jahn, G., Oehme, A., Krems, J.F., & Gelau, C. (2005). Peripheral detection as a workload measure in driving: Effects of traffic complexity and route guidance system use in a driving study. *Transportation Research Part F: Traffic Psychology and Behaviour 8*, 255-275.

Merat, N., Johansson, E., Engstrom, J., Chin, E., Nathan, F., & Victor, T. (2006). *Specification of a secondary task to be used in safety assessment of IVIS*. AIDE Deliverable 2.2.3. European Commision, IST-1-507674-IP.

Miura, T. (1986). Coping with situational demands: A study of eye movements and peripheral vision performance. In A.G. Gale, I.D. Brown, C.M. Haselgrave, P. Smith, & S. Taylor (Eds.), *Vision in Vehicles-II*. Amsterdam: Elsevier.

Neale, V.L., Dingus, T.A., Klauer, S.G., Sudweeks, J. & Goodman, M. (2005). An overview of the 100-car naturalistic study and findings. *Proceedings 19th Enhances Safety Vehicles Conference*. Washington, D.C.: Paper number 05-0400.

Verwey, W.B. (1993b). *Driver workload as a function of road situation, age, traffic density and route familiarity*. Soesterberg, Netherlands: TNO report IZF 1993 C-11.

Verwey, W.B. (2000). On-line driver workload estimation: Effects of road situation and age on secondary task measures. *Ergonomics 43(2)*, 187-209.

Zeitlin, L. (1998). Micromodel for objective estimation of driver mental workload from task data. *Transportation Research Record 1631*, 28–34.

Zijlstra, F.R. (1993). *Efficiency in work behavior. A design approach for modern tools. PhD Thesis*. Delft University of Technology. Delft, The Netherlands: Delft University Press.

Technologies to support socially connected journeys: Designing to encourage user acceptance and utilisation

Sarah Sharples, David Golightly, Caroline Leygue, Claire O'Malley, James Goulding, & Ben Bedwell
Human Factors Research Group and Horizon Digital Research
University of Nottingham, Nottingham
UK

Abstract

Financial, practical and environmental incentives are increasingly encouraging people to seek opportunities to share journeys taken by car. Location-enabled mobile technologies provide opportunities for real-time dynamic communications to support transport decision making in general and car sharing in particular. Theories such as the Technology Acceptance Model (TAM, Venkatesh & Davis, 2000) and Theory of Planned Behaviour (TPB, Ajzen, 1991) have provided insight into the factors that can affect user behaviour with technologies in a range of contexts. This paper presents the emerging priorities from a series of human factors studies that were conducted to investigate user attitudes to and requirements for technologies to support car sharing. Methods applied included: user interviews, mobile diary studies and evaluations of prototype dynamic car and taxi share technologies. A model to inform the design of technologies to support socially connected travel, that highlights issues of privacy, security, flexibility, planning and social context of use is presented.

Introduction

With increased fuel costs and long-term concerns about the impact of automotive travel on the environment, the concept of car sharing, or car pooling, is of increasing interest and value. Much car sharing happens on an ad-hoc and informal basis, where friends, families or colleagues coordinate journey plans. However, formal systems including paper based notice boards to facilitate matching, and, more recently, web based 'buddying' systems (e.g. liftshare) have also been used to enable people to plan shared journeys with other system users. Increasing prevalence of ubiquitous, location based and mobile computing, in particular smart phone technologies, provide an opportunity to extend the functionality of such systems to enable more ad-hoc arrangements of car sharing, and also introduce the potential to link such systems with other social networking systems such as Facebook.

The use and design of technologies to support such applications needs careful thought. Previous research, such as the Technology Acceptance Model (TAM) (Venkatesh & Davis, 2000) has demonstrated that a number of factors combine to influence the behaviour of users with technologies. More generally, theories such as the Theory of Planned Behaviour (TPB, Ajzen, 1991) have considered the influences on an individual when selecting a behaviour or action.

A research project to examine the potential for ubiquitous technologies to support socially connected travel was conducted as part of the transport research theme within the Horizon digital hub funded by Research Council UK's Digital Economy programme. The aim of this research initiative, based at the University of Nottingham, is to develop understanding of the potential of ubiquitous digital technology, looking at the challenges of providing a new generation of personal applications that use the traces that are left behind whenever mobile, internet and other digital technologies are used. There is a particular emphasis on approaches that respect personal privacy and enrich, rather than supplant, the wide range of social interactions that people engage in at work and at home. This paper provides an overview of this research, and considers some of the emergent themes that were identified from the different methodological approaches applied.

Previous work into socially connected travel

Shared travel as a concept has existed for many years – indeed Woodworth & Behnke (2006) describe 'Jitneys', an early type of ride-giving service in Los Angeles, that "for a slight additional charge, … would deviate from their main route to deliver passengers to their homes, particularly in inclement weather". This is an early example of a hybrid of what would now be termed car sharing or taxi sharing. Car pooling has been a particularly prevalent activity in the US and Canada, (Jacobson & King, 2008; Morency, 2007) and technologies have emerged to support car sharing – for example, Morency (2007) identified five separate matching tools to support car sharing in Montreal. However, there are clear cultural differences between uptake of car share in different countries.

Previous research into car sharing technologies have identified user requirements (Kowshik et al., 1993) and issues such as usability generally (Levosky & Greenberg, 2001), as an important consideration. Specific technology factors include: the ability to finesse arrangements and build trust through chat-like communication (Brereton and Ghelewat, 2010); the use of multiple platforms for any given system (web, mobile, shared displays) to increase exposure to the service (Wash et al., 2005; Dailey et al., 1999); and effective human support within the organisation (Buliung et al., 2010). In addition to factors associated with implementation and design, the concept of sharing a journey, probably with someone who the user has not previously met, also presents barriers to the success of such schemes or systems, with factors such as social awkwardness (Laurier et al., 2001), safety and trust (Chaube et al., 2010) and perceived loss of control, mastery and status often experienced with car usage (Steg, 2005; Gardner & Abraham, 2007) having been highlighted as issues of particular importance. Therefore, it is critical that the design or requirements specification of any such technology considers both the requirements from a technological perspective *in addition* to an understanding of the factors that can influence an individual's decision to change the way in which they complete a journey.

Theoretical perspectives

Two well-established theoretical models can be considered in relation to technologies to support socially connected travel. The first is the Technology Acceptance Model (TAM, Davis, 1993). This model proposes that the concepts of perceived usefulness and perceived ease of use combine to influence a users 'intention to use' a technology, and the model was later extended (Venkatesh & Davis, 2000) to provide more explanation as to the influences on the perceived usefulness element, including voluntariness, experience, subjective norm, image, job relevance, output quality and result demonstrability. This model has been applied to a range of technologies, many of which (e.g. Venkatesh & Davis, 2000) have been desktop-type applications used in a work context.

A second theory of relevance is the Theory of Planned Behaviour (TPB, Ajzen, 1991). This theory describes the combined impact of the attitude towards a behaviour, the subjective norm and perceived behavioural control on intention, which then influences behaviour. This theory has been considered in relation to behaviours such as donating money to charity, personal health behaviours or electoral voting decisions, amongst many others (Azjen, 1991) and was considered in the development of TAM (Venkatesh & Davis, 2000)

These two theories provide a useful basis for the understanding of technology to support socially connected travel. The TAM emphasises the role not only of technology design (perceived ease of use) but also the way in which the technology is implemented, including factors such as the perceived relevance of the technology to an individual's job, the perceived similarity between the technology and required interactions and other types of systems already used (subjective norm) and the extent to which a person is required or chooses to use a system (voluntariness). However, TAM has been primarily applied within work contexts, so it is interesting to see how its elements transfer to the social and collaborative nature of car sharing technologies. The TPB emphasises the roles of attitudes and control on behavioural choice, but does not easily denote the influence of design of systems such as might be used in a mobile, location-based technology to support car share. Potentially therefore, the context of technology to support socially-connected travel may be one that is explained by considering elements of both of these models, in addition to other emergent themes.

Methodological approach

To investigate the requirements for technology to support socially connected travel a number of small-scale studies were conducted, applying methods including prototype design, interviews with potential car sharers, online surveys and a mobile phone based diary study, where user journeys were recorded and users were asked to report whether they had shared journeys, and, if not, whether they would have been amenable to sharing in the future. Car sharing and Taxi sharing were used as example concepts to elicit user requirements for technologies to support socially connected journeys. The aims of the overall programme of work were:

- To investigate the viability of technology enabled socially connected travel

- To understand the influencing characteristics of the technology (including privacy, data presentation)
- To analyse the contextual factors that may mediate the effectiveness of technology enabled socially connected travel.

Table 1 presents a brief description of the studies conducted within the Horizon socially connected journey project, and Figure 1 illustrates screenshots of some of the prototype demonstrators that were developed to explore the possible technologies to support socially connected travel, demonstrating the types of information that people might use in real time shared journey planning.

Figure 1. Screenshots of car share and taxi share prototype demonstrators developed

Emergent themes

A number of themes emerged from the overall data set. All of the above data sets informed the themes identified below, with quotations particularly taken from the car share attitude and user interface requirements interviews. These themes are briefly summarised below.

Participants reported *limited membership of formal car share schemes*. However, during the interviews conducted there were clear examples of previous *informal car sharing*; in fact, some participants who did not initially describe themselves as 'car sharers', when probed, provided examples of sharing journeys with family members, work colleagues or friends in a range of contexts including work meetings, leisure trips and providing transport for children. This finding reflects previous research (Morency, 2007).

Table 1. Summary of studies conducted with Socially Connected Travel Project

	Aim	Brief statement of findings
Car share attitudes interview	Interviews with potential car sharers about attitudes	Flexibility is the main barrier to car share. Social, costs and environmental impact are all seen as positive outcomes of car share.
GPS journey sharing	Identify opportunities for journey sharing from users' current travel habits by monitoring journeys using GPS mobile diary app	Over 400 journeys recorded, with the majority being car-based and/or work orientated. Participants indicated there were many more that they were happy to share, there was only one instance of a driver in a single-occupancy vehicle trip indicating they were willing to be a passenger for a shared journey
Car share motivations questionnaire	Understand values and motivations to car share and individual characteristics	Social networking and environmental impact are key perceived benefits of car share. People like the social company of others in their car. Financial motivations don't always come first.
Car share stakeholders focus group	Capture of organizational and technology provider perspective on management of car share schemes	Car sharing is an organisational necessity in many new built offices. Technology is only an facilitator
User Interface requirements interview	Paper-based concepts used as probes to elicit requirements for user interaction in socially connected journey technologies	Privacy and safety influence interface requirements for input, output and verification of shared journey partners. People are uncomfortable with a complete absence of planning before a journey (needs to not be completely ad-hoc). Time (rather than route) is a major factor in determining desirability of share.
Car share meeting demonstrator	Using demonstrator as prompt to discuss issues around technology enabled car sharing	Maps are not the automatically best way to represent a journey. People liked using text messages as communication tools. "less is more"
Common meeting point travel demonstrator	Demonstrator implemented to allow journey sharing at a conference	People prefer to pre plan where need for commitment, and need for feedback and confirmation during journey
Taxi share requirements survey and interviews	Interviews with conference attendees on taxi sharing concept	Appeal of concept as perceived as a safe and anonymous way to travel (taxi driver as 'neutral')
Taxi share demonstrator	Technical demonstrator of taxi sharing concept	Need for human operator as part of system to negotiate details of costs
Online matching technology questionnaire	Survey of members of an organization where car share 'buddy' scheme being introduced	Organisational setting was important for supporting journey share. People did not have clear mental models of pool of matches available and associated privacy issues.
Incentives survey	Survey of motivators and incentives for potential journey sharers	People can be categorised into different groups according to distinct categories of barriers and incentives. "one size does not fit all"
Multimodal interaction requirements focus group	Focus group on different technology requirements at different journey stages using paper prototypes	There are differences in technology requirements at the journey planning and during journey stages.

The studies above were conducted over a period of approximately 12 months.

Motivators to car share included *cost savings*, *social networking*, *prioritised parking* and *'sharing the stress'* of driving. It is interesting to note that most data, including the formal survey of motivations, did not find that direct financial incentives (e.g. building up financial credits or receiving a direct financial bonus) were particularly powerful motivators. Discussions with organisational stakeholders (Golightly et al.,

2010) indicated that for one organisation where monetary incentives were provided, these were not motivators that had a strong impact on car sharing behaviour. During the organisational stakeholder interviews it was also noted that participation in and support for car sharing provided benefits to organisations who wished to demonstrate *Corporate Responsibility* and meet CO_2 *targets*.

A major barrier to car sharing was a perceived *lack of flexibility*. For example, one respondent in the diary interviews reported *"I would like the car so you can go when you want to go"*. This desire for flexibility and immediacy suggests a potential opportunity for ad-hoc car sharing, such as the system promoted by Avego (www.avego.com), where users can indicate their wish for a lift and a driver can be immediately alerted of a possible share opportunity. However, this conflicts with the reported *importance of planning*. Many interviewees reported that situations in which they wished to car share were return journeys, and that they would be particularly nervous about not having a guaranteed lift home. One participant reported *"I think it would be better to pre-arrange before travelling for the way back. Because you wouldn't want to turn up and find out that no-one's logged into the system... and you've got no-one to go back with"*.

The way in which a socially connected travel scheme is implemented is critical to its success. This may require *personal contact* from a scheme manager, or specific *marketing* via a trusted route (e.g. the official web site of a local football team to support supporters sharing lifts to a match). Illustrations of the incentives is also critical, and it is important to acknowledge that the incentives may be different for different individuals (e.g. some may prefer to have someone to talk to during a journey, whereas others may not be persuaded by increased social contact but may wish to save money or have prioritised parking) and also for the same individual in different circumstances (e.g. the same person may not wish to share their journey to work but may be happy to share when travelling to a music festival).

In terms of the design of any technology to support socially connected travel, three key themes emerge: *Security*, *Information content*, and *Usability*. It is important to note that of these three, security is an absolutely critical consideration and a significant barrier to many, demonstrated by the interview comment: *"If there's an element of doubt anywhere along the line that someone's security could be compromised, I think people would leave it alone"*. Therefore the information that is shared, and the stage at which the information is shared, is an important design consideration. However, it was acknowledged by many participants in the studies that if you wished to have information about another sharer, you would need to be prepared to share that information yourself. Therefore careful consideration needs to be given to the use of nicknames, sharing of address information and links between a socially connected travel system and other social media (e.g. Facebook). Participants had particular preferences associated with the level of control they had over selected sharers. A further issue associated with information content is that the finding of Chatterton (2009) was reinforced, in that representation of CO_2 saving was not a key requirement. In addition, the predicted time taken for the journey was generally prioritised over specific map or route information. This has implications for the nature of communication and type of device – it is possible for example to communicate predicted time information via a simple text message, whereas map or route information would require a more sophisticated smartphone or tablet display.

Finally, usability is critical, particularly in terms of the efficiency of use of the system and the clarity of information displayed, but is a facilitator or catalyst for car sharing rather than a fundamental requirement – in reality, many of the other social, personal and organisational factors will play a much more dominant role than the design of the system.

These findings contributed to the development of a model to demonstrate the factors influencing use of socially connected travel systems.

Socially Connected Travel Influences and Adoption (SCoTIA)

The data collected from the programme of work has enabled the development of the Horizon model of influences and sustainers of socially connected travel (HISSCoT). This model illustrates the role of *personal*, *contextual* and *technological* factors in both initiating socially connected travel and, critically, ensuring that it is maintained to form behaviour, attitude or habit change.

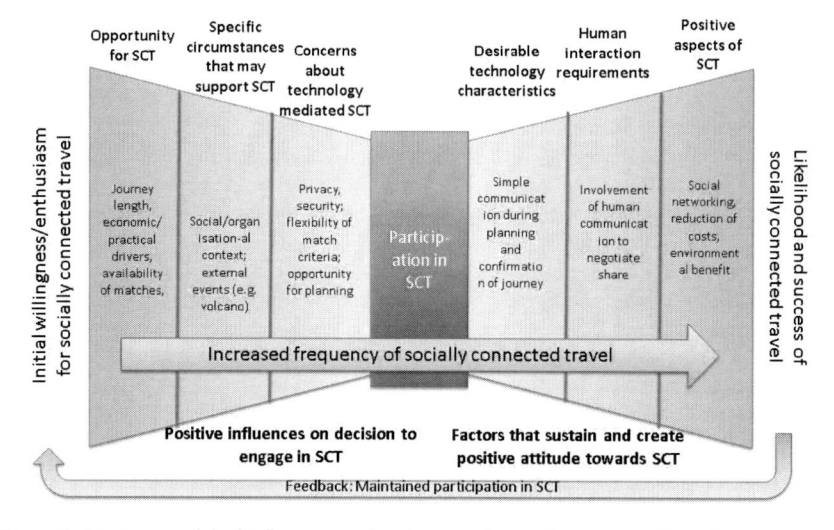

Figure 2. Horizon model of influences and sustainers of socially connected travel

The left-hand side of the model (Figure 2) describes the factors that have been identified as critical for a user to initiate participation in socially-connected travel. Firstly, the user begins with an *initial level of willingness or enthusiasm for socially connected travel*. Influences on this willingness or enthusiasm may be attitudes towards the environment or financial circumstances, as well as preferences for social interaction during travel. For these individuals the *opportunity for socially connected travel* may then be encountered. Typical determinants of such opportunities that emerged from data included journey length, specific economic or practical drivers (e.g. a journey that would be particularly costly or where it would be particularly hard to find a parking place) and the perceived availability of potential matches. Once such opportunities are present, the social or organisational context (e.g. specific company incentives to encourage shared travel) or one-off external factors (e.g. such as was seen in the aftermath of the volcano eruption that affected flights in April 2010) that produce *specific circumstances that may support socially connected travel*. This is then followed by the specific *concerns about technology-mediated socially connected travel* that may exist, and included issues associated with privacy and security, flexibility of match criteria (e.g. wishing to share with a non-smoker) and opportunity for planning vs. ad-hoc systems.

Critically, the research results demonstrate that these factors combine to act as a "filter" – it is only when ALL of these elements are favourable that a user attempts participation in socially connected travel. This point in the model can be described as the point of 'intention' as described in the Theory of Planned Behaviour. Attitude towards the behaviour and perceived behaviour control were also highlighted in the data obtained from the research programme. The role of the subjective norm was less clear, but there is some suggestion from other cultures where there is a greater critical mass of socially connected travel participants that this would also play a role. Much of the data collected has therefore elaborated upon the contributors to the three elements that feed into the point of intention in the TPB.

The role of technology design only emerges in the HISSCoT model at the stage of 'concerns about technology-mediated SCT'. At this point the technology has a role in communicating issues such as the way in which personal data is stored and shared and identifying and presenting matches in an appropriate form, as well as allowing advanced planning of journeys where desired. The stage described as 'Intention to use' in the Technology Acceptance Model (TAM), which refers to the intention to use the technology itself, is therefore encompassed by several elements of the HISSCoT model, and needs to be considered in the personal, situational and organisational contexts identified in the left-hand side of the model.

Once a user has decided to participate in socially connected travel there are several factors that can make this process easier (facilitators) or faster (catalysts). The first of these are the *desirable technology characteristics* highlighted such as the use of simple communication (rather than complex graphical displays for example) during the process of planning and confirming a journey. The second element that is a key influence is the *human interaction requirements,* denoting the desire for human communication to communicate the share. This could be actual communication by telephone or direct message, or could be mediated by the technology, in which case that technology designer has a role in ensuring that the communication has characteristics that suggest it is 'human' e.g. in the language used, the use of pictures of people or speech bubbles associated with messages etc. It is at this point, once a person has shared a journey, that the *positive aspects of socially connected travel*, such as social networking, reduction of travel costs and actual environmental benefit have an impact. These elements on the right-hand side of the model can be considered as elements that encourage the 'habit' of socially connected travel to be formed.

Therefore, this 'habit' could be articulated as either the 'behaviour' element of TPB or the 'usage behaviour' element of TAM. The data collected here suggests that whilst both TAM and TPB have important contributions in terms of explaining and predicting likelihood of socially connected travel,. However, in isolation, neither of them cover all of the elements that the data that was collected in this research programme have highlighted as of importance.

Conclusions

Ubiquitous technologies offer potential to support socially connected travel. However, human factors research needs to fundamentally address the challenge of designing technology to influence change in transport and travel behaviour; technology on its own cannot drive that change. As has been found in the context of car sharing, transport and travel decisions are rooted in habit and attitude – both in terms of current travel behaviour, and in terms of social attitudes or habits when people are being asked to adopt new behaviours that might be seen as either uncomfortable or even risky. Therefore a journey is viewed not as an end in itself but as a function that serves higher order domain goals (work, leisure) with related domain constraints (safety, comfort, cost). Such a contextual approach, using tools such as Cognitive Work Analysis (e.g. Millen et al., 2011; Stanton et al., submitted) may be powerful to fully describe the context and motivations for adopting travel change, and the decisions involved.

New forms of technology may be able to encourage new forms of travel behaviour such as socially connected travel. The research programme presented here has found some potential for change, but has demonstrated that the importance of flexibility is critical for users. As well as technology design considerations, concepts such as automation, workload, situation awareness, physical interaction design and embedding the application within a socio-technical context are also important. If technologically-enabled solutions to socially-connected travel are to be realised, then human factors has a central role to play in involving users, and the factors that constrain, shape and motivate their travel choices, in the design and implementation process.

Acknowledgements

This work is funded by Horizon Digital Economy Research Institute (RCUK Grant No. EP/G065802/1)

References

Ajzen, I. (1991) The theory of planned behavior. *Organisational Behavior and Human Decision Processes*, 50, 179-211.
Brereton, M. & Ghelewat, S. (2010). Designing for participation in local social ridesharing networks - grass roots prototyping of IT systems. In *Proceedings of PDC '10,* Sydney, Australia. (pp. 199-202). New York, NY: ACM Press
Buliung, R.N., Soltys, K., Bui, R., Habel, C., & Lanyon, R., (2010). Catching a ride on the information super-highway: toward an understanding of internet-based carpool formation and use. *Transportation, 37,* 849–873.

Chatterton, T.J., Coutler, A., Musselwhite, C., Lyons, G., & Clegg, S. (2009) Understanding how transport choices are affected by the environment and health: Views expressed in a study on the use of carbon calculators. *Public Health, 123,* 45-49.

Chaube, V. Kavanaugh, A.L., & Perez-Quinones, M.A. (2010). Leveraging Social Networks to Embed Trust in Rideshare Programs. In *Proceedings of 43rd Hawaii International Conference on System Sciences.* (pp. 1-8). IEEE

Dailey, D.J., Loseff, D., & Meyers, D. (1999) Seattle smart traveler: dynamic ridematching on the World Wide Web. *Transportation Research Part C, 7,* 17–32.

Davis, F.D. (1993) User acceptance of information technology: system characteristics, user perceptions and behavioral impacts. *International Journal of Man-Machine Studies, 38,* 475-487.

Gardner, B., & Abraham., C. (2007). What drives car use? A grounded theory analysis of commuters' reasons for driving. *Transportation Research Part F, 10,* 187–200

Golightly, D., Sharples, S., Irune, A., Leygue, C., Cranwell, J. & O'Malley, C. (2010) User and Organisational needs for adhoc car sharing. In *Proceedings of Digital Futures '10.* Nottingham, November, 2010

Jacobson, S. H., & King, D. M. (2008). Fuel saving and ridesharing in the US: motivations, limitations, and opportunities. *Transportation Research Part D,* 14(1), 14-21.

Levofsky, A. & Greenberg, A. (2001). Organized Dynamic Ridesharing: The potential Environmental Benefits and the Opportunity for Advancing the Concept. Available from http://ridesharechoices.scripts.mit.edu/home/wp-content/papers/GreenburgLevofsky-OrganizedDynamicRidesharing.pdf (accessed 29.09.2011)

Laurier, E., Lorimer, H., Brown, B., Jones, O., Juhlin, O., Noble, A., Perry, M., Pica, D., Sormani, P., Strebel, I., Swan, L., Taylor, A., Watts, L., & Weilenmann, A. (2008) Driving and passengering: notes on the ordinary organisation of car travel. *Mobilities, 3,* 1-23.

Millen, L., Edwards, T., Golightly, D. Sharples, S., Wilson, J.R., & Kirwan, B. (2011) Systems Change in Transport Control: Applications of Cognitive Work Analysis. *The International Journal of Aviation Psychology 21,* 62-84

Morency, C. (2007) The ambivalence of ridesharing. *Transportation, 34,* 239–253

Stanton, N. A., McIlroy, R. C., Preston, J. M., & Ryan, B (submitted) Following the CWA train of thought: Exploring barriers and constraints to modal shift to rail transport.

Steg, L. (2005). Instrumental, symbolic and affective motives for car use. *Transportation Research Part A, 39,* 147-162.

Venkatesh, V. & Davis, F.D. (2000) A Theoretical Extension of the Technology Acceptance Model: Four Longitudinal Field Studies. *Management Science, 46,* 186-204.

Wash, R., Hemphill, L., & Resnick, P. (2005). Design decisions in the RideNow project. In *Proceedings of the 2005 international ACM SIGGROUP conference on Supporting group work,* New York, NY: ACM press.

Woodworth, P. & Benhke, R.W. (2006) <u>Smart Jitney/Community-Enhanced Transit Systems</u>. In *Proceedings of Bus and Paratransit Conference.* Washington DC: American Public Transport Association.

Acknowledgement to reviewers

We'd like to express our gratitude to the following colleagues who helped to review and in this way improve the quality of the contributions in this book:

Jerome Bourbousson, Université de Nantes, France
Beatrice Feuerberg, Egis Avia, Toulouse, France
Michaela Heese, Austro Control GmbH, Vienna, Austria
Bob Hockey, the University of Sheffield, UK
Jettie Hoonhout, Philips Research, Eindhoven, the Netherlands
Stig Johnson, SINTEF, Trondheim, Norway
Lena Kecklund, MTO Safety AB, Stockholm, Sweden
Ben Lewis-Evans, University of Groningen, the Netherlands
Ben Mulder, University of Groningen, the Netherlands
Ilit Oppenheim, Ben Gurion University of the Negev, Beer Sheva, Israel
Paul Schepers, Rijkswaterstaat, Ministry of Infrastructure and the Environment, Delft, the Netherlands
Frank Steyvers, University of Groningen, the Netherlands
Danielle Verstegen, Maastricht University, the Netherlands
Ellen Wilschut, TNO, Soesterberg, the Netherlands

In D. de Waard, N. Merat, A.H. Jamson, Y. Barnard, and O.M.J. Carsten (Eds.) (2012). *Human Factors of Systems and Technology* (pp. 397). Maastricht, the Netherlands: Shaker Publishing.